Micromovement in Orthopaedics

Micromovement in Orthopaedics

EDITED BY

ALAN R. TURNER-SMITH

Senior Lecturer in Rehabilitation Engineering,
Department of Medical Engineering and Physics,
King's College, London
(Formerly Director, Oxford Orthopaedic Engineering Centre)

CLARENDON PRESS · OXFORD
1993

Oxford University Press, Walton Street, Oxford OX2 6DP
Oxford New York Toronto
Delhi Bombay Calcutta Madras Karachi
Kuala Lumpur Singapore Hong Kong Tokyo
Nairobi Dar es Salaam Cape Town
Melbourne Auckland Madrid
and associated companies in
Berlin Ibadan

Oxford is a trade mark of Oxford University Press

Published in the United States
by Oxford University Press Inc., New York

A catalogue record for this book is available from the British Library

Library of Congress Cataloging in Publication Data
Micromovement in orthopaedics/edited by Alan R. Turner-Smith.
Includes bibliographical references.
1. Bones–Pathophysiology–Congresses. 2. Bones–Mechanical properties–Congresses. 3. Bones–Adaptation–Congresses. 4. Bones–Movements–Congresses. 5. Biomechanics–Congresses.
I. Turner-Smith, Alan R.
[DNLM: 1. Movement–physiology–congresses. 2. Fractures–congresses. 3. Musculoskeletal System physiology congresses. 4. Joint Prosthesis–congresses. WE 103 M626 1993]
RD732.M48 1993 616.7'1–dc20 93–18460
ISBN 0-19-262306-0 (hb)

Set by Cotswold Typesetting Ltd from disk supplied by editor
Printed in Great Britain by
Bookcraft (Bath) Ltd,
Midsomer Norton, Avon

Preface

This book is a selection of contributions to a conference hosted by the Oxford Orthopaedic Engineering Centre in April 1992. Clinicians and scientists met together to explore the clinical implications of micromovement and to evaluate methods for its measurement. The record of the meeting provides a review of current thinking about the medical and engineering aspects of micromovement in orthopaedics.

The musculoskeletal system undergoes constant micromovement in response to mechanical influences. Micromovements are thought to have an important influence on the healing of fractures, the development of arthritis and osteoporosis, and the body's response to prosthetic implants. At one magnitude, micromovement has been shown to promote the healing of fractures, but at a higher magnitude it inhibits healing. Micromovement of a total joint replacement can lead to bone resorption and subsequent loosening of the prosthesis. These phenomena and their influence on bone formation have excited the interest of orthopaedic surgeons and engineers.

Two particular concerns emerged in the meeting. The first was the extent to which theory, laboratory measurements, and early *in-vivo* measurements relate to each other and to final clinical outcome. For example, the time-scale from prosthetic joint design to final clinical judgement is more than 10 years, so establishment of these relationships is of enormous importance in improving patient care and reducing costs of health provision. The second concern was the role played by physiological feedback in the control of load during healing of the musculoskeletal system. For example, the healing of bone as it adjusts to a loading regime is under the control both of the clinician—through the choice of stabilization of the fracture, instructions for weight-bearing, and the method of rehabilitative treatment—and of the patient, through proprioceptive feedback.

We are only beginning to appreciate the interaction between external control of micromovement, physiological control of micromovement, and the influence of micromovement on bone growth. Both bone repair and implant support will benefit when the action of micromovement in these clinical conditions is understood.

Concerning the structure of this book: in clinical science a natural cycle exists that moves from clinical observation, through measurement, to formulation of theory, and from the macroscopic to the microscopic. The articles have been arranged into sections to follow this general pattern, rather than preserving their original order of presentation. One of the most enjoyable features of the meeting was the discussion after each paper and during the two workshops on the clinical themes of fracture repair and prosthetic joint performance. Where debate added usefully to the full articles it has been recorded. As it proved impossible to identify all the speakers individually I simply have to acknowledge the debt owed to all those who took part.

I would like to thank Howmedica International which sponsored the meeting, the Nuffield Orthopaedic Centre which provided the venue, the contributors for their efforts in preparing the manuscripts, and the staff of the Oxford Orthopaedic Engineering Centre and the Girdlestone Memorial Library for their suggestions and help in checking proofs. In particular, a large measure of the credit for this book rests with Barbara Marks who administered the meeting, and typed and checked all the manuscripts.

Oxford A.R.T.-S.
February 1993

Contents

Contributors

Arokoski, J., Department of Anatomy, University of Kuopio, Finland
Asada, K., Osaka City University Medical School, Department of Orthopaedics, Japan
Ascenzi, A., Department of Human Biopathology, 'La Sapienza' University, Rome, Italy
Aspden, R. M., Department of Orthopaedics, University of Aberdeen, Scotland
Boyd, R., Bone and Joint Center, Henry Ford Hospital, Detroit, USA
Boyde, A., Department of Anatomy and Developmental Biology, University College, London, England
Brooks, R., Princess Elizabeth Orthopaedic Hospital, Exeter, England
Bulstrode, C. J., Nuffield Department of Orthopaedic Surgery, University of Oxford, England
Callaghan, J. J., Department of Orthopaedics, University of Iowa Hospital, Iowa, USA
Carlson, Å., Department of Orthopaedics, Malmo General Hospital, Sweden
Choudhry, R. R., The Royal Infirmary, Leicester, England
Collier, R. J., Electronic Engineering Laboratories, University of Kent, Canterbury, England
Craigen, M. A. C., Accident and Orthopaedic Division, Royal Infirmary, Glasgow, Scotland
Cunningham, J., School of Engineering and Applied Science, University of Durham, England
Derbyshire, B., Department of Mechanical Engineering, University of Leeds, England
Derbyshire, J., Department of Radiology, Nuffield Orthopaedic Centre, Oxford, England
Donarski, R. J., Electronic Engineering Labotatories, University of Kent, Canterbury, England
Draper, E. R. C., Department of Orthopaedic Surgery, Hammersmith Hospital, London, England
Edwards, J., Department of Orthopaedic Mechanics, University of Salford, England
Evans, M., Oxford Orthopaedic Engineering Centre, University of Oxford, England
Fenner, J. A. K., Department of Clinical Physics and Bio-engineering, Royal Infirmary, Glasgow, Scotland
Figueiredo, U. M., Comparative Orthopaedic Research Unit, University of Bristol, England
Forrest, A. K., Centre for Robotics and Automated Systems, Imperial College of Science, Technology and Medicine, London, England

Francis, M., Nuffield Department of Orthopaedic Surgery, University of Oxford, England

Freeman, M. A. R., The Royal London Hospital Medical College, Bone and Joint Research Unit, England

Fulghum, C. S., Duke University, Durham, USA

Fyhrie, D., Bone and Joint Center, Henry Ford Hospital, USA

Gardner, T. N., Oxford Orthopaedic Engineering Centre, University of Oxford, England

Gie, G. A., Princess Elizabeth Orthopaedic Hospital, Exeter, England

Glisson, R. R., Duke University, Durham, USA

Goldspink, G., Unit of Veterinary Molecular and Cellular Biology, Royal Veterinary College, London, England

Goodship, A. E., Comparative Orthopaedic Research Unit, University of Bristol, England

Hale, R., Princess Elizabeth Orthopaedic Hospital, Exeter, England

Heinegård, D., Department of Physiological Chemistry, University of Lund, Sweden

Helminen, H. J., Department of Anatomy, University of Kuopio, Finland

Hooper, R. M., School of Engineering, University of Exeter, England

Houston, W. J. B., Department of Orthodontics, U. M. D. S., London, England

Hua, J., Biomedical Engineering Department, Institute of Orthopaedics, Stanmore, England

Hughes, S. P. F., Department of Orthopaedic Surgery, Hammersmith Hospital, London, England

Indrekvam, K., Haukeland Hospital, University of Bergen, Norway

Johnston, I., Department of Engineering Science, University of Oxford, England

Jurvelin, J. S., Department of Clinical Physiology, Kuopio University Hospital, Finland

Kay, P., Department of Orthopaedic Surgery, University of Manchester, England

Kenwright, J., Nuffield Orthopaedic Centre, Oxford, England

Kiviranta, I., Department of Surgery, Kuopio University Hospital, Finland

Kyberd, P. J., Oxford Orthopaedic Engineering Centre, University of Oxford, England

Langeland, N., Haukeland Hospital, University of Bergen, Norway

Larsson, T., Department of Physiological Chemistry, University of Lund, Sweden

Lay, A., Electronic Engineering Laboratories, University of Kent, Canterbury, England

Laycock, D. C., Department of Orthopaedic Mechanics, University of Salford, England

Lee, A. J. C., School of Engineering, University of Exeter, England

Ling, R. S. M., School of Engineering, University of Exeter, England

Liossis, G., Centre for Robotics and Automated Systems, Imperial College of Science, Technology and Medicine, London, England

McCarthy, I. D., Department of Orthopaedic Surgery, Hammersmith Hospital, London, England

McDonald, F., Department of Oral Biology, United Medical and Dental Schools, London, England

Murray, D. W., Nuffield Department of Orthopaedic Surgery, University of Oxford, England

Norrdin, R., Comparative Orthopaedic Research Unit, University of Bristol, England

O'Connor, B., Director of Clinical Studies, Robert Jones and Agnes Hunt Orthopaedic Hospital, Oswestry, England

O'Connor, J. J., Department of Engineering Science, University of Oxford, England

Önsten, I., Department of Orthopaedics, Malmo General Hospital, Sweden.

Pan, H. Q., Bone and Joint Center, Henry Ford Hospital, Detroit, USA

Radin, E. L., Bone and Joint Center, Henry Ford Hospital, Detroit, USA

Räsänen, T., Department of Anatomy, University of Kuopio, Finland

Richardson, J. B., University Department of Orthopaedic Surgery, Glenfield General Hospital, Leicester, England

Rushton, N., University of Cambridge Orthopaedic Research Unit, England

Ryd, L., Department of Orthopaedics, University Hospital, Lund, Sweden

Saamanen, A.-M., Department of Anatomy, University of Kuopio, Finland

Sakane, H., Department of Orthopaedics, Osaka City University Medical School, Japan

Sanzén, L., Department of Orthopaedics, Malmo General Hospital, Sweden

Simpson, A. H. R. W., Nuffield Orthopaedic Centre, Oxford, England

Stother, I. G., Bio-engineering Unit, University of Strathclyde, Glasgow, Scotland

Stoyle, T. F., Glenfield General Hospital, Leicester, England

Strachan, R. K., Department of Orthopaedic Surgery, Hammersmith Hospital, London, England

Stranne, S. K., University of Iowa, USA

Taktak, A. F. G., Department of Orthopaedic Mechanics, University of Salford, England

Tammi, M., Department of Anatomy, University of Kuopio, Finland

Toksvig-Larsen, S., Department of Orthopaedics, University Hospital, Lund, Sweden

Turner-Smith, A. R., Department of Medical Engineering and Physics, King's College, England

Vaarnamo, A., Department of Anatomy, University of Kuopio, Finland
van Audekercke, R., Katholieke Universiteit, Leuven, Belgium
van der Perre, G., Katholieke Universiteit, Leuven, Belgium
Vander Sloten, J., Katholieke Universiteit, Leuven, Belgium
Walker, P. S., Department of Biomedical Engineering, Royal National Orthopaedic Hospital, Stanmore, England
Wallace, A. L., Department of Orthopaedic Surgery, University of Edinburgh, Scotland
Watkins, P. E., Comparative Orthopaedic Research Unit, University of Bristol, England
Wilkinson, R., Department of Clinical Physics and Bio-engineering, Royal Infirmary, Glasgow, Scotland
Williams P. E., Department of Applied Biology, University of Hull, England
Worley, A. J., Electronic Engineering Laboratories, University of Kent, Canterbury, England
Yang, K. H., Bioengineering Center, Wayne State University, USA
Yoshida, K., Osaka City University Medical School, Department of Orthopaedics, Japan

Section 1—Fractures

1 Fracture stiffness as a measure of fracture healing

J. B. RICHARDSON, J. L. CUNNINGHAM, J. KENWRIGHT, AND B. O'CONNOR

Introduction

In the days before radiography, Arbuthnott Lane (1914) undertook an early attempt to define fracture healing by examining the earning capacity of manual labourers following a fracture of the tibia. He found an average fall to 70 per cent of previous earnings in patients managed conservatively. This drove his desire to develop the management of tibial fractures by internal fixation. His enthusiasm for a mechanistic approach to the problem was taken up by Lambotte in Belgium (Lambotte 1907). Danis was in turn the student of Lambotte, and developed further the idea of 'primary healing' by rigid internal fixation. These principles of internal fixation developed by Lane then became the foundation that Muller used to form the AO group. Tracing back the history of 'primary healing' would appear to originate in Lane's development of a relatively objective measure of healing. The fact that a Scotsman should base the measure of success of a healing fracture on its financial implications may be considered predictable by some!

Since that time many have sought an objective measure of healing. In 1942 Page called for a standard definition of union which could be applied in all studies. Manual and radiological assessment of fracture healing were held to be unreliable, and this was confirmed by Mathew when he invited surgeons to test a fracture model of controlled stiffness. They were unable to detect bending movements, in a model, of less than 3° (Mathew *et al.* 1974).

The difficulties and disagreements in radiological diagnosis of union were tested by Nicholls who concluded that 'the orthopaedist or radiologist is not very reliable at determining early osseous union' (Nicholls *et al.* 1979). The most widely used definition of healing is the point at which a patient is able to weight-bear independently, but this is not a precise measure.

Complications have the disadvantage of occurring intermittently in fracture healing. Healing time as an end point does take the presence of complications into account to the extent that they delay independent weight-bearing.

Dynamic bone scintigraphy has been shown to predict delayed fracture union, but is not recommended as a routine procedure (McDougall and

Keeling 1988), although it can predict nonunion with a specificity of 90 per cent and a sensitivity of 70 per cent (Smith *et al.* 1987).

Rymaszewski (1984) examined the use of a load sensor ('flexigage') applied to the subcutaneous border of the tibia beneath the plaster cast in 24 patients. The gauge measured movements across the fracture, and these were seen to fall as healing progressed. Objective comparisons were found to be difficult in the preliminary study, and the gauge caused problems of pressure necrosis against the skin.

Sound transmission in the assessment of fracture healing

The application of sound transmission in the diagnosis of fractures was first reported in 1903 by Andrews who listened to the transmission of sound from a tuning fork applied to a distal boney part with a stethoscope proximal to the fracture. In 1972 Abendschein reported the use of ultrasound as a measure of returning stiffness in healing bone using 100 kHz sound measured across guinea-pig femoral fractures. The elastic modulus of a substance can be determined from the speed at which it conducts sound, but in practice on patients the soft tissue interface is variable and reduces the accuracy of the method (Abendschein and Hyatt 1972). The following equation relates tissue elastic moduli, speed of sound transmission, and tissue density:

$$E = v^2 D,$$

for tissue of modulus of elasticity E, sound conduction velocity, v, and mass density D.

Spectral analysis

The resonant frequency of a body is related to its length, and the determination of this frequency might be largely independent of soft tissue problems. A method of observing fracture stiffness healing based on a spectral analysis of bone vibration was reported by Doemland *et al.* (1986). The normal tibia was used for comparison, and the progress of union observed from a pattern of vibration returning to normal. The disadvantages of this approach are that it is difficult to quantify in terms more objective than a comparison to the normal leg, and that methods of internal or external fixation interfere with the measurements.

Fracture stiffness

Resistance to stress is a fundamental property of a material, and can be derived from the elasticity, or stiffness, of a known structure. Fracture stiffness may be the appropriate mechanical measure of returning function during fracture healing as, according to both Frost (1964) and Perren and Rahn (1980), stiffness is the chief function of the skeleton, rather than strength. The gradual change in tissue during healing can be characterized by its stiffness, changing from a Young's modulus of 0.05 MN/m^2 for granulation tissue to 20 000 MN/m^2 for mature bone, representing a 400 000-fold change (Perren and Rahn 1980). Measurements of fracture stiffness have a potential contribution as a functional measure of healing in two areas: monitoring progress in an individual patient, and comparisons between patients. The principal demands on the measuring system will be for accuracy in the matter of comparison, and for precision of stiffness measurement in the assessment of an individual patient's progress. (Accuracy describes how close a measurement is to the true value. Precision reflects how close repeated measurements may be to each other, irrespective of how far the group of results is from the true value.) The clinician needs information to allow safe removal of splintage at an early date. The method of measurement evidently cannot be one that directly measures the strength of the bone as to do so would involve its fracture!

Henry *et al.* (1968) used stiffness as a measure of healing in an investigation of rabbit fracture fixation, and found that a return to normal stiffness correlated with about half of normal strength. They found fracture stiffness a more consistent observation than strength and proposed its use as the measure of return of function. Brighton and Krebs (1972) made use of this measure in animal studies, and the curve in Fig. 1.1 also demonstrates the later fall in stiffness to normal following 'remodelling' of excess callus. Four studies on patients allow application of this healing curve to tibial fracture healing in patients.

Jernberger in 1970 undertook an extensive study of fracture stiffness, using a strain-gauged beam fixed to the patient's tibia by temporary bone screws at every out-patient visit. This rather invasive approach did give very accurate results and also information on the normal stiffness of intact tibiae. He

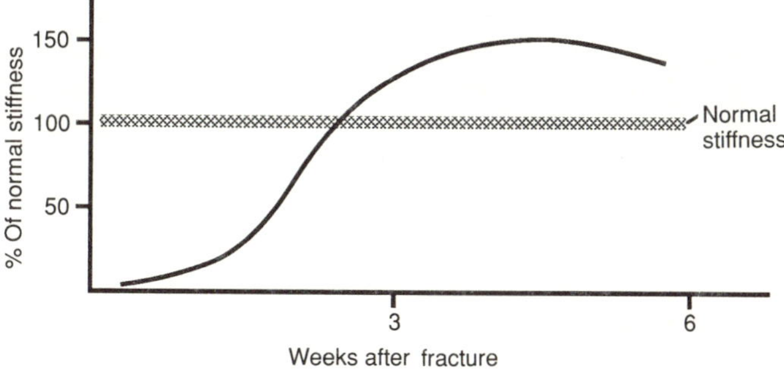

Fig. 1.1 Progress of stiffness with healing in the rabbit fibula (after Brighton and Krebs 1972. Reproduced with permission from *Surgery, Gynecology and Obstetrics*).

reported a progressive rise in fracture stiffness during healing which was generally of an exponential type, corresponding to the first half of Fig. 1.1. Intact tibiae in volunteers were measured and a remarkably narrow range of stiffness was found, 55–60 N m in men. He did not recommend a stiffness level at which to recommend independent weight-bearing, but the final stiffness recorded allowed him to gauge that patients were apparently 'healed' at stiffnesses of 7.5–10.5 N m/degree (Fig. 1.2).

Jírgensen (1972) devized one of the simplest methods of fracture stiffness measurement, using a dial gauge to measure the amount of bending during application of a fixed torque to the distal tibia in patients in external fixation. He had an overall error of 13 per cent with the method. The deflection fell with increasing stiffness, and by calculation the deflection can be interpreted as a stiffness to allow comparison with other workers in Fig. 1.2. He found that the rate of increase in fracture stiffness in his patients correlated with their clinical progress.

The use of strain gauges bonded to a patient's fixator does achieve good precision of measurement, that is repeatability, which has been the aim of Burny *et al.* (1978). His group demonstrated patterns of healing by plotting the fall in fixator loads as the fracture heals. They found patterns of healing that demonstrated refracture during healing, but these results cannot be calculated into objective or comparable figures—the curves are of the rate of fall of strain in the fixator column as healing proceeds.

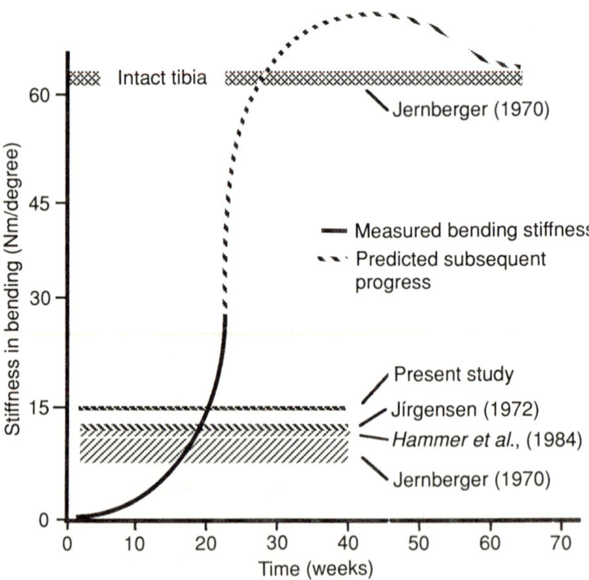

Fig. 1.2 Progress of fracture stiffness during fracture healing in the human tibia.

Hammer *et al.* (1984) developed a system of measuring fracture stiffness with the aim of defining 'healing', using a radiographic method. The patient lay on a radiographic table with the knee supported and the leg straight. An anteroposterior radiograph was taken, and then repeated after a fixed load was applied to the foot in the coronal plane. A system of subtracting the two images allowed angulation at the fracture site to be calculated, and the fracture stiffness to be calculated from known applied moments. Errors of 20 per cent were quoted, and a safe level of stiffness was chosen to be a minimum of 12 N m/degree, although with their method, the actual average was somewhat higher.

The Oxford group developed a strain gauge transducer for the fixator column which is removable (Evans 1984; Evans *et al.* 1988). This allows calibration of the transducer. Tanner (1985) developed a method of calculating an objective fracture stiffness from the fixator column strain by modelling the fixator stiffness, and assuming tight bone screws (Churches *et al.* 1985). By applying the calculation of fracture stiffness as a routine in the mechanical testing of fractures, the many factors that vary between patients could be assessed (Richardson *et al.* 1984).

Each series' results were examined, using results calculated in terms of Newton-metres per degree. The direction of each test and other parameters such as the range of bending over which a stiffness was calculated was seen to differ between the various methods. The resulting curve seen in Fig. 1.2 does however allow a valuable comparison of these different methods of fracture stiffness measurement and also a useful mechanical model of the progress of fracture healing.

An objective definition of healing

What is the evidence for 15 N m/degree to be the definition of 'healing'? We need a reasonable test that this level generally equates to the healed condition. The most relevant test may be whether a healing tibia is sufficiently strong above this level. Refracture can be a significant problem following external fixation, affecting almost 3 per cent of patients in de Bastiani's series (de Bastiani *et al.* 1984). The relatively rigid method of external fixation may contribute to little external callus forming, or perhaps simply the use of external fixation in more severe fractures may lead to difficulties in determining healing from radiographs. Refracture can be defined as pain or deformity occurring at a fracture site after splintage has been removed. It is evidently important for the patient that this complication is kept to a minimum, and it can provide a benchmark against which to test a definition of fracture healing.

Data have been gathered on over 200 patients with tibial fractures. In half of these patients, the fixator was removed on clinical grounds, although fracture stiffness measurements were taken. In the other half, the fixators were removed when fracture stiffness had reached 15 N m/degree. In the first group there were seven refractures whereas in the subsequent group there were none (Richardson *et al.* 1992). Figure 1.3 demonstrates that the refractures in group 1 were all at a stiffness of less than 15 N m/degree. This would indicate that this level is not only generally acceptable as a definition of healing but that using it does reduce the incidence of refracture.

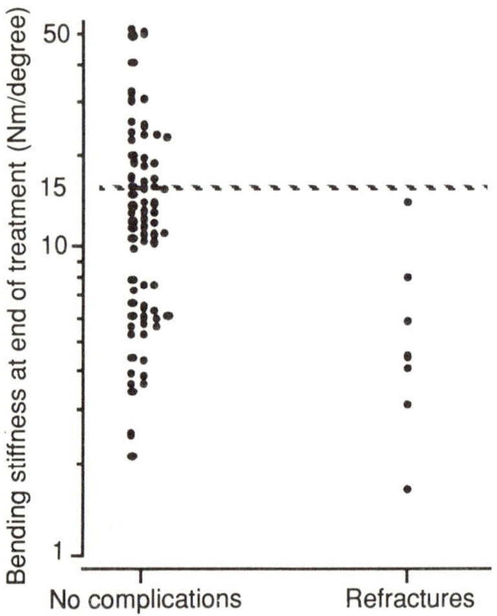

Fig. 1.3 Bending stiffness at end of treatment and incidence of refractures.

Methods of measuring fracture stiffness

Surface measurements of fracture stiffness are possible, but may have low accuracy (see Chapter 17, p. 140). Measurement from fixator pins is more accurate, assuming the pins are tight in relation to the stiffness of the measuring device. The disadvantage of measurements made with stiff external fixators in place is the possibility of over-estimating stiffness in the presence of loose pins. The authors have developed a system based on a flexible goniometer developed by Nicol which is capable of accurate measurements with very loose bone screws (Fig. 1.4). It is in use in several countries and is available from Orthofix or can be built from simple components. It is best used as a routine in the fracture clinic where a measurement can be obtained in 6 min. Once stiffness has risen to 6 N m/degree the time to healing can be predicted to within a few weeks, which is a benefit to patients. One advantage of measuring fracture stiffness is that patients can understand the progress of their fracture more readily and take part in decisions to intervene if fracture healing is slow.

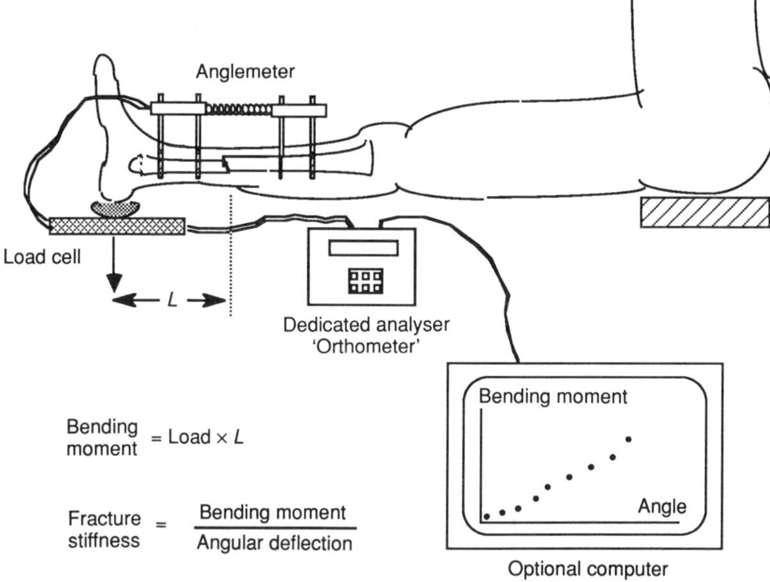

Fig. 1.4 Fracture stiffness measurement equipment.

If a patient were particularly heavy, or a fracture angulated, then higher levels of stiffness might be appropriate prior to removal of the fixator. This does not prevent one using the time to reach 15 N m/degree as a measure of healing in a clinical trial. It is also necessary to see external callus evident on X-ray since it is possible to have 'brittle healing' where the stiffness of a fracture might eventually rise by direct healing without the strength of external callus.

An *et al.* (1988) predicted that fracture stiffness measurements can only be made up to 25 per cent of normal stiffness. Fortunately the period for which the fracture remains protected in these studies was found to be only this first 25 per cent of the curve. The strength of healing bone was found by Henry *et al.* (1968) to be some 50 per cent of that of intact bone of similar stiffness, and in practice some patients suffer refracture from less than severe injuries several months after protection has been removed. Fracture stiffness measurement of even a low precision would however appear to offer protection against refracture. It also provides the ability to compare different methods of fracture management (Kenwright *et al.* 1991).

Debate in the meeting

The author (JBR) admitted that the apparent success of choosing one particular value for fracture stiffness at which the fixator could be removed might be due to the fact that the particular value chosen was a very conservative one. He emphasized, however, that he did not wish anyone to be narrow-minded about one particular figure. As well as stiffness, the quality of callus, the weight of the patient, their activity level, and the alignment of the fracture all affect strength and so should be incorporated into the clinical decision for fixator removal.

In each of the two groups many different types of treatment had been used. This was not a prospective, randomized study. The end-point in the clinical decision group was in no way a 'gold standard' but, as variable as it was, it was all the author had to measure his quantitative method against.

Early stiffness measurements are distorted by muscle activity which adds an unquantified force during stiffness measurement. So although bending stiffness is a valuable test at later stages of healing, it is not useful for quantifying early healing.

2 The vascular response to micromovement in experimental fractures

A. L. WALLACE, E. R. C. DRAPER, R. K. STRACHAN, I. D. MCCARTHY, AND S. P. F. HUGHES

Introduction

The definitive management of fractures of long bones remains a controversial issue in orthopaedic surgery. While it has been recognized for many years that the rate and pattern of healing may be influenced by the mechanical conditions to which the fragments are subject (Ashhurst 1986), it has become evident that there are other factors such as the prevailing vascular supply (Cavadias and Trueta 1965), adjacent soft tissue damage (Count-Brown and McQueen 1987), neural and humoral effects (Aro *et al.* 1985; Bidner *et al.* 1990), changes in local oxygen tension (Heppenstall *et al.* 1976), and electrical potential (Gross and Williams 1982) which may also be important.

None the less, the idea that the rate of fracture healing might be accelerated by mechanical means is an attractive concept. As a consequence of observations of fractures in experimental animals held with devices of differing rigidity, Perren (1979) proposed the hypothesis that the sequential differentiation of tissues from haematoma to new bone in the fracture gap was dependent on the progressive reduction of interfragmentary movements. Although this theory has never been validated, it has prompted a number of investigators to attempt to define the optimal magnitude, timing, and orientation of relative displacements (commonly expressed as strain) for the healing of diaphyseal fractures.

It has been argued that union by bridging external callus provides the most rapid restoration of structural integrity in long bones (McKibbin 1978), and for this reason many studies have manipulated the mechanical environment of the fracture using external fixation (Williams *et al.* 1987), intramedullary nails (Molster and Gjerdet 1984), or flexible plates (Claes 1989). Goodship and Kenwright (1985) showed that ovine tibial osteotomies held with sliding clamps recovered stiffness more rapidly than those held with rigid clamps. More recently, Egger *et al.* (1991) found that 'dynamization', or conversion of an external fixator on the canine tibia from a rigid mode to intermittent axial compression, at 7 days resulted in superior torsional strength at 6 weeks.

Other workers have disputed the importance of the mechanical environment (Gilbert *et al.* 1989), although there is now some clinical evidence that micromotion can alter the rate of healing in human tibial fractures (Kenwright *et al.* 1991).

The other factor generally recognized to be of significance is the state of the prevailing blood supply. The effect of vascular damage (which is common after trauma but which is generally minimal in experimental osteotomies), in relation to the effect of the mechanical environment, has not been defined. In recent studies performed in our laboratory, the authors have shown that in the early phase of healing the periosteal collateral circulation is predominant (Strachan *et al.* 1990) and that obliteration of this network has a profound effect on early healing, even in the presence of micromovement (Wallace *et al.* 1991*b*).

The purpose of this study was to define the effect of micromovement upon the early phase of healing of an ovine tibial osteotomy (i) in a well-vascularized environment and (ii) in a devascularized environment, in order to provide guidelines for the appropriate clinical management of long bone fractures, with special reference to high-energy trauma.

Materials and methods

Instrumented fixation system

An instrumented, unilateral external fixator was designed which would allow prescription of different axial stiffnesses and measurement of axial loads carried by the device. This has been described in detail elsewhere (Wallace *et al.* 1991*b*). Displacement at the osteotomy site due to an axial load was determined by varying the material used for the spring component. Displacement due to pin bending was reduced by increasing the diameter of standard 110 mm long tapered self-tapping pins (Orthofix srl, Verona, Italy) from 6 mm to 10 mm. Each fixator and pin assembly was calibrated under axial load *in vitro* using wooden fragments and a dial gauge in the gap; the response of the transducer was found to be essentially linear and error in the measurement system was estimated to be less than 5 per cent.

Experimental groups

Skeletally-mature 3-year old female Scottish blackface sheep weighing 45–65 kg were randomly allocated to three groups. The axial stiffness chosen for the standard group (Group S, $n=13$) and the devascularized group (Group D, $n=13$) was 240 N/mm, achieved with a silicone rubber block, and was approximately equivalent to the stiffness of a bilateral Hoffmann configuration applied to a human tibia (Kempson and Campbell 1981). The stiffness chosen for the rigid group (Group R, $n = 10$) was 460 N/mm, or nearly twice that of Group S, and was achieved with a stainless steel block. In each group approximately half the animals were assessed at 2 weeks, which has been shown to be the peak of blood flow after osteotomy (McCarthy and Hughes 1984) and at 6 weeks, which appears to be the stage of maximum callus in similar models (Markel *et al.* 1990).

Surgical procedures

Each sheep was anaesthetized with a halothane/nitrous oxide mixture and intubated. The fixator was applied in the craniocaudal plane at an offset distance of 60 mm from the right tibia by means of the six pins which were inserted after pre-drilling with a 4.8 mm bit. A longitudinal incision was then made over the medial subcutaneous border of the right tibia and deepened down to the periosteum, which was incized transversely. An osteotomy was made with a Gigli saw, protecting the soft tissues and leaving a 2-mm gap which was checked with a feeler gauge. In Groups S and R the wound was closed in layers and the pins tightened using a calibrated torque wrench.

In Group D, the periosteum was stripped circumferentially for 20 mm proximally and distally from the osteotomy site, and a 40 mm silicone rubber sleeve was inserted over the fragment ends, between the cortex and surrounding muscle. This procedure was intended to prevent the revascularization of the cortex from the periosteum and its muscular anastomoses, leaving the intracortical or medullary networks as the only possible sources of revascularization. After each procedure, the animals recovered and were housed in large pens in an indoor facility.

In-vivo *measurement of applied loads*

Pre-operatively, each sheep underwent a week of training on an instrumented treadmill which allowed measurement of the vertical component of the ground reaction force (GRF). Post-operatively, tests were carried out on a weekly basis. Pins were tightened to predetermined thresholds before each test. During the test, which typically recorded between 50 and 100 individual steps in about 3 min, simultaneous measurement of the GRF and axial load in the fixator bar were fed to a microcomputer via a strain gauge amplifier.

An editing program allowed rejection of abnormal data such as when the animal placed the opposite limb on the force plate, which was easily identified by visual assessment of the curves for GRF and axial load for each step. The primary measurements made were of fixator axial load, GRF, and stance phase time. Parameters calculated were axial displacement (using the *in-vitro* axial stiffness of each fixator system), which was expressed as strain by dividing displacement by the original gap size of 2 mm; and normalized GRF, which accounted for variation due to weight differences between animals.

Measurement of regional blood flow

Radiolabelled microspheres were used to determine blood flow in bone and soft tissue by the reference organ technique (Gross *et al.* 1981). Approximately 4.5 million microspheres (New England Nuclear Ltd, Southampton, UK) of L5 1 0.1 µm diameter and labelled with ^{57}Co (half-life 270 days) were injected over 30 s into the left ventricle. A reference sample was obtained from the brachial catheter for 2 min, after which the animal was killed with an intraventricular dose of potassium chloride.

Mechanical testing

The hindlimbs were amputated through the knee joints and the tibiae cleaned of major soft tissue attachments. The ends of the right and left tibiae were mounted in cups containing rapid setting Woods' metal, and tested in torsion within 2 h of death. A custom-built testing machine was used which loaded the bones to failure at a rate of 360°/min in a clockwise direction. Applied torque and angular displacement were measured, giving maximum torque, torsional stiffness and energy absorbed to failure for each bone.

Quantitative microradiography

At the osteotomy site, a transverse section measuring 150 µm thick was cut 1 mm proximal and distal to the gap using a tensioned annular diamond edged microtome and polished on ground glass to about 120 µm. These undecalcified sections were stored in 70 per cent alcohol before being placed on Kodak 4415 high-resolution X-ray plates, and exposed in a Faxitron 43805 X-ray machine (Hewlett-Packard, Vinten Instruments, Weybridge, Surrey) for 5 min at 20 kV and 1 mA tube current, 30 cm from the X-ray source.

The microradiographs were quantified using a computerized image analysis system (Magiscan, Joyce-Loebl Ltd, Gateshead, UK). Under low power magnification, images of the illuminated cross-sections were captured by a video camera and reproduced on a high-resolution colour monitor. Measurements of endosteal, cortical, and periosteal area and perimeter were calibrated by a scaling factor and expressed in square millimetres using GENIAS software on an IBM microcomputer. In order to account for inherent differences in original bone dimensions between animals, area measurements were normalized as a fraction of the outer cortical perimeter, which was assumed to remain constant in each sheep during the experiment. The porosity of endosteal, cortical and periosteal mineralized tissue was calculated and expressed as a percentage of the relative region of interest.

Statistical analysis

At each stage, the means in all three groups were compared using a one-way analysis of variance. If a significant difference was found, the unpaired Student's *t* test was applied, either comparing Group S with Group R, or Group S with Group D. Group S essentially acted as a control for both groups, and therefore no comparisons could be made directly between Group R and Group D because of inherent differences in mechanical environment or blood supply as a consequence of the experimental design. A difference was regarded as being significant when $p<0.05$, and standard deviations are quoted.

Results

In-vivo *loads, displacement, and strain*

Analysis of variance revealed significant differences in normalized ground reaction force in both the early and late phase of the experiment, which is demonstrated in Fig. 2.1. At seven days after osteotomy, the rigidly held group (Group R) were exerting higher loads (33.2 ± 4.7 per cent body weight) than Group S (24.8 ± 10.8 per cent body weight), but thereafter the two groups were not different. There was no difference between Group S and the devascularized

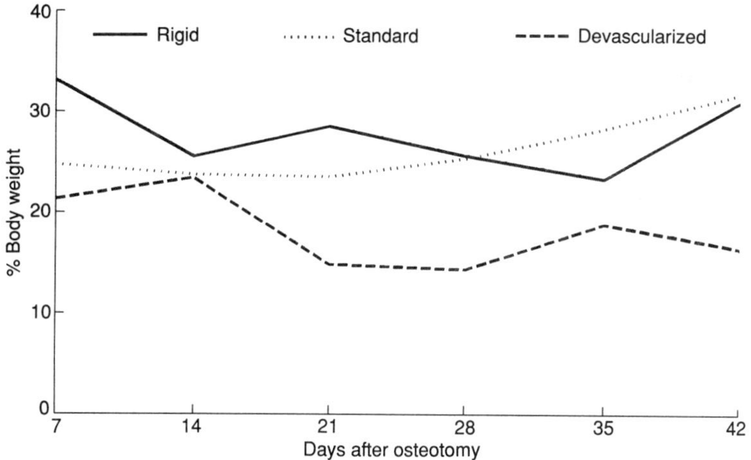

Fig. 2.1 Ground reaction force, expressed as a percentage of the pre-operative body weight in newtons.

group until 21 days after osteotomy, but for the remainder of the experimental period, Group D limped progressively, so that by 42 days the GRF was only about half that measured in Group S.

The fixator axial loads are given in Table 2.1, and expressed as axial gap strain in Fig. 2.2. Again, significant differences were found in the early and late experimental periods. Despite a 92 per cent increase in axial fixation stiffness, the sheep in Group R increased their weight-bearing so that by 14 days the mean axial strain of 40 per cent (equivalent to 0.81 ± 0.25 mm) was only 25 per cent lower than that calculated in Group S ($p<0.05$). During the following weeks, as the osteotomy healed, axial loads and displacements in the well-vascularized groups decreased in a fairly linear fashion at a strain loss rate of

Table 2.1 Fixator axial load (N)

Post-operative day	Rigid	Standard	Devascularized
7	438 (±144)[a]	297 (±92)[a]	270 (±132)
14	371 (±115)	314 (±128)	273 (±148)
21	259 (±143)	169 (±98)	169 (±83)
28	107 (±51)	87 (±86)	143 (±59)
35	29 (±18)	52 (±52)[b]	178 (±104)[b]
42	28 (±32)	23 (±20)[c]	155 (±95)[c]

Corresponding superscripts refer to significant differences ($p<0.05$).

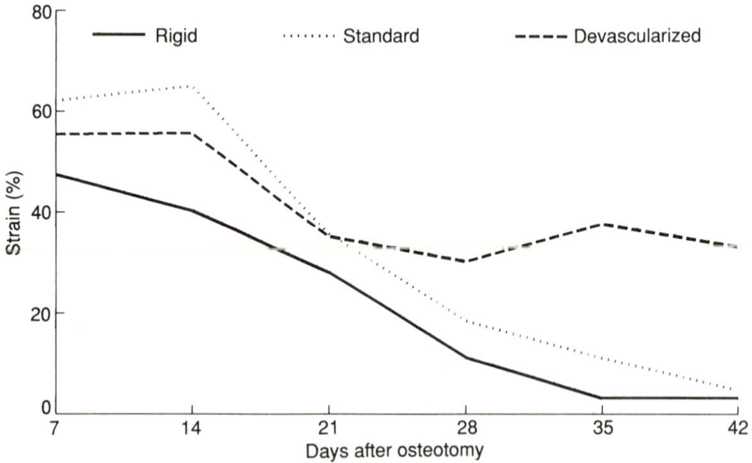

Fig. 2.2 Axial interfragmentary gap strain. Osteotomy displacement expressed as a percentage of the unloaded osteotomy gap width (2 mm).

about 1.8 per cent per day, reaching low values of less than 30 N by 42 days. In contrast, the devascularized group showed no net reduction in axial load, displacement, or strain in the latter half of the experiment. At 42 days after osteotomy, the fixator was carrying a mean axial load of 155 ± 95 N, equivalent to a displacement of 0.66 ± 0.41 mm.

Osteotomy site blood flow

In the well-vascularized groups, both cortical and medullary flows were significantly higher than those in the contralateral tibia, measured by the paired t test ($p<0.01$). Analysis of counts showed that 100–150 microspheres per tissue sample were obtained with the administered dose, which has been shown to give results in bone with an accuracy of within 10 per cent (Li *et al.* 1989). However at 14 days, corticomedullary flow in Group S was about 4 times higher (Fig. 2.3) than in the rigidly held group ($p<0.05$) although, by 42 days, flows had returned toward contralateral values and there was no significant difference. In the devascularized group, cortical flow at 14 days was low, at baseline values of 1.70 ± 1.65 ml/min/100 g, despite an increase in medullary flow equivalent to that in Group S. By 42 days, cortical flow had recovered to similar levels to those measured in the other groups, but the massive rise in medullary flow persisted ($p<0.01$).

Mechanical testing

The torsional properties of the osteotomized and intact contralateral bones are summarized in Table 2.2. It can be seen that although slightly higher values

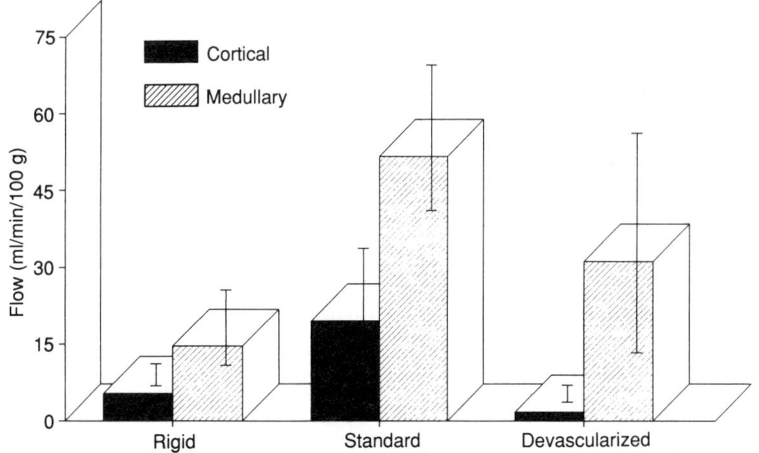

Fig. 2.3 Regional blood flow at the osteotomy site at 14 days after osteotomy.

Table 2.2 Torsional properties

	Rigid	Standard	Devascularized
(i) Maximum torque (N m)			
Osteotomy—14 days	1.17 (±0.41)	1.41 (±0.37)	NR
Osteotomy—42 days	12.72 (±4.08)	21.20 (±15.18)	0.35 (±0.47)
Control tibia—14 days	68.88 (±7.61)	73.53 (±15.97)	60.77 (±4.84)
Control tibia—42 days	60.70 (±9.77)	66.41 (±11.97)	59.54 (±3.10)
(ii) Torsional stiffness (N m/degree)			
Osteotomy—14 days	0.06 (±0.03)	0.07 (±0.02)	NR
Osteotomy—42 days	0.91 (±016)	1.05 (±0.28)	0.01 (±0.02)
Control tibia—14 days	2.00 (±0.68)	2.62 (±0.58)	1.85 (±0.18)
Control tibia—42 days	1.98 (±0.33)	2.14 (±0.24)	1.98 (±0.10)

NR = not recordable.

were recorded in Group S than in Group R at 14 and 42 days, there was no statistically significant difference between them. By 42 days about 50 per cent of stiffness and 25 per cent strength had been recovered in the well-vascularized groups. However, in the devascularized group no mechanical link was demonstrable at 14 days, and by 42 days torsional properties were approximately equivalent to the 14-day values of the well-vascularized groups, indicating a substantial delay in recovery.

Microradiography

Qualitatively, it appeared that at 42 days the cortex in the devascularized osteotomies was undergoing extensive endosteal resorption, and this was confirmed by the higher endosteal area and lower cortical area (Table 2.3) than that in Group S. Comparing the well-vascularized osteotomies, the rigid group had a significantly smaller total and periosteal cross-sectional area, and less cortical porosity, than the semi-rigid group (Fig. 2.4, $p<0.05$).

Discussion

In the field of research into basic mechanisms of fracture healing, there have been two main approaches to defining the mechanical environment at the osteotomy or fracture site. In simple terms, these may be considered as either 'active' where the fragments are loaded or displaced by a powered device to pre-set values (Goodship and Kenwright 1985; Cheal *et al.* 1991), or 'passive' where the loads and resulting displacements are a function of both the

Table 2.3 Microradiographic measurements

	Rigid	Standard	Devascularized
Perimeter (mm)			
Periosteal	79.02 ($\pm$4.56)[a]	97.28 ($\pm$24.11)[a]	—
Cortical	50.60 ($\pm$1.80)	51.35 ($\pm$3.25)	49.75 ($\pm$1.52)
Endosteal	27.77 ($\pm$3.12)	27.48 ($\pm$1.77)[b]	32.33 ($\pm$4.63)[b]
Normalized area (/CP)			
Total	7.85 ($\pm$0.99)	8.48 ($\pm$1.63)	—
Periosteal	4.15 ($\pm$1.02)	4.74 ($\pm$1.52)	—
Cortical	2.56 ($\pm$0.19)	2.66 ($\pm$0.22)[c]	2.06 ($\pm$0.39)[c]
Endosteal	1.14 ($\pm$0.21)	1.08 ($\pm$0.09)[d]	1.47 ($\pm$0.37)[d]
Porosity (%)			
Periosteal	44.08 ($\pm$8.12)	41.64 ($\pm$8.99)	—
Cortical	8.55 ($\pm$3.25)[e]	12.22 ($\pm$6.65)[ef]	1.12 ($\pm$0.77)[f]
Endosteal	56.49 ($\pm$12.99)	43.33 ($\pm$12.62)	—

CP = cortical perimeter. Corresponding superscripts refer to significant differences ($p < 0.05$).

mechanical characteristics of the fixator and the activity of the animal itself (Gilbert *et al.* 1989).

For this study, the latter type of arrangement was selected because the aim was to prescribe an initial environment (such as would be done in the clinical management of a long bone fracture) and monitor both the mechanical and haemodynamic response of the healing tissues, with and without a devascularizing insult. Had an active system been selected, it would have been difficult to measure the 'physiological' reduction of strain within the extremes of the vascular environments, and to measure the recovery of limb function.

It became apparent, however, that despite the 92 per cent increase in axial fixation stiffness selected for the rigid group, the animals were able to increase the external loads on the right hindlimb in the first 2 weeks after osteotomy to the extent that the calculated axial strains were much closer to those of the standard group than had been expected, and indeed appeared to converge in the late phase of the experiment. Although less spectacular than predicted, the early disparity in strain environments was associated with a significant difference in the vascular response when measured at 14 days. With a 25 per cent difference in axial strain, a fourfold difference in cortical and medullary blood flow was observed, and this suggests that the vascular system at the osteotomy site is very sensitive to mechanical stimuli.

It is not clear from this experiment whether the rise in flow represents simply greater flow in normally-perfused vessels, recruitment of functionally 'resting' capillary beds (Hughes *et al.* 1978) or *de novo* angiogenesis. The mechanism by

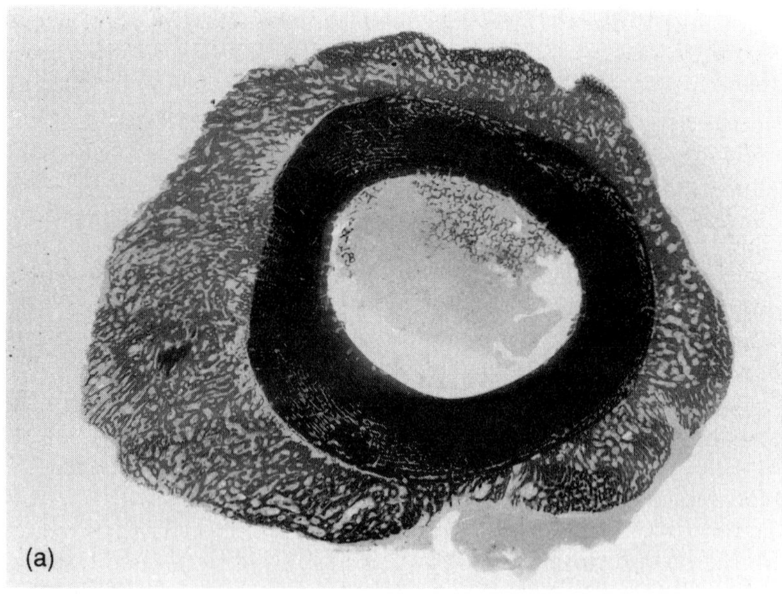

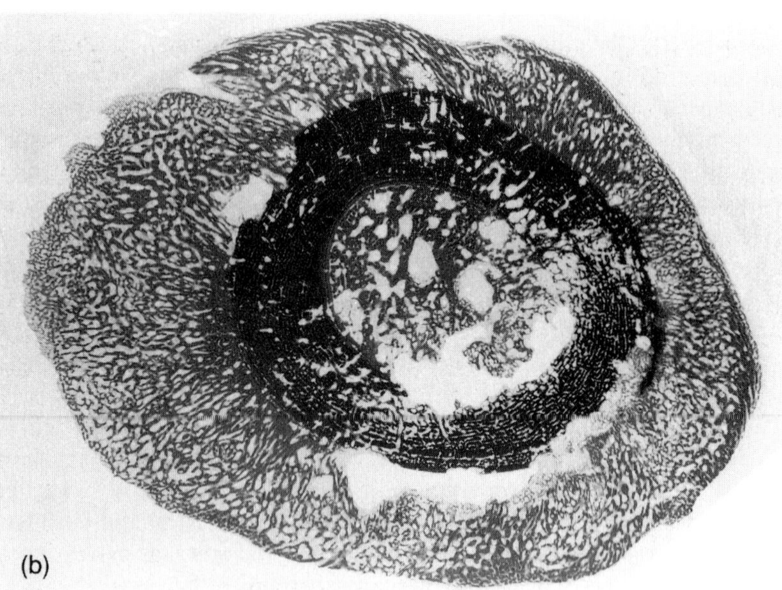

Fig. 2.4 Examples of cross-sectional microradiographs from (a) rigid osteotomy (fixator axial stiffness = 460 N/mm), and (b) standard osteotomy (fixator axial stiffness = 240 N/mm).

which the response occurs is also unknown, but may either be a fundamental property of the vascular endothelium itself (Hudlicka and Tyler 1986), or more likely, a consequence of local release of vasodilatory or angiogenic mediators from differentiating cells within the osteotomy gap (Mohler *et al.* 1990; Wallace *et al.* 1991*a*).

From the results in the well-vascularized groups in the present study it appears that the vascular response is also linked to the quantity and distribution of new bone formation, as suggested by Trueta (1963). In Group S, the higher blood flows seem to be associated with a greater cross-sectional area of callus, although there was no statistical difference in mechanical properties between the two groups. This lends support to the concept that when the vascular reserve is optimal, the formation and distribution of new bone is proportional to the magnitude of the early axial strain, and that once formed the periosteal callus fronts are able to reduce the gap strain at a constant rate, as measured in this study. The result is the achievement of a common structural 'end-point' by initially differing pathways.

If, however, the periosteal collateral route is blocked, cortical flow remains low in the face of relatively high gap strains. The evidence from this study is that although the medullary vascular response is impressive there is considerable endosteal resorption, and formation of medullary callus is slow, contributing little to the restoration of the structural properties of the bone by 42 days.

The clinical relevance of this work is that the effect on healing produced by variation of the mechanical environment is related to, and seems to be dependent upon, the vascular response. This inter-relationship means that in severely-devascularized fractures the potential influence of the device chosen for fixation will initially depend on the rate of periosteal vascular recovery. Once this has occurred, however, the vascular response is quantitatively related to the degree of micromovement or strain, and subsequently to the magnitude and distribution of osteogenesis. The specific cellular and molecular mechanisms have yet to be elicited, but there would appear to be considerable potential for the therapeutic use of angiogenic factors, within a controlled mechanical environment, in the revascularization of fractures with extensive soft tissue injury.

Acknowledgements

This work was supported by a grant from Action Research for the Crippled Child. The authors would particularly like to thank Mr R. H. Fleming and Mrs B. C. Wyatt for their technical expertise and assistance in the conduct of the experiments.

Debate in the meeting

The authors selected their two fixation stiffnesses to represent the limits of normal clinical practice, however, it was suggested that uniaxial external fixators in common use were very much less stiff. Maybe there is an upper limit to strain which would be limited by the biological system. It is commonly held that if callus is not initiated early on it is difficult to stimulate its formation later. It may be possible, therefore, to use a flexible frame early-on to allow the biological system to adjust micromovement to an optimal value, and to use a stiffer frame later in order to protect the pin/bone interface?

The role of the vascular system cannot be dissociated from the complex morphometric changes required for mechanical stability. Although the literature concentrates on the medullary dominance of vascularity in normal bone and the periosteal dominance in the early phase after fracture, the role of intercortical circulation and the changes in it during healing are not so well known. Maybe longitudinal vascular channels are not established until late in healing, once all remodelling has taken place across the fracture site.

Although it was the axial strain which was recorded at the macroscopic level, this may bear no relationship to the strain in cells between the fragments where high shear forces can exist.

3 Transient dynamic loading in a rabbit osteotomy model

K. H. Yang, R. Boyd, D. Fyhrie, H. Q. Pan, and E. L. Radin

Introduction

Although angular deformities are associated with osteoarthrosis (Ory 1964), not all patients with genu/varum (knock knees) or genu/valgum (bow legs) develop clinical symptoms. The additional factors which promote clinical osteoarthrosis remain unknown. Suspecting that transient dynamic loading may be a contributing factor, Radin *et al.* (1990) compared the difference in axial acceleration of normal and valgus proximal tibias of the same rabbit.

Method

The author created, under general anaesthesia and sterile conditions, 30° valgus osteotomies of the proximal tibia just distal to the tibial tuberosity in 4 kg (adult) white New Zealand rabbits. The osteotomy was fixed with a plate and screws. The rabbits were exercized daily beginning 3 weeks post-operatively. Axial accelerations of both tibias were measured at the end of $4\frac{1}{2}$ months. Two miniature Entran EGA-125 *g* accelerometers were mounted on aluminium plates. Two rubber bands were used to fasten the accelerometer mounts to the tibias just below the knees, as shown in Fig. 3.1. The rabbits were placed on a matted floor and allowed to move freely. The acceleration data was filtered at 1000 Hz and digitized on a Tektronix 2510 digitizer at a sampling rate of 10 000/s for each channel. Data were recorded when three consecutive steps were observed. A 500 Hz Butterworth digital filter was used to process the data of the middle step.

Results

In all animals treated, a peak acceleration was present on the experimental side which was always higher than on the control side, see Fig. 3.2. The average peak acceleration for the experimental limbs was 8.5 *g*, as compared with the control limb average of 3.3 *g*; $p < 0.0001$. The duration of the transient peak was 12–18 ms. These peak accelerations occurred at push-off.

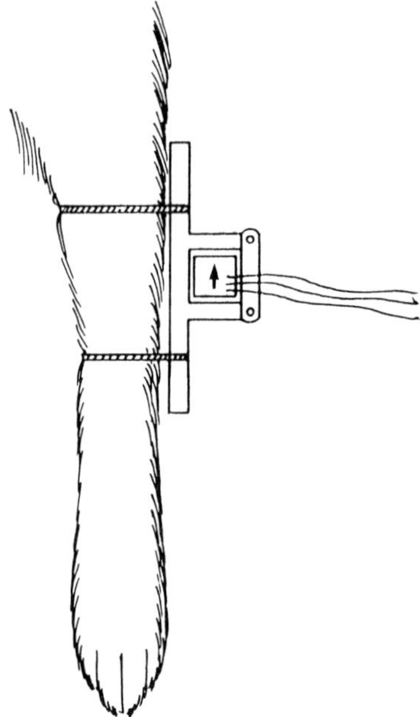

Fig. 3.1 Accelerometer attached to the lateral leg.

Discussion

The acceleration measurements in this study showed that the experimental legs were subjected to transient dynamic loading at push-off and at the end of stance phase. This phenomenon was not present on the control side. It has been established that rapidly applied loading is detrimental to joints. Rabbits ambulate by bipedal hopping. Along with the well-known lateral displacement of the point of application of the resultant interarticular force in angular knees, we have found an element of transient dynamic loading. These data suggest that dynamic loading may play a role in osteoarthrosis following angulation deformity of the knee.

The authors hypothesize that lateral translation of the tibia momentarily subluxes the femur as it slides down the inclined tibial plateau and that muscle action at the time of push-off, a critical phase in hopping gait, is responsible for the transient dynamic loading (Fig. 3.3). It is also interesting that the duration of the peak loading is of the same magnitude as used in a repetitive impulsive loading apparatus which creates osteoarthrosis in rabbits (Simon *et al.* 1972).

In summary, a significant transient peak acceleration was found following valgus osteotomy in rabbits at the end of stance phase. Since it is established that 30° valgus osteotomy leads to gonarthrosis in rabbits, it is hypothesized

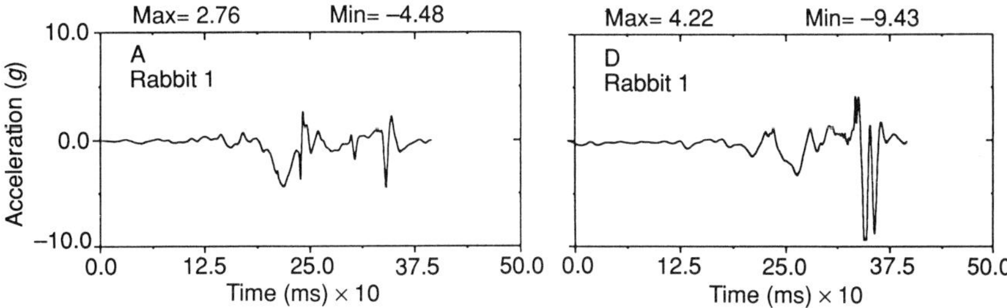

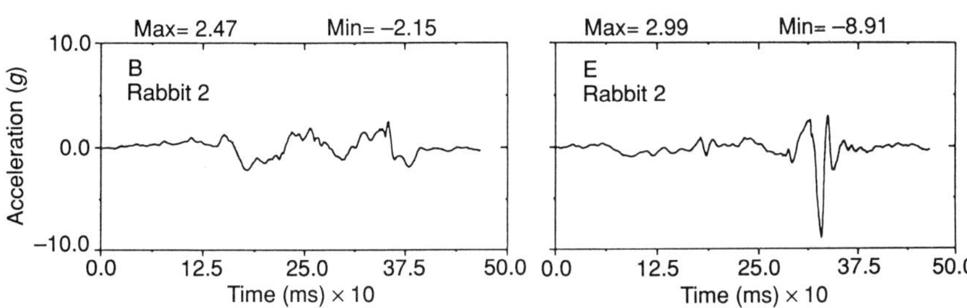

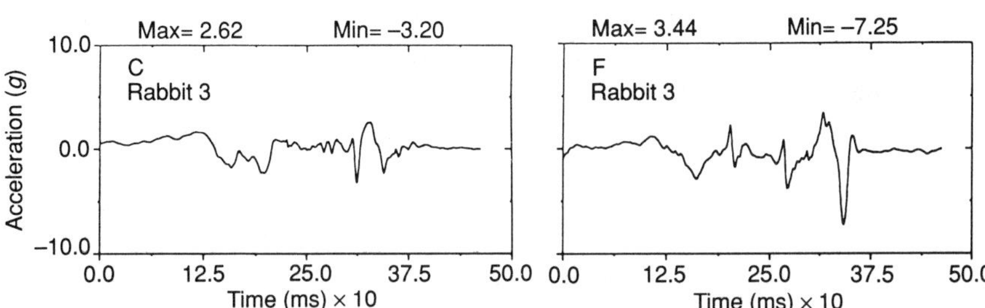

Fig. 3.2 Representative acceleration from three rabbits with tibia angulation. Data obtained from control legs are shown on the left. Note that maximum acceleration in the negative direction always occurs at the end of stance phase for the osteotomized legs (right).

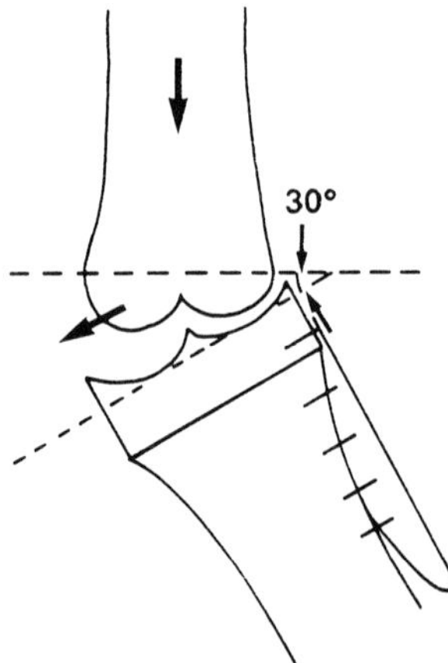

Fig. 3.3 Proposed explanation of the results. A momentary subluxation of the femur sliding down the 30° inclination on the tibial plateaux causes the impulsive load.

that this transient peak acceleration is a factor in the subsequent joint deterioration.

Acknowledgement

This study was supported by NIH Grant No. 27127.

4 The healing of tibial plateau fractures. A Roentgen stereophotogrammetric study

L. Ryd and S. Toksvig-Larsen

Introduction

In the treatment of fractures, recent research suggests that immobilization is counter-productive to the preservation of cartilage integrity and that motion exercises should be started as soon as possible (Salter 1989). Fractures through the lateral tibial condyle are caused by tremendous vertical forces, which have a pronounced crushing effect on the cartilage (Kennedy and Bailey 1968) making early motion exercises all the more indicated in the treatment of this fracture. Conversely, however, inferior results are correlated to a remaining depression of the joint surface (Rasmussen 1971; Lansinger *et al.* 1986) and proper restoration of articular congruency is important. There is thus a conflict of interest between early mobilization to benefit the cartilage and sufficient immobilization to benefit proper fracture healing.

At the authors' department, knees that are stable in the extended position are treated conservatively, while unstable ones are operated on with open reduction, bone transplant, and internal fixation. We report here on five cases, thus treated, in which the depressed fragment was marked with tantalum markers to allow the study of possible migration using Roentgen stereophotogrammetric analysis (RSA) in the post-operative period.

Patients and methods

Five knees were studied, belonging to the 'split-depressed' type (Fig. 4.1(a,b)) according to Rasmussen (1971), or corresponding to the Type II fracture according to Schatzker *et al.* (1979). The operations were performed through an antero-lateral approach where the vertical split fracture was developed to gain access to the interior of the lateral condyle. After elevation of the depressed fragment to the proper level, bone chips from the ipsilateral iliac crest were impacted in the void left in the condyle beneath the elevated fragment. The split fragment was subsequently replaced and secured by one or two AO spongious bone lag screws. Where possible, the lateral meniscus was

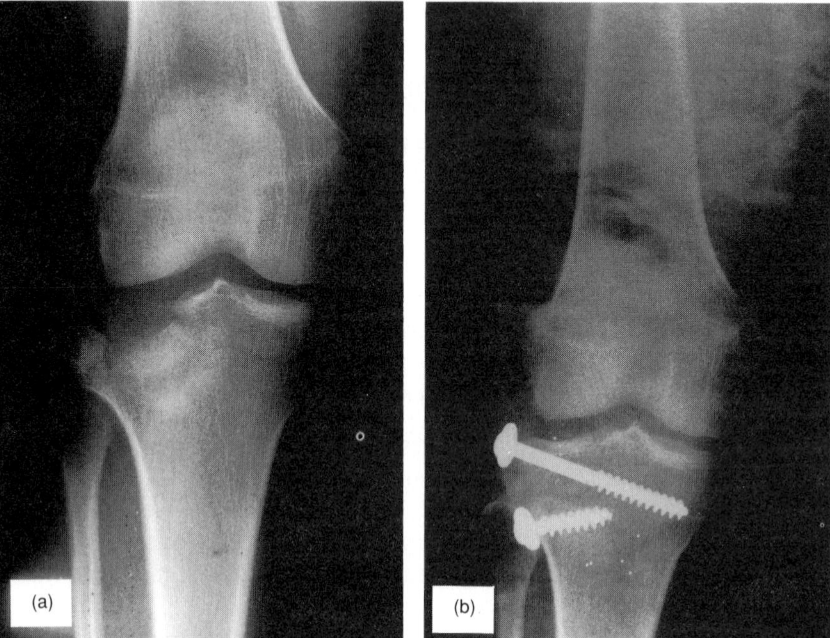

Fig. 4.1 Antero-posterior view of a severely depressed lateral tibial plateau fracture (Case 2) (a) before and (b) after operative reduction. Note tantalum markers in the depressed fragment and reference markers in the tibial methaphysis.

sutured. Post-operatively, all patients were treated with a whole-leg plaster for 4 weeks after which physiotherapy without plaster was initiated. The patients were not allowed to bear weight before 12 weeks. Post-operative radiographs showed the reduction to be close to anatomical in all cases.

At the operation, 3–4 tantalum bone markers were inserted in the depressed fragment and 3–4 more in the surrounding tibial methaphysis. Post-operatively, and at 1, 3, 6, and 12 months, stereoradiograms were obtained according to Selvik (1974). The two radiographs comprising a stereo pair, were digitized using a high precision tablet and the x- and y-coordinates were fed into an IBM AT clone microcomputer. Using two software packages, the three-dimensional coordinates of the patients markers were obtained and subsequently the migration of the markers in the fracture segment relative to the tibial methaphysis was obtained using rigid-body kinematics. Motion of the fracture segment was expressed as rotation about and translation along the three cardinal axes coinciding with the main planes of the body. As a simplistic way to denote the magnitude of the motion, the total three-dimensional vector motion of the marker which moved the most, the maximum total point motion (MTPM) was given (Ryd 1986). The accuracy of RSA applied to the knee has previously been determined to be 0.2 mm/0.3° (Ryd 1986). Conventional radiographs were obtained in the AP and lateral projections at the same time as the stereoexaminations. Care was taken to position the central beam through the centre of the knee on all occasions. A consistent film-focus distance of about 100 cm was used.

Results

All segments migrated during the healing phase. This migration (MTPM) occurred almost exclusively during the first month (Fig. 4.2). The migration was rotatory in all cases. With the exception of Case 3, whose segment rotated 30° about the vertical axis, all rotations were rather small. In three cases (2, 3 and 4) the elevated segment subsided back down into the tibial methaphysis 0.9–2.8 mm (Table 4.1). Again, this subsidence occurred almost exclusively during the first month. Small and erratic translations in other directions were recorded.

Discussion

The 'split-depressed' fractures studied here constitute a substantial part, 25–40 per cent, of all tibia plateau fractures (Rasmussen 1971; Lachiewicz and Funcik 1990; Vandenberghe *et al.* 1990; Blokker *et al.* 1984). While the treatment of plateau fractures in general is controversial, this subgroup raises

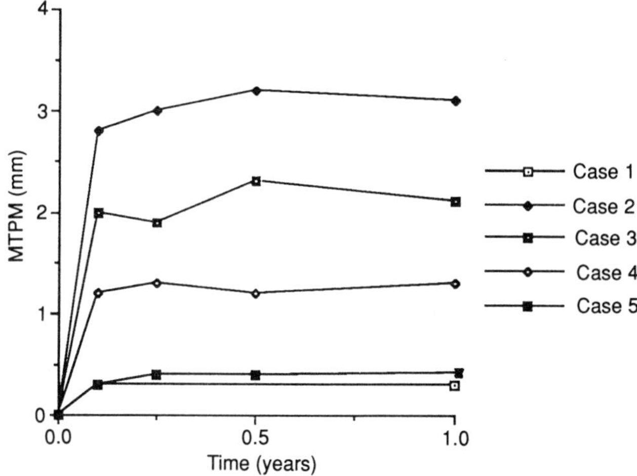

Fig. 4.2 Diagram of the migration of the depressed fragment in five cases with fractures through the lateral tibial plateau.

Table 4.1

	Rotation			Translations		
Case	*X*	*Y*	*Z*	*X*	*Y*	*Z*
1	−0.5	0.9	2.8	0	0	0
2	3.3	−1.3	−6.8	−0.7	−2.8	−0.3
3	−2.8	−30.2	3.2	−0.8	−1.5	0
4	3.1	−3.3	3.7	0	−0.9	0.6
5	−0.9	1.6	1.9	0	0	0

The *x* axis is transverse; + sign indicates forward tilt and lateral translation, *y* axis is vertical; + sign indicates 'toe-out' rotation and upward translation, *z* axis is sagittal; + sign indicates tilt towards the centre of the knee and anterior translation.

little controversy and authors with a conservative approach admit to the need for reduction of the depressed fragment, be that by open or percutaneous means (Duwelius and Connolly 1988; Apley 1956; DeCoster *et al.* 1988).

The objective of this investigation was to assess the degree of stability of this semi-rigid type of osteosynthesis regarding re-dislocation of the elevated fracture segment. The results clearly show that by the time the plaster was discarded at 1 month, the fractures were stable and well adapted to support the strains caused by physiotherapy. Also, at the start of weight-bearing at 3 months the fractures had consolidated. In summary, the post-operative treatment was a safe one, which did not jeopardize healing, despite the fact that the time of immobilization was rather on the short side. Also, the clinical results were satisfactory. In three cases, however, a minute amount of re-dislocation of the depressed fragments occurred during the first month. Albeit small, these re-dislocations show that the stability of the osteosynthesis is critical. The method of osteosynthesis used in this study was conservative and additional stability could have been achieved by a condylar plate. The advantage of using only screws was less surgical trauma, as the split fragment did not have to be denuded of its periosteal attachment with ensuing risk of osteo-necrosis (Chaix *et al.* 1982; Duparc and Cavagna 1987; Duwelius and Connolly 1988) and pseudarthrosis (Blokker *et al.* 1984). Also, adverse effects on post-operative range of motion from excessive soft tissue damage was avoided.

With depressions exceeding 4–5 mm, lack of achieving and maintaining reduction has been shown to be the one most decisive factor linked to the development of secondary osteoarthrosis (OA) (Blokker *et al.* 1984; Vandenberghe *et al.* 1990). Such OA is often compatible with reasonable results within the study periods reported (Apley 1956) but has been linked to bad results in other studies (DeCoster *et al.* 1988; Vandenberghe *et al.* 1990;

Lachiewicz and Funcik 1990) and is a cause for concern in the long term (Duparc and Cavagna 1987). Apley (1956) has already stated that early motion might be the best way to make the damaged articular surfaces mould one another. Salter (1989) later showed the beneficial effect on cartilage of motion, though for another reason. The authors regard these results as encouraging regarding a more aggressive post-operative approach. Immediate motion exercises in a hinged brace (Vandenberghe *et al.* 1990; DeCoster *et al.* 1988) and weight-bearing after 1 month may be warranted. Such patients would have to be monitored carefully.

Debate in the meeting

How well did the articular cartilage perform after the operation? Were there any signs of continued degeneration?—There were no signs, but it is too short a period to judge with certainty. If the bone is devascularized perhaps it will not soften and collapse. Is it possible that some of these fragments of bone are devascularized due to the injury and that after a year they may revascularize, soften and collapse? Dead bone is very hard and appears normal on X-ray. The bone from the iliac crest is certainly devascularized and certainly dead at the time of application but the authors were confident that later collapse would not occur.

It would be useful to continue this experiment to the second and third year. It might even be that patients should be advized not to weight bear at this late stage. Could not RSA or even uniplanar measurements be taken to resolve this question?—Yes!

5 The influence of micromovement in experimental leg-lengthening

U. M. FIGUEIREDO, P. E. WATKINS, AND A. E. GOODSHIP

Introduction

Leg length discrepancy may develop as a result of many pathological conditions, congenital or acquired, and it represents a significant orthopaedic problem. A single bone or the entire limb may be affected. Correction of the inequality may be achieved by lengthening the short bone, shortening the longer bone, or by a combination of both. However, the idea of surgically shortening the normal bone is not generally accepted by the patients themselves. It has been suggested that some action should be taken to compensate for inequality greater than 2 cm (Kenwright 1989).

Distraction has, for a long time, been considered with some reservation due to the various complications associated with the procedure (Coleman and Scott 1991). These include delay in bone healing associated with the use of external fixation (Kenwright 1989).

Over time, new technical developments and biological concepts related to bone lengthening have been published. Stimulation of bone growth, diaphyseal osteotomy, methaphyseal corticotomy, percutaneous osteoclasia, callotasis, and chondrodiastasis, have been developed, but the most appropriate technique(s) have still to be identified. Gradual bone lengthening by callus distraction has been the preferred procedure in many circumstances. (Ilizarov 1989; Paley *et al.* 1989; Kawamura *et al.* 1981; Aldegheri *et al.* 1989.)

The concept of bone healing being influenced by the mechanical environment is well established (Kenwright and Goodship 1989). However, techniques for leg-lengthening have to a large degree been developed from a clinical basis with sometimes little, if any, consideration of the underlying biomechanical environment of the bone (Paterson 1990). It is now well recognized that bone is a mechanically sensitive tissue, both when intact (Lanyon 1987) and when disrupted (Goodship and Kenwright 1985). This has led to the application of controlled interfragmentary micromovement to accelerate fracture healing (Goodship and Kenwright 1985). These studies have shown clearly that the magnitude, direction, and timing of the applied load are all critical factors in influencing osteogenesis (Goodship *et al.* 1989; Kenwright and Goodship 1989).

The aim of the present research was to apply this technique for stimulating osteogenesis to the problem of experimental leg-lengthening. The hypothesis to be tested was that micromovement applied simultaneously with distraction would enhance the quantity and quality of new bone, compared with distraction alone.

Materials and methods

Experimental animals

Twelve young, but skeletally mature, Welsh ewes were used in this study. Their ages ranged from 2 years and 6 months to 6 years (mean 4 years), and their body weight ranged from 58 kg to 72 kg (mean 65 kg).

Experimental design

All animals were trained to walk over a force plate. A 3 mm tibial osteotomy was performed, and the bone fragments were supported using an Oxford Mark 2 external fixator. One week after surgery, distraction was started and continued for a period of 5 weeks. The animals were randomly divided into two equal groups:

Group A—static distraction
Group B—pulsed distraction.

After distraction, the animals were allowed to weight-bear for a further period of 6 weeks before euthanasia. Bone lengthening and healing were assessed during the 12 weeks after surgery, and post-mortem.

Surgical procedure

General anaesthesia was induced by intravenous sodium thiopentone (Intraval). Following endotracheal intubation, anaesthesia was maintained by a mixture of oxygen, nitrous oxide, and halothane delivered by a semi-closed circuit. An oesophago-ruminal tube was introduced to prevent the development of ruminal tympany. The animal was placed in dorsal recumbency and the right leg was prepared for surgery by shaving the hair and then cleansing the skin with povidine–iodine solution and hibitane.

The tibia was approached by a longitudinal incision performed on the cranial aspect of the leg, extending from 2 cm proximal to the tibial tubercle to 2 cm proximal to the tibiotarsal joint. The proximal annular ligament of the hock joint was severed to improve exposure. The belly of the peroneus tersius was retracted laterally and the tibia exposed. Using an engineered template, holes were made in the bone and careful tapping was followed by the insertion of 6 mm diameter screws, into both cortices of the tibia. An Oxford Mark 2 external fixator was attached to the screws inserted in the bone. The midpoint of the bone, between the two innermost screws was located by reference to the template, and the periosteum was cut transversely at the level of the osteotomy site. The neurovascular bundle was protected and the soft tissue retracted for the introduction of the Gigli saw. The bone was cut transversely following the orientation of the mark on the periosteum. The osteotomy site was enlarged by distracting the distal fragment so producing a 3 mm gap. This was maintained by a metal spacer attached to the fixator between the two innermost turrets.

The wound was closed by suturing the fascia using 2-0 dexon and the skin using 3-0 nylon. Dressing was applied using cotton wool and crepe bandage. All animals received post-operative analgesic (buprenorphine, temgesic). In addition they received prophylactic antibiotic (ampicillin) for 5 days consecutively.

Post-operative care

The animals were kept in a recovery room immediately after the operation. They were allowed to freely weight-bear after recovering from anaesthesia and were housed in loose boxes.

The Oxford Mark 2 external fixator

The fixator is a rigid unilateral frame composed of a square section base 230 mm long, joined via six clamping turrets to the six screws. The turrets are held rigidly onto the bar by means of screws and back plates. During the procedure of distraction the rigid back plates were replaced by sliding plates.

Distraction regime

The sheep were supported in a special chair, and the actuator from a hydraulic servo-controlled materials testing machine (DARTEC) was attached to the fixator.

In both groups, 0.8 mm of distraction was applied twice daily (Fig. 5.1). For animals in the static group (Group A) distraction was applied by linear extension over a period of 500 s. Animals in the pulsed group (Group B) received a superimposed cyclical micromovement regime, comprising 250 cycles at a frequency of 0.5 Hz, with a pulse amplitude of 0.08 mm and a maximum rate of movement of 40 mm per second. In both groups the maximum permitted force for distraction was 500 N. During each period of distraction, both the peak force and the energy required to produce distraction were measured.

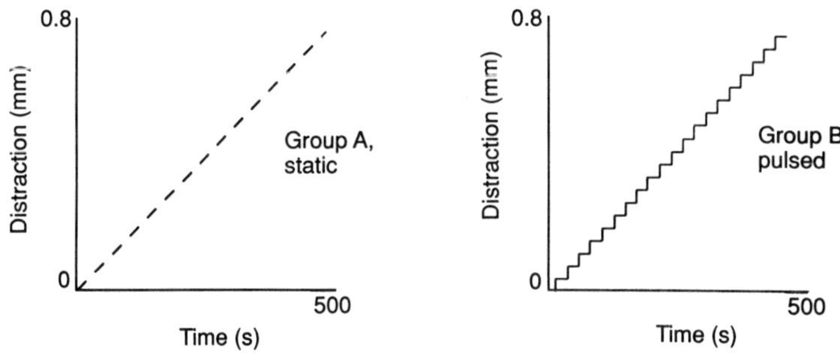

Fig. 5.1 Group A—distraction of 0.8 mm on each occasion by linear extension. Group B—distraction of 0.8 mm with superimposed micromovement regime.

In both groups distraction commenced 1 week after surgery and continued for 5 weeks. The animals were then allowed a further period of 6 weeks to permit consolidation of the new bone.

Assessment of healing and lengthening

Radiography

Standard radiographs of the osteotomy were taken every 2 weeks under general anaesthesia and the distance between the bone ends measured on the radiographs to evaluate the elongation achieved.

Bone mineral content (BMC)

Dual source photon absorption densitometry (Watkins *et al.* 1986) was used to measure BMC around the osteotomy at weeks 2, 4, 6, 8, 10, and 12 post-surgery. Scans were made in two axes at the osteotomy site.

Longitudinal scan. One scan was made along the axis of the tibia, passing from the proximal fragment, through the gap, into the distal fragment.

Transverse scan. Three scans were made along the cranio-caudal axis of the bone, at the level of the gap, through the proximal fragment at the edge of the gap, and through the distal fragment also at the edge of the gap.

Limb function

Restoration of limb function was assessed by measuring the ground reaction force acting through the operated limb during locomotion, using a Kistler force plate.

Post-mortem

Euthanasia was carried out at 12 weeks. Both tibiae were removed and measured using a vernier to quantify the amount of lengthening achieved. Both tibiae of the same animal were compared. For mechanical testing, each end of the bone specimen was embedded in a block of dental cement for fixation and mounted in the testing device with the long axis of the bone coinciding with the rotational axis of the testing machine. Torsional stiffness was evaluated by applying a torque of 16 N m to the tibiae and measuring the resultant angular displacement. The torsional stiffness of the operated tibia was expressed as a percentage of the stiffness of the contralateral (control) tibia from the same animal.

Sequential fluorchrome bone labelling was used at regular intervals to monitor new bone formation. Intravenous injection of tetracycline was given on the second and fourth weeks and alizarine on the eighth and tenth weeks post-operatively, in order to establish the time of bone deposition. Specimens removed from the new bone formation site were examined histologically.

Results

Radiography

Callus was initially visible 2 weeks after surgery. Distraction was seen to be greatest between the second and fourth weeks and it was greater in those animals receiving static distraction (Fig. 5.2). The mean elongation of the gap at the end of 6 weeks was 14.5 mm ($\pm$4.3 mm) for the pulsed group and 18.3 mm ($\pm$5.7 mm) for the static group.

The pattern and amount of bone formation showed little difference between the two groups. Consolidation was achieved by week 12 in 11 animals. One animal in Group B, sustained a fracture through the callus on removal of the fixator after euthanasia.

Bone mineral content

Both groups showed an increase in BMC throughout the healing process but there was no significant difference between the groups.

Limb function

As distraction progressed, the weight-bearing through the operated leg diminished, but once full distraction had been achieved, after 6 weeks, it increased gradually. However, by week 12 it had not returned to the pre-

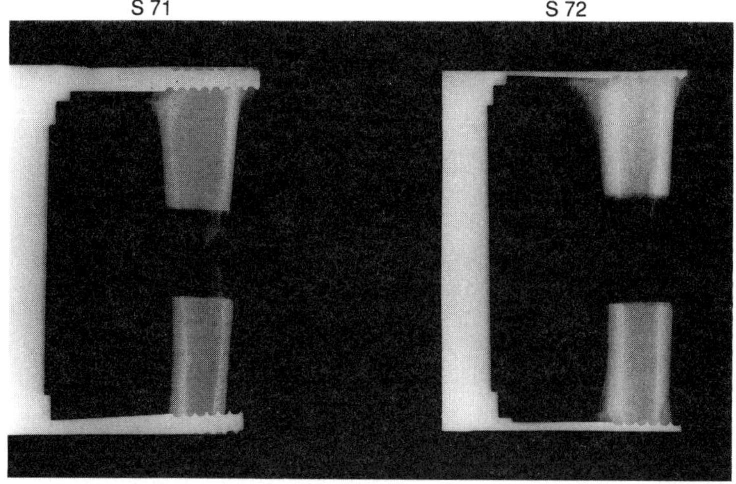

Fig. 5.2 Radiographs at 6 weeks post-operation. S71 (pulsed), S72 (static). Distraction was greater in the static group.

operative levels. This residual dysfunction may be attributed either to the acquired leg length discrepancy and/or to continuing consolidation.

Peak force and energy

There was no significant difference between the groups with respect to the peak force generated during distraction. Animals in the pulsed group showed a significantly greater rate of increase of energy required for distraction compared with those in the static group, as shown by regressional analysis ($p<0.05$).

Post-mortem

The overall length of the tibia was measured and the mean increase in the operated leg was 12.1 mm for the pulsed group and 16.8 mm for the static group. All tibiae were subjected to torsional stiffness evaluation. Mean torsional stiffness was greater for the pulsed group; however, the difference between the two groups was not statistically significant according to a matched pair *t* test (Fig. 5.3).

There was no significant difference between the groups regarding the fluorchrome distribution. Microradiographs showed significantly greater endosteal new bone formation ($p<0.022$) in the static group but there was no difference in the pulsed group comparatively.

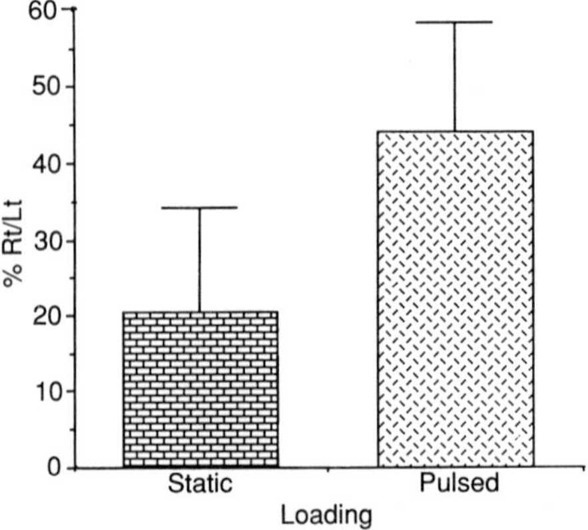

Fig. 5.3 Mean torsional stiffness (+ standard error) was greater for the pulsed group, compared with the static group.

Discussion

In this pilot experimental study, many of the criteria for assessing healing showed no significant difference between the two treatment groups. In the pulsed group, energy for distraction showed a significantly greater rate of increase compared with the static group. This taken along with the results of torsional stiffness measurements, would indicate that micromovement stimulates osteogenesis and strengthening during distraction. However, the X-ray measurements suggest that this may restrict the amount of distraction which is finally achieved.

The timing of micromovement in a distraction regime appears to be critical; to achieve satisfactory lengthening it may be argued that micromovement

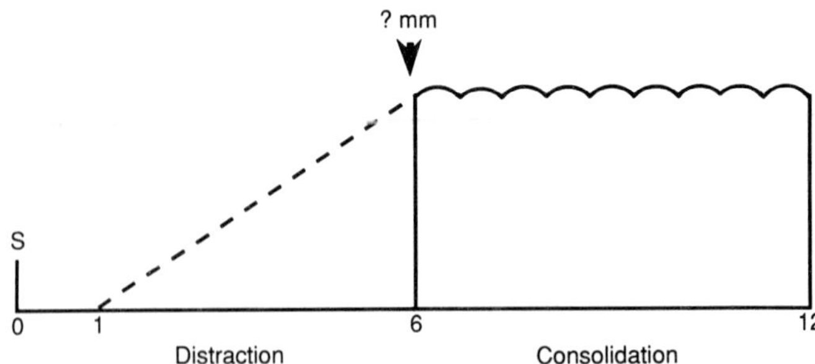

Fig. 5.4 In order to achieve satisfactory length it is suggested that micromovement should be applied once full distraction has been achieved, thereby enhancing the consolidation of the regenerate bone.

should be applied once full distraction has been achieved, thereby enhancing the consolidation of the regenerate bone (Fig. 5.4). This line of investigation warrants further examination.

Acknowledgement

This project was supported by 'Conselho Nacional De Pesquisa' Cnpq, Brazilian Government.

Debate in the meeting

With such a small magnitude of micromovement, the meeting wondered whether there was a genuine difference between the groups at all. It is easier to demonstrate that there *is* a difference between the two groups than it is to prove that there is *no* difference, and no effect due to micromotion. It is very difficult to control the precise mechanical environment which animals will impose on their fracture, and in point of fact the 500 steps of distraction per day can themselves be considered a micromovement of sorts. The paper was defended as part of a pilot study in which a range of mechanical regimes were being studied. It was thought that it might be possible to find a first cyclic regime which would inhibit bone healing, and so give more time for the distraction of soft tissues, and then apply a second, different, osteogenic micromovement to stimulate consolidation.

6 Controlling the load-bearing movement at externally-fixated tibial fractures

T. N. Gardner and M. Evans

Introduction

During the last decade or two, it has become more common to use unilateral external fixators for stabilizing difficult tibial fractures (comminuted or open) during healing, where perhaps the traditional method of plaster-casting has resulted in nonunion. In the early stages of healing, the stiffness of the fixator framework determines the mechanical environment at the fracture site where the new bone material develops. The influence of this environment on healing is discussed by Kenwright and Goodship (1989). 'Primary healing' arises from a fairly rigid environment, for example with internal fixation. Here, there is an absence of callus and because of the considerable remodelling of devitalized bone, the development of normal bone strength is delayed. 'Secondary healing' is initiated by a mechanical environment of less rigidity, as with unilateral external fixation. Here the fracture is first stabilized by the formation of callus as a bridging structure, which is then followed by a remodelling phase that replaces woven and lamella bone. Therefore, since the mechanical environment influences the process by which the fracture is repaired, it follows that the magnitude and direction of strains at the fracture site may determine the effectiveness of healing. This work therefore addresses the question of whether those fixators currently available restrict strain at the fracture site to within levels that are conducive to an efficient healing process.

Method

A displacement transducer has been developed that uses magnetic field 'Hall effect' devices to measure fracture movement for six degrees of freedom. It is clamped externally to fixator bone screws across the fracture and measures the 3-dimensional movement of the bottom bone fragment with reference to the top. The movement is measured over a period of time, during normal patient function. Figure 6.1 shows load and displacement curves in a 5-s walking period, for a patient fitted with a 'blue' monotube fixator (Howmedica International) at 18 weeks post-fracture. The load curve indicates the vertical reaction measured at a force plate as the patient walks over. Displacements are

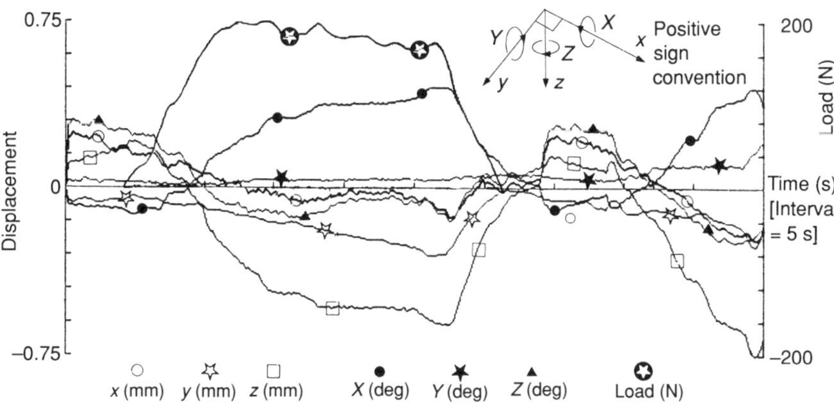

Fig. 6.1 Displacement and load curves vs time during walking.

recorded in six degrees of freedom and comprise three orthogonal directions (x, y, z mm), with rotation about each (X°, Y°, Z°). Direction z is measured along the longitudinal axis of the bone, while x and y are approximately medial and anterior, respectively. The fracture is nonunited and the bone fragments are end-bearing, having displaced towards each other by approximately 0.75 mm, to close the fracture gap (z). Therefore axial strain is around 100 per cent. Also, there is a 0.5° peak rotation about the x axis due to bending of the main bar of the fixator, since it is eccentrically loaded by the force through the longitudinal axis of the tibia. All unilateral fixator frames are susceptible to this form of displacement. In addition to linear axial displacement there is shear movement in line with the bone screws, peaking at around 0.3 mm. In view of the magnitude of axial fracture strain permitted by the fixator frame, and the angular displacement it imposes at the fracture, the question arises—are fixator frames more flexible than they need to be?

Axial stiffness was measured for six different fixator frames in a geometric configuration typical in clinical practice. Stiffness here is the compressive force required along the axis of the bone to compress the fracture site by 1 mm, where resistance is provided only from the fixator/screw framework with no contribution from the material at the fracture. Reading from the left in Fig. 6.2, the fixators are red (How-R) and blue (How-B) monotubes (Howmedica International), Lazo-rolling rod (Hoffmann/Howmedica International), Modulsystem (Orthofix), Dynabrace (Oxford), and Unifix (AO). Numbers contained in the fixator abbreviations refer to the use of either 5- or 6-mm diameter bone screws.

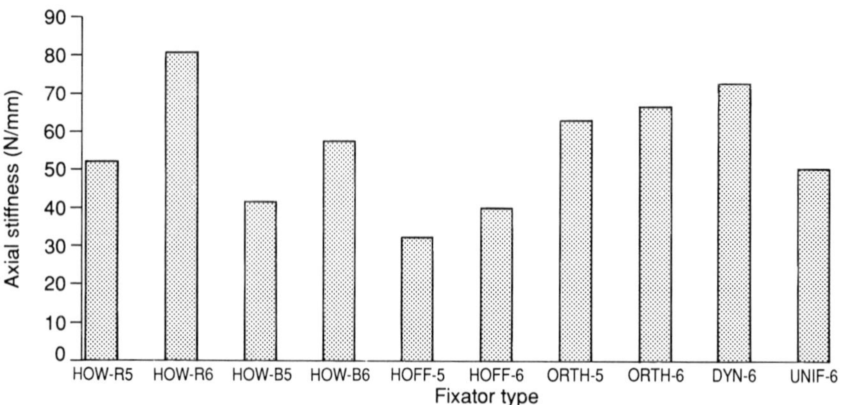

Fig. 6.2 Axial stiffness tests on commercially available fixators.

Results

The frame stiffness provided by the blue monotube with 6-mm screws is approximately 60 N/mm and this is midway between the upper and lower bounds of the 6-mm systems. For the 200 N peak load in Fig. 6.1, this means that the fixator frame, when not assisted by the fracture material, will allow a compressional displacement of 3.3 mm at the fracture site. Since actual displacement is less than 0.75 mm, further compression is avoided through the support provided by the end-bearing bone fragments. It is, therefore, the developing bone at the fracture site that resists almost all the self-weight loading on the tibia. The nature of this load sharing disparity is illustrated in Fig. 6.3, which predicts the percentage load on the fracture as a function of the

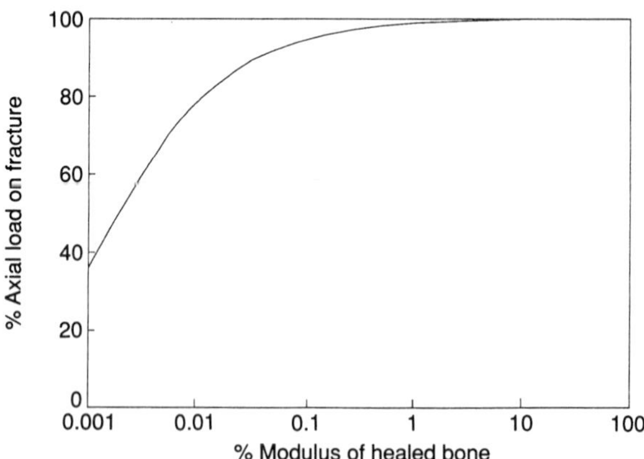

Fig. 6.3 Axial load bearing of healing bone.

elastic modulus of the developing bone material. Here the average axial stiffness of the six fixator frames is used and it is assumed for healed bone that the elastic modulus in compression is 28 GPa and that the cross-sectional area and gap length at the fracture are 300 mm^2 and 3 mm, respectively. It can be seen that 80 per cent of the load on the tibia is taken by the fracture material before the modulus of this tissue has achieved 0.01 per cent of its healed condition.

Discussion

What is the effect of the level of strain suggested by Fig. 6.1 and the disparity in load-bearing suggested by Fig. 6.3? Kenwright and Goodship (1989) show in Fig. 6.4 that this level of strain, and therefore load-bearing, may be detrimental to the healing process. For their animal trials, 3-mm gaps were created by standardized mid-diaphyseal tibial osteotomy in three groups of six sheep. A control group was fixated fairly rigidly, while the other two groups were axially dynamized by 0.5 mm and 2 mm for 17 min at 1 Hz and fracture healing progress was assessed by dual photon absorption densitometry and stiffness tests. Significant improvement in the rate of mineralization and stiffening was demonstrated in the group displaced by 0.5 mm (17 per cent axial strain at the osteotomy), but a reduced rate was demonstrated in the group displaced by 2 mm (67 per cent strain). Clearly, the level of strain in the early stages of fracture healing should be sufficient to stimulate tissue development but should not exceed a level where strain inhibits the healing process. Although it is not known why the 67 per cent strain has slowed development, an interesting parallel may be drawn from the testing of nonbiological specimens. Here, detrimental effects are caused by allowing

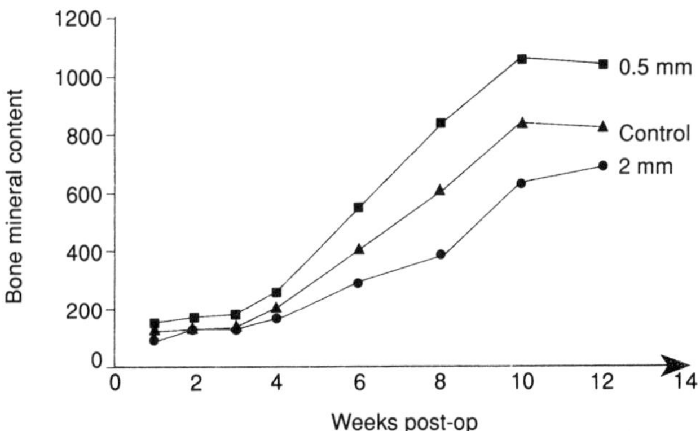

Fig. 6.4 Bone mineral content for three regimes of controlled axial micromovement (Kenwright and Goodship 1989).

strain to exceed the 'elastic limit' of the material, thereby causing irreversible change. This change could be considered as damage for biological specimens that have a similar stress–strain response.

Since the fracture site material will be increasing in stiffness as healing progresses, the levels of strain that will be beneficial to development are likely to decrease. Figure 6.5 illustrates why this is probably the case. The idealized load–displacement curves for this material at 8, 12, 16, and 20 weeks post-operation have been simplified to an initially linear-elastic, then a plastic behaviour. This profile is similar to the idealized stress–strain curve for compression, illustrated by Carter and Spengler (1978). It can be seen that the maximum elastic displacement occurs at the limit of the linear-elastic response (the elastic limit). Therefore, if maximum displacement is not exceeded during the loading cycle, the material will recover its initial condition when load is removed. Thus it may be possible to confine the axial displacement of the fracture material, during self-weight loading, to within the limit of linear elasticity on each curve (at each stage in the healing process). This may provide a strain environment that stimulates the rapid development of new bone while avoiding the levels of strain that cause plastic deformations and may inhibit healing. Figure 6.6 shows a diagram of the probable relationship between the increasing stiffness, from Evans *et al.* (1988), and the maximum strains from Perren and Cordey (1977) as healing progresses.

There are two difficulties in adopting the above procedure for fracture management. The first is the method of predicting safe displacement at stages during healing. The second is developing a fixator framework that can be pre-set to control maximum axial displacement at the fracture within the required safe limits of strain (while stabilizing against shear (x, y) and angular displacements (X, Y, Z)). Accurate prediction of the safe displacement is probably impossible since the initial fracture gap is usually nonuniform in its extent and material properties, which means that the value of strain cannot be

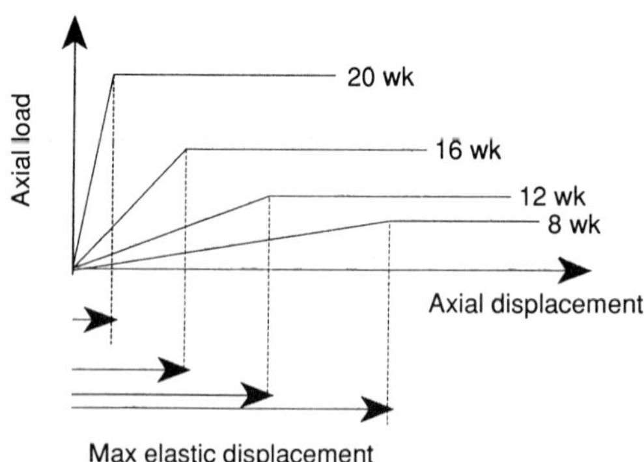

Fig. 6.5 Idealized load–displacement characteristics during bone healing.

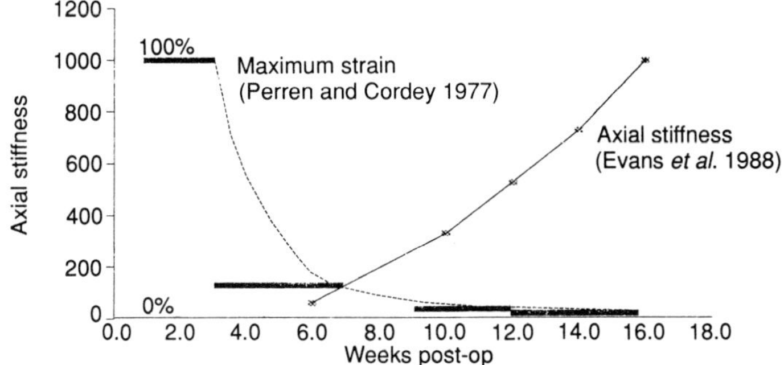

Fig. 6.6 Developing stiffness and maximum strain during bone healing.

found because it is a function of initial length. Here, it should be possible to measure the effective fracture gap by measuring displacement at the fracture when axial loading causes the fragments to end-bear. The maximum strains provided by Perren and Cordey (1977) may then be used to supply the required safe displacements, within the elastic limit of the material, for the appropriate stage in the healing process.

A fixator framework that can control axial displacement within adjustable limits requires a rigid base frame. Frames that compress and angulate at the fracture site under load-bearing do not allow the clinician to control strain. Here, strain is largely a function of the patient's willingness to weight-bear on the fracture. Therefore, patients who are not tentative in weight-bearing are likely to cause the higher levels of strain that inhibit, rather than enhance, healing and patients who are tentative weight-bearers may not cause sufficient strain to stimulate healing. Although developing bone may respond to the greater challenge initiated by damage due to inhibitive strains, at some threshold of strain the rate and extent of damage may exceed its capacity to heal. Since the repair process is a prolonged period anyway, the effect of extending the process and increasing the occurrence of nonunion may not be apparent in such circumstances.

Since none of the fixator frames tested provided the regime of stiffness needed, two blue monotube fixators were connected to the same set of four bone screws to provide a virtually rigid base frame on which the required level of axial displacement might be superimposed. Figure 6.7 shows the arrangement used where the tubing represents the two bone fragments each side of the fracture site. The combination of the two fixators provides a frame that has negligible flexibility in angular mode (X, Y, Z) and shear mode (x, y), and that is virtually rigid in axial mode (z). Since both fixator bars may be set to provide a spring-loaded telescoping action under axial load (adjustable between 0 and 3 mm), the frames may be used to provide a range of maximum axial displacements at the fracture to accommodate the reducing safe tissue

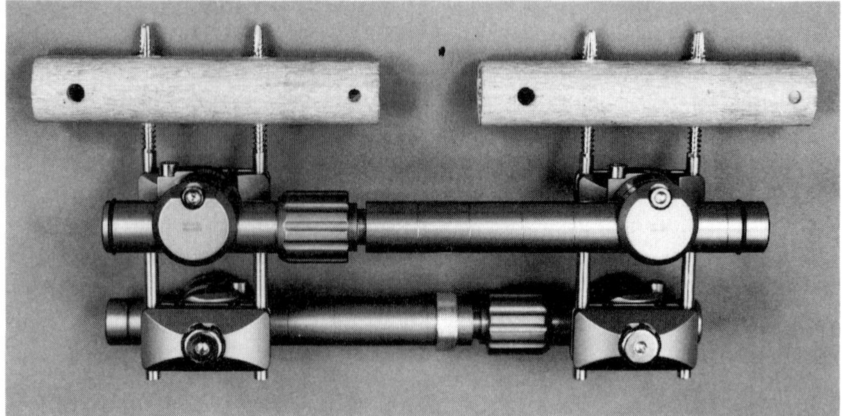

Fig. 6.7 Double 'Bi-tube' fixator arrangement.

strains throughout the healing period. Therefore, the magnitude of maximum axial strain allowed at the fracture may be fixed by the clinician at regular intervals, and remains unaffected by the patient's willingness to weight-bear on the limb. Tentative weight-bearing patients are accommodated by low spring pressures which ensure that maximum strains are reached for low levels of weight-bearing, providing that full recovery of gap size is achieved on unloading; more willing weight-bearing patients cannot strain the fracture beyond the maximum fixed by the clinician.

The clinical trials now under way at the Oxford Orthopaedic Engineering Centre will incorporate an investigation of the double fixator system. Load–displacement performance during walking will be used to determine whether this approach to fracture management is likely to be successful.

Acknowledgements

This work has been funded by Howmedica International.

Debate in the meeting

The double tube frame is very stiff in all six degrees of freedom and allows a limit of axial displacement to be set by the clinician. The single tube will provide a strain at the fracture site, which is a function largely of the weight that the patient is prepared to put on it. Some patients who are very tentative in their weight bearing, put next to no load on, and produce next to no strain. Other patients over-strain their fractures by causing 100 per cent strain, allowing the fragments to end-bear across the fracture. In this case the mechanical environment of the fracture is not being controlled by the fixation but rather by the patient, who is determining the level of strain there. The double fixator would allow the clinician to determine strain rather than the patient.

Why is it an advantage to be able to adjust the amount of axial displacement?—During healing the bone is stiffening, so the level of strain should be reduced at the fracture site over the healing period. If strain is controlled to only one value, at some point as the material stiffens it will be strained beyond its elastic limit and be damaged. A reasonable degree of challenge will promote bone healing but a stage will come when the rate of damage will exceed the capacity of the bone to repair.

7 Workshop—fracture repair

Chaired by A. E. Goodship

A theme that has been running through the preceding contributions is the interaction of biology and mechanics, and now we consider the principles of biomechanics in relation to fracture healing. We understand the biology of hard tissue repair in basic terms: engineers understand the mechanics of fixation frames, surgeons understand the clinical problems, biologists understand interactions at the cellular level. But conversely, surgeons do not know what they want from their mechanical frames, biologists are not quite sure how their models relate to exact clinical situations, and engineers have been unable to produce a system to deliver accurate mechanical control of micromovement at the fracture site.

Gardner presented an engineer's point of view of the mechanical environment for fracture healing (Chapter 6, p. 40). He showed that an external fixator could be provided which could control the magnitude of strain at the fracture site to within the elastic limit of the interfragmentary material at all stages of its development, from initial callus formation to final consolidation. Although this seems sensible, some thought that rather than a fixation device, perhaps the biological system should be allowed to control the strain that is imposed at each stage of healing.

A simple mechanical model of healing bone may not be appropriate because strains occur across a very heterogeneous group of tissues. If rigid control theory were applied to healing in plaster casts, none would heal at all because the evidence in the literature is that fractures within casts move 2–3 mm during the first 2–3 weeks of healing.

If a fracture responds to strain this presumably means that there is an optimum strain to promote fracture healing. With rigid frames, the ground reaction force is high. With more flexible frames, the ground reaction is smaller initially and gradually builds up. Results from the Bristol laboratory indicate that if stimulation is imposed early it will provide for better healing than if the fracture is initially protected and the load-bearing is brought in later. Micromovement is delayed to 7 days after surgery purely as a feature of the experimental model in order to reduce risk of soft tissue damage. It may be that an even earlier start for micromovement would be advantageous.

An analysis of patients in Oxford who had micromovement imposed within 2 weeks of fracture showed that they did better than those receiving micromovement later than 2 weeks. There is no evidence that this movement does the soft tissues any harm, but nevertheless, if patients have had serious injury, micromovement is not inflicted until later, when it is given as part of a regime of physiotherapy. A common anecdotal observation is that lively, outgoing patients who insist on taking exercise usually do well, whereas cautious patients who will not self-load have delayed healing. It is this second group who would benefit from having more highly controlled micromovement.

There may be a neurological feedback loop from the fracture that is modulated by higher levels. It would be interesting to know whether modulation of this overall behaviour pattern would indeed change fracture repair rates. Is the failure of a fracture to unite in a minority of patients associated with a failure of the biological feedback mechanism, imposing either too great or too small a strain? If it is, it could be that it is in this group alone that the clinician should take over control of the strain regime at the fracture. If there is to be interference with the feedback loop then a measure of union is needed as well as a measure of displacement at the fracture. Some measures have been presented in the preceding contributions. Strain may not always be an appropriate measure. In some fractures there is a significant gap of perhaps 2 mm in which many results seem to indicate that a strain of about 15–30 per cent, will stimulate healing. However, there are other patients in whom the fractures are in apposition, with fragment end-bearing rather than micromovement or strain.

Are we in the position where we can say we know what the osteogenic strain in a gap is? Or should we take each case individually and look at the configuration and orientation of the fracture line and work out from the patient's weight and their likely psychological status what kind of fixation and stiffness we should apply?—The latter is probably not possible because of the heterogeneous nature of the tissues and fixation system, but neither is it necessary in practice as long as some other aspect of healing can be measured. The best measure of healing is the restoration of mechanical integrity. How is healing to be assessed? What is the end-point?

Some suggest that stiffness correlates with strength, others deny that it does. In the end the patient's question is 'How soon is it safe to use my leg again?' There may be a relationship between stiffness and structural strength that is adequate for normal clinical use but there is by no means a one-to-one correlation between the two. When the stiffness of a healing fracture, as measured by a four point bending test, equals that of the normal contralateral bone there is still a vast difference in strength between the two.

It is probably important to distinguish material properties from structural properties. Bone structure seems to arrange itself to best cope with the magnitude and orientation of the forces that are applied to it, in a similar way to the application of Wolff's Law for intact bone. Are the resonance and acoustic techniques independent of the structural arrangement? These measurements often indicate that stiffness increases rapidly in one plane but not in another so perhaps we should be measuring mechanical parameters during a normal physiological activity such as walking. The single measure of axial stiffness is not a particularly useful measure of healing as stiffness increases exponentially during the period in which a fixator is applied. Resistance to bending or some other plane of movement may be indicative of healing.

The material properties of callus may not be homogeneous, quite apart from the shape being variable. Just as the mechanical environment of the whole fracture influences its healing, the local microscopic mechanical environment may determine which tissue type will exist in each region of callus. There is a complex interaction between tissue type and mechanical properties which, if it were understood, would enable optimum control of the healing process. A number of different monitoring techniques seem to be emerging now as alternatives to radiographs that can measure a mechanically-related property.

There is evidence that osteoblasts can be stimulated to produce extra mineral and extra matrix proteins. Callus formation is very efficient. At every point in the healing process the system appears to be examining itself to see whether there is sufficient callus in order to produce more if needed. It is unlikely that a pill (biochemical control alone) will ever be able to control such a system. There is no point in inducing bone unless it exists in an environment in which the mechanical factors will allow it to persist. Fractures are always going to have to have a mechanical input, and modifying that mechanical input will always make a difference to healing.

Is it still a viable proposition to accelerate fracture healing by mechanical means? Is it even unnecessary, because the main clinical problem is those patients who hardly heal at all, rather than those who heal at a normal rate? Even if we cannot yet accelerate the natural process of fracture repair, however, we are on the way to reducing its inhibition.

References to Section 1

Abendschein, W. F. and Hyatt, G. W. (1972). Ultrasonics and physical properties of healing bones. *Journal of Trauma*, **12**, 297–301. *Abstracts of the European Society of Bioengineers*.

Aldegheri, R., Renzi-Brivio, L., and Agostini, S. (1989). The callotasis method of limb lengthening. *Clinical Orthopaedics and Related Research*, **241**, 137–46.

An, K.-N., Kasman, R. A., and Chao, E. Y. S. (1988). Theoretical analysis of fracture healing monitoring with external fixators. *Engineering in Medicine*, **17**, 11–15.

Andrews, A. H. (1903). Tuning fork and the stethoscope in the diagnosis of fractures. *Chicago Medical Recorder*, **24**,182–6.

Apley, A. G. (1956). Fractures of the lateral tibial condyle treated by skeletal traction and early mobilization. *Journal of Bone and Joint Surgery*, **38B**, 699–708.

Aro, H., Eehola, E., and Aho, A. J. (1985). Development of nonunions in the rat fibula after removal of periosteal neural mechanoreceptors. *Clinical Orthopaedics and Related Research*, **199**, 292–9.

Ashhurst, D. E. (1986). The influence of mechanical conditions on the healing of experimental fractures in the rabbit a microscopical study. *Philosophical Transactions of the Royal Society, London*, **B313**, 271–302.

Bastiani, G. de, Aldegheri R., and Brivio, L. R. (1984). The treatment of fractures with a dynamic axial fixator. *Journal of Bone and Joint Surgery*, **66B**, 538–45.

Bidner, S. M., Rubins, I. M., Desjardins, J. V., Zukor, D. J., and Goltzman, D. (1990). Evidence for a humoral mechanism for enhanced osteogenesis after head injury. *Journal of Bone and Joint Surgery*, **72A**, 1144–9.

Blokker, C. P., Rorabeck, C. H., and Bourne, R. B. (1984). Tibial plateau fractures. *Clinical Orthopaedics and Related Research*, **182**, 193–9.

Bonnin, J. G. (1942). Time of union in the fractured tibia. *British Medical Journal*, **1**, 440.

Brighton, C. T. and Krebs, A. G. (1972). Oxygen tension of non-union of fractured femurs. *Surgery, Gynecology and Obstetrics*, **135**, 379–85.

Burny, F., Bourgois, R., Dankerwolcke, M., and Moulart, F. (1978). Utilisation clinique de jauge de constrainte—situation actuelle et perspectures d'avenir. *Acta Orthopaedica Belgica*, **44**, 895–920.

Carter, D. R. and Spengler, D. M. (1978). Mechanical properties and composition of cortical bone. *Clinical Orthopaedics and Related Research*, **135**, 192–217.

Cavadias, A. X. and Trueta, J. (1965). An experimental study of the vascular contribution to the callus of fracture. *Surgery, Gynecology and Obstetrics*, **20**, 731–47.

Chaix, O., Herman, S., Cohen, P., LeBalch, T., and Lemare, J.-P. (1982). Osteosynthese par plaque epiphysaire dans les fractures des plateaux tibeaux. *Revue de Chirurgie et Reparatrice de L'appareil Moteur*, **68**, 189–97.

Chao, E. Y., Kasman, R. A., and An, K. N. (1982). Rigidity and stress analyses of external fracture fixation devices: a theoretical approach. *Journal of Biomechanics*, **15**, 971–84.

Cheal, E. J., Mansmann, K. A., DiGioia, A. M., Hayes, W. C., and Perren, S. M.

(1991). Role of interfragmentary strain in fracture healing: ovine model of a healing osteotomy. *Journal of Orthopaedic Research*, **9**, 131–42.

Churches, A. E., Tanner, K. E., and Harris, J. D. (1985). The Oxford external fixator: fixator stiffness and the effects of bone pin loosening. *Engineering in Medicine*, **14**, 3–11.

Claes, L. (1989). The mechanical and morphological properties of bone beneath internal fixation plates of differing rigidity. *Journal of Orthopaedic Research*, **7**, 170–7.

Clifford, R. P., Lyons, T. J., and Webb, J. K. (1987). Complications of external fixation of open fractures of the tibia. *Injury*, **18**, 174–6.

Coleman, S. S. and Scott, S. M. (1991). Biology and technology of limb lengthening. *Clinical Orthopaedics and Related Research*, **264**, 76–83.

Count-Brown, C. M. and McQueen, M. M. (1987). Compartment syndrome delays tibial union. *Acta Orthopaedica Scandinavica*, **58**, 249–52

Danis, R. (1949). *Theorie et pratique de l'ostéosynthèse*. Masson et Cie, Paris.

DeCoster, T. A., Nepola, J. V., and El-Khoury, G. Y. (1988). Cast-brace treatment of proximal tibia fractures: a 10-year follow-up study. *Clinical Orthopaedics and Related Research*, **231**, 196–204.

Doemland, H. H., Jacobs, R. R., Spence, J., and Roberts, F. G. (1986). Assessment of fracture healing by spectral analysis. *Journal of Medical Engineering and Technology*, **10**, 180–7.

Duparc, J. and Cavagna, R. R. (1987). Resultat du operatoire des fractures de plateaux tibiaux (à propos de 110 cas). *International Orthopaedics*, **11**, 205–13.

Duwelius, P. J. and Connolly, J. F. (1988). Closed reduction of tibial plateau fractures. *Clinical Orthopaedics and Related Research*, **230**, 116–26.

Edholm, P., Hammer, R., Hammerby, S., and Lindholm, B. (1984). The stability of union in tibial shaft fracture: its measurement by a non-invasive method. *Archives of Orthopaedic and Traumatic Surgery*, **102**, 242–7.

Egger, E. L., Gottsauner-Wolf, F., Aro, H. T., Markel, M. D., and Chao, E. Y. S. (1991). Effect of dynamisation after one week on healing of a delayed union model. *Transactions of the Orthopaedic Research Society*, **15**, 413–14.

Evans, M. (1984). Fracture monitoring transducer. *Oxford Orthopaedic Engineering Centre Annual Report*, **11**, 64–6.

Evans, M., Kenwright, J., and Cunningham, J. L. (1988). Design and performance of a fracture monitoring transducer. *Journal of Biomechanical Engineering*, **10**, 64–9.

Frost, H. M. (1964). *The laws of bone structure*. Charles C. Thomas, Illinois.

Gilbert, J. A., Dahners, L. E., and Atkinson, M. A. (1989). The effect of external fixation stiffness on early healing of transverse osteotomies. *Journal of Orthopaedic Research*, **7**, 389–97.

Goodship. A. E. and Kenwright, J. (1985). The influence of induced micromovement upon the healing of experimental tibial fractures. *Journal of Bone and Joint Surgery*, **67B**, 650–5.

Goodship, A. E., Adams, M. A. and Kenwright, J. (1989). The influence of mechanical stimulation on different stages of fracture healing. *Transaction of the 35th Annual Meeting of the Orthopaedic Research Society*, p. 592.

Gross, D. and Williams, W. S. (1982). Streaming potential and the electromechanical response of physiologically-moist bone. *Journal of Biomechanics*, **15**, 277–95.

Gross, P. M., Marcus, M. L., and Heistad, D. D. (1981). Measurement of blood flow to bone and marrow in experimental animals by means of the microsphere technique. *Journal of Bone and Joint Surgery*, **63A**, 1028–31.

Hammer, R., Edholm, P., and Lindholm, B. (1984). Stability of union after tibial shaft fracture. *Journal of Bone and Joint Surgery*, **66B**, 529–34.

Henry, A. N., Freeman, M. A. R., and Swanson, S. A. V. (1968). Studies on the mechanical properties of healing experimental fractures. *Proceedings of the Royal Society of Medicine*, **61**, 40–4.

Heppenstall, R. B., Goodwin, C. W., and Brighton, C. T. (1976). Fracture healing in the presence of chronic hypoxia. *Journal of Bone and Joint Surgery*, **58A**, 1153–6.

Hudlicka, O. and Tyler, K. R. (1986). *Angiogenesis: the growth of the vascular system*. Academic Press, London.

Hughes, S., Khan, R., Davies, R., and Lavender, P. (1978). The uptake by the canine tibia of the bone-scanning agent Tc-99m MDP. *Journal of Bone and Joint Surgery*, **60B**, 579–82.

Ilizarov,G. (1989). Experimental studies of bone elongation. In *External fixation and functional bracing* (ed. R. Coombs, S. Green, and A. Sarmiento), pp. 375–9. Orthotext.

Jernberger, A. (1970). Measurement of stability of tibial fractures. A mechanical method. *Acta Orthopaedica Scandinavica (Suppl.)*, **135**.

Jírgensen, T. E. (1972). Measurement of stability of crural fractures of the tibia treated with Hoffman osteotaxis. *Acta Orthopaedica Scandinavica (Suppl.)*, **43**, 207–18, 264–79, 280–91.

Kawamura, B., Hosono, S., and Takahashi, T. (1981). The principles and techniques of limb lengthening. *International Orthopaedics*, **5**, 69–83.

Kempson, G. E. and Campbell, D. (1981). The comparative stiffness of external fixation frames. *Injury*, **12**, 297–304.

Kennedy, J. C. and Bailey, W. H. (1968). Experimental tibial-plateau fractures. Studies of the mechanism and a classification. *Journal of Bone and Joint Surgery*, **50A**, 1522–34.

Kenwright, J. (1989). Leg-length inequality. *Current Orthopaedics*, **3**, 176–82.

Kenwright, J. and Goodship, A. E. (1989). Controlled mechanical stimulation in the treatment of tibial fractures. *Clinical Orthopaedics and Related Research*, **241**, 36–47.

Kenwright, J., Richardson, J. B., Cunningham, J. L., White, S. H., Goodship, A. E., Adams, M. A. *et al.* (1991). Axial movement and tibial fractures. *Journal of Bone and Joint Surgery*, **73B**, 654–9.

Lachiewicz, P. F. and Funcik, T. (1990). Factors influencing the results of open reduction and internal fixation of tibial plateau fractures. *Clinical Orthopaedics and Related Research*, **259**, 210–15.

Lambotte, A. (1907). *L'intervention operatoire dans les fractures récentes et anciennes*. Henri Lamertin, Bruxelles.

Lane, W. A. (1914). *The operative treatment of fractures* (2nd edn). The Medical Publishing Company, London.

Lansinger, O., Bergman, B., Korner L., and Andersson, G. B. J. (1986). Tibial condylar fractures. *Journal of Bone and Joint Surgery*, **68A**, 13–19.

Lanyon, L. E. (1987). Functional strain in bone tissue as an objective and contributing stimulus for adaptive bone remodelling. *Journal of Biomechanics*, **20**, 1083–93.

Li, G., Bronk, J. T., and Kelly, P. J. (1989). Canine bone blood flow estimated with microspheres. *Journal of Orthopaedic Research*, **7**, 61–7.

Linden,W. van der and Larson, K. (1975). Plate fixation versus conservative treatment of tibial shaft fractures. *Journal of Bone and Joint Surgery*, **57A**, 321–7.

Lindsay, M. K. and Howes, E. L. (1931). The breaking strength of healing fractures. *Journal of Bone and Joint Surgery*, **13**, 491.

Markel, M. D., Wilkenheiser, M. A., and Chao, E. Y. (1990). A study of fracture callus material properties relationship to the torsional strength of bone. *Journal of Orthopaedic Research*, **8**, 843–50.

Mathew, L. S., Kaufer, H., and Sonsdegard D. A. (1974). Manual sensing of fracture stability. *Acta Orthopaedica Scandinavica*, **45**, 373–81.

McCarthy, I. D. and Hughes, S. P. F. (1984). Extraction of Tc-99m methylene diphosphonate as a function of bone blood flow. In *Bone circulation* (ed. J. Arlet, R. P. Ficat, and D. S. Hungerford), pp. 167–70. Williams and Wilkins, Baltimore.

McDougall, I. R. and Keeling, C. A. (1988). Complications of fractures and their healing. *Seminars in Nuclear Medicine*, **18**, 113–25.

McKibbin, B. (1978). The biology of fracture healing in long bones. *Journal of Bone and Joint Surgery*, **60B**, 150–62.

Mohler, D. G., Cohen, A. M., Fehnel, J. M., Lane, J. M., and Tomin, A. (1990). Effect of basic fibroblast growth factor on angiogenesis and calcification of rat femoral defect. *Transactions of the Orthopaedic Research Society*, **15**, 380.

Molster, A. O. and Gjerdet, N. R. (1984). Effects of instability on fracture healing in the rat. *Acta Orthopaedica Scandinavica*, **55**, 342–6.

Nicholls, P. J., Beng, E., Bliven, F. E. Jr, and Kling, J. M. (1979). X-ray diagnosis of healing fractures in rabbits. *Clinical Orthopaedics and Related Research*, **142**, 234–6.

Ory, M. (1964). Des influences mécaniques dans l'apperition et le dévelopement des manifestations dégénératives du genou. *Journal Belge de Medicine Physique*, **19**, 103–20.

Page, C.M. (1942). Time to union in the fractured tibia. *British Medical Journal*, **1**, 305.

Paley, D., Catagni, M., Argnani, F., Villa, A., Benedetti, G. B., and Cattaneo, R. (1989). Ilizarov treatment of tibial non-union with bone loss. *Clinical Orthopaedics and Related Research*, **241**, 146–65.

Paterson, D. (1990). Leg-lengthening procedures: a historical review. *Clinical Orthopaedics and Related Research*, **250**, 27–33.

Perren, S. M. (1979). Physical and biological aspects of fracture healing with special reference to internal fixation. *Clinical Orthopaedics and Related Research*, **138**, 175–96.

Perren, S. M. and Cordey, J. (1977). Die Gewebsdifferenzierung in der Frakturheilung. Monatsschrift. *Unfallheilkunde*, **80**, 161–4.

Perren, S. M. and Rahn, B. A. (1980). Biomechanics of fracture healing. *Canadian Journal of Surgery*, **23**, 228–32.

Radin, E. L., Burr, D. B., Fyhrie, D., Brown, T. D., and Boyd, R. D. (1990). Characteristics of joint loading as it applies to osteoarthritis. In *Symposium on biomechanics of diarthrodial joints*, Vol. 1 (ed. V. C. Mow, S. L.-Y. Woo, and T. Radcliffe), pp. 437–51. Springer-Verlag, New York.

Rasmussen, P. S. (1971) A functional approach to evaluation and treatment of tibial condylar fractures. Thesis. Goteborg University, Sweden.

Richardson, J., Evans, M., Tew, A. I., and Harris J. D. (1984). Fracture stiffness measurement. *Oxford Orthopaedic Engineering Centre Annual Report*, **11**, 69–72.

Richardson, J. B., Kenwright, J., and Cunningham, J. L. (1992). Fracture stiffness in the assessment and management of tibial fractures. *Journal of Biomechanics*, **25**, 75–9.

Ryd, L. (1986). Micromotion in knee arthroplasty. A Roentgen stereophotogrammetric analysis of tibial component fixation. *Acta Orthopaedica Scandinavica*, **51** (*Suppl*. 220).

Rymaszewski, L. A. (1984). Preliminary clinical investigation into the use of 'flexigage' as a sensor to monitor tibial fracture healing. Thesis submitted to the University of Strathclyde, Glasgow for the Degree of MSc in the Department of Physiology and Pharmacology.

Salter, R. B. (1989). The biologic concept of continuous passive motion of synovial joints. The first 18 years of basic research and its clinical application. *European Surgical Research*, **242**, 12–25.

Schatzker, J., McBroom, R., and Bruce, D. (1979). The tibial plateau fracture. The Toronto experience. *Clinical Orthopaedics and Related Research*, **138**, 94–104.

Selvik, G. (1974). A Roentgen stereophotogrammetric system for the study of the kinematics of the skeletal systems. Thesis. University of Lund, Sweden. Reprint in: (1989). *Acta Orthopaedica Scandinavica*, **60** (*Suppl*. 232).

Simon, L. R., Radin, E. L., Paul, I. L. and Rose, R. M. (1972). Response of joints to impact loading, II: *in vivo* behaviour of subchondral bone. *Journal of Biomechanics*, **5**, 267–72.

Smith, M. A., Jones, E. A., Strachan, R. K., Nicoll, J. J., Best, J. J., Tothill, P. *et al.* (1987). Prediction of fracture healing in the tibia by quantitative radionucleotide imaging. *Journal of Bone and Joint Surgery*, **69B**, 441–7.

Strachan, R. K., McCarthy, I. D., Fleming, R. H., and Hughes, S. P. F. (1990). The role of the tibial nutrient artery. *Journal of Bone and Joint Surgery*, **72B**, 391–4.

Tanner, K. E. (1985). The design of a fracture movement transducer. Thesis submitted for the degree of D.Phil. University of Oxford.

Trueta, J. (1963). The role of the vessels in osteogenesis. *Journal of Bone and Joint Surgery*, **45B**, 402–18.

Vandenberghe, D., Cuypers, L., Rombouts, L., Van Dooren, J., and Rombouts, J. (1990). Internal fixation of tibial plateau fractures using the AO instrumentation. *Acta Orthopaedica Belgica*, **56**(2), 431–42.

Wallace, A. L., McLaughlin, B., Weiss, J. B., and Hughes, S. P. F. (1991*a*). Endothelial cell stimulating angiogenesis factor in patients with tibial fractures. *Injury*, **22**, 375–6.

Wallace, A. L., Draper E. R. C., Strachan, R. K., McCarthy, I. D., and Hughes, S. P. F. (1991*b*). The effect of devascularisation upon early bone healing in dynamic external fixation. *Journal of Bone and Joint Surgery*, **73B**, 819–25.

Watkins, P. E., Kelly, D. J., Leendertz, J. A., Rigby, H. S., and Goodship, A. E. (1986). The application of photon densitometry to fracture healing (Abstract). *British Journal of Radiology*, **45**, 819.

Wllliams, E. A., Rand, J. A., An, K. N., Chao, E. Y. S., and Kelly, P. J. (1987). The early healing of tibial osteotomies stabilized with one-plane or two-plane external fixation. *Journal of Bone and Joint Surgery*, **69A**, 355–65.

Section 2—Prosthetic Joints

8 Migration patterns in uncemented ceramic THR. An eight year follow-up

T. F. Stoyle and R. R. Choudhry

Introduction

After the initial success of THR, surgeons became aware in the mid-seventies of failures. Subsequently, papers such as Gruen *et al.* (1979) and Dorr (1983) were published which drew attention to the relatively poor results of cemented THR in young active and heavy patients. These poor results were attributed to the wear rate of metal on polyethylene articulations, to adverse tissue reactions, and to cement failure. These problems required a solution if an operation was to be offered to the not inconsiderable number of young patients requiring arthroplasty. Various options were considered, among them, elimination of either or both the polyethylene and the cement, improvement in cementing techniques, and the development of different materials for joint replacement. It was decided to investigate the results of uncemented arthroplasty with alumina ceramic articular surfaces.

The material

A group of patients operated upon between November 1979 and December 1985 were studied. Only patients undergoing primary hip replacement with the Mittelmeier (Autophor) prosthesis, without cement and with both articular surfaces of alumina ceramic, were included. The patients were consecutive and unselected except by age. No patient older than 61 years was included. There were 91 patients with 104 hips in the trial. The patients were finally assessed at the end of December 1989. The follow up rate was 93.6 per cent and the mean follow up time was 76.8 months (range from 120 months to 48 months). The patients' ages ranged from 17 to 61 years and the average was approximately 40 years. There were 58 hips in 51 females and 36 hips in 31 males. The subjects' weight varied from 39 to 118 kg, with an average of 69 kg.

The method

The patients were assessed clinically and radiologically. For the radiological assessment X-rays were taken at the outset, after operation, and annually.

That variations in X-ray technique lead to variations in measured parameters is well recognized. This is due mainly to differing magnifications with different tube–object distances and spherical error due to differences in the positioning of the patient. Corrections for these potential errors were made when calculating prosthesis migration and adaptive changes in the bone.

With regard to the acetabular prosthesis, migration was measured by changes in the position of the centre of the prosthetic head of femur. Variations in magnification were corrected using a factor derived from the difference in the size of the femoral head as projected on the X-ray and the known diameter of the inserted prosthesis. Variations due to spherical error were derived from an analysis of rotation of the pelvis around the horizontal axis (average 6.6°) and around the vertical axis (average 4.0°).

In summary, the error in determining the intersection of the 'tear-drop' line and the vertical was probably 0.5 mm with a similar error in the head centre point. Calculation of magnification was subject to an error of 1.0 mm. The spherical error gave a maximum error of 2.6 mm. Thus the sum of these errors was assumed to give a maximum of 4.5 mm. Only a displacement in the head centre of 5.0 mm or more was regarded as significant.

For the femoral component, lines horizontal to the vertical axis of the prosthesis were drawn at the superior and inferior poles. The magnification factor was calculated from the relationship between the known length of the inserted prosthesis and the length as measured on the X-ray. Three further horizontal lines were drawn, the first at the lowest point of the underside of the collar of the prosthesis, the second at the proximal cut end of the calcar and the third at the lower point of the intersection of the lesser trochanter and the medial femoral cortex. Thus it was possible to obtain corrected data with regard to prosthesis migration. All measurements were made in millimetres. The presence of a pedestal at the tip of the prosthesis was recorded as absent, partial, or complete.

The degree of apposition of the prosthesis to the internal surface of the femur was analysed and five groups were identified: (1) solid—no space between the inner cortex and the implant; (2) tight—1 mm distance in two zones; (3) loose—1–2 mm distance in two or more zones; (4) very loose—>2 mm in three or more zones; (5) three point fixation—when the tip of the prosthesis touches the medial cortex and the upper lateral surface is in contact with the lateral cortex.

The results

The serial X-rays were analysed and the following factors recorded: the acetabular insertion angle; the position of the head centre; the distance between the underside of the prosthesis collar and the cut surface of the calcar;

the distance between the collar and the lesser trochanter; the presence of porosis of the proximal femur and femoral fit as described above. These findings were correlated to the clinical findings with particular reference to the case which required re-operation.

The natural history of subsidence was explored. All 94 hips available for study were found to have subsided. The subsidence ranged from 0.6 to 14 mm (mean 4.2 ± 3.0 mm). All the hips subsided between 0.2 and 6.0 mm in the first year. During the second year, 10 hips stabilized and stopped subsiding. In the third year 36 more hips stabilized. In summary, 82 per cent subsided less than 2 mm in the first year, 94 per cent subsided less than 2 mm in the second year, and 92 per cent did not subside after the third year, see Fig. 8.1.

The relationship between the group of prostheses in which the collar was calcar-bearing and the group in which it was not calcar-bearing was of interest. There were 57 hips in the noncalcar-bearing group and 37 in the calcar-bearing group in which the collar of the prosthesis was well seated on the calcar. The average gap between the collar and the calcar was 3.4 ± 3.5 mm (range 0.8 to 14.2 mm). The mean subsidence in the well seated group was 4.9 ± 3.6 mm while the non-calcar-bearing group subsided on average 3.8 ± 2.4 mm.

Subsidence was examined against femoral fit. Sadly, only 52 per cent of the prostheses were found to be in the solid and tight categories. Regression analysis showed a close relationship between fit and subsidence.

Sixteen of the 94 hips required revision: 12.7 per cent (12 hips) had aseptic loosening. Nine of the patients had pain, 13 had a femoral fit categorized as loose or worse, 12 had a partial or complete pedestal, and 10 subsided 4.0 mm or more.

Discussion

Subsidence in cemented implants has been extensively reported, but a review of the English literature reveals few studies of uncemented hips (Engh 1983; Kattapuram 1990). All of these papers reviewed small numbers of prostheses for short periods. Loudon and Charnley (1980) showed a relationship between subsidence and the shape of the femoral stem in the Charnley low friction arthroplasty and in a further paper (1986) he described two patterns of subsidence in cemented low friction arthroplasty. The first was a predominant group, which he called 'stable subsidence', in which there was a reduction in subsidence with time, and the second a lesser group, in which 'progressive subsidence' was associated with loosening. In other reviews of cemented arthroplasty Chafetz *et al.* (1985) concluded that subsidence was inevitable and desirable being a sign of successful load transmission. Ling (1986) believed that distal movement of the femoral prosthesis was essential to the successful

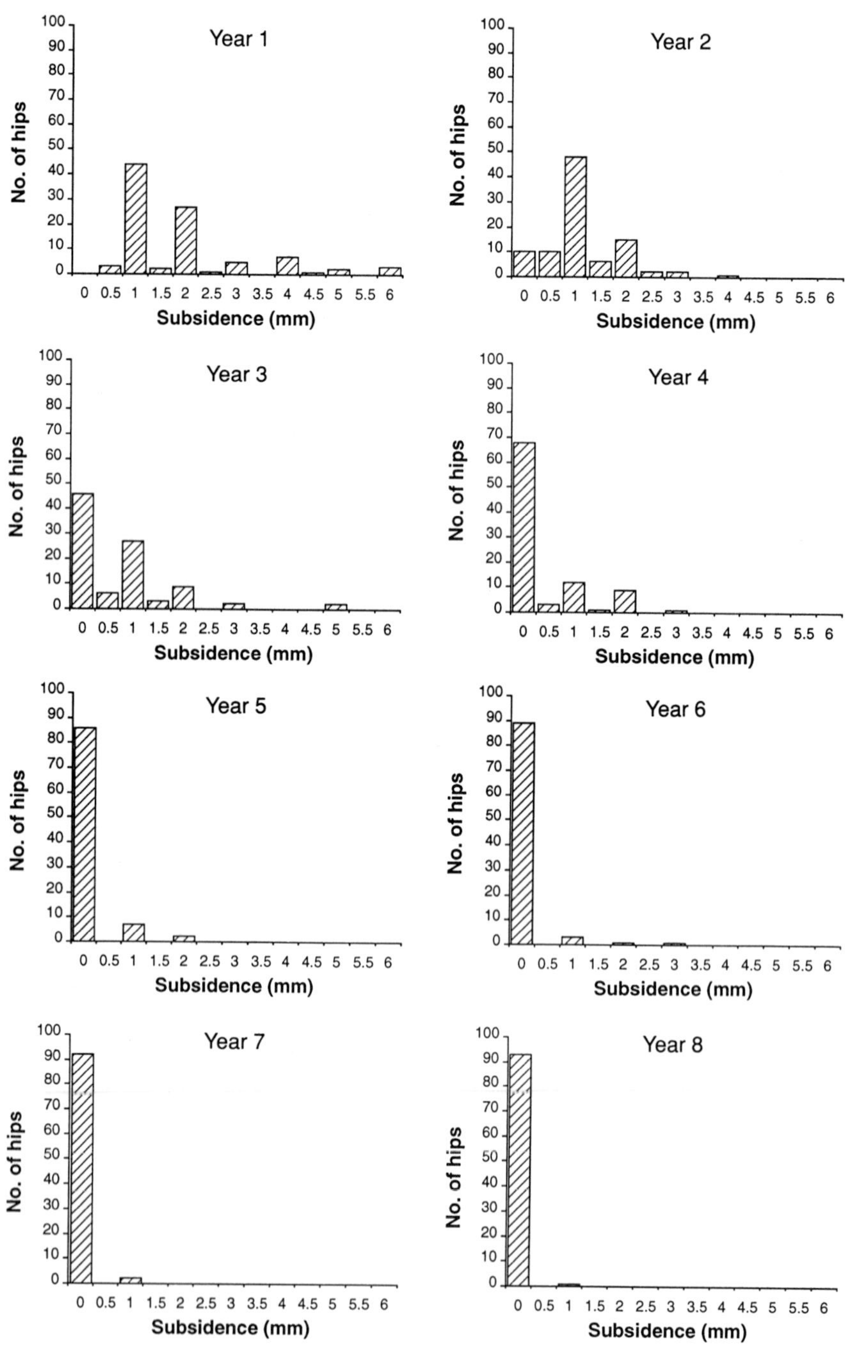

Fig. 8.1 Distribution of subsidence over 8 years.

engagement of the taper and weight transmission. Among the papers on uncemented hips, Salzer (1984) reported subsidence despite the presence of a collar and that subsidence seldom continued with time. Amstutz *et al.* (1989) opined that a collar would prevent an exact fit and lead to toggle. Kattapuram (1990) reporting on the porous-coated anatomic stem, observed that subsidence of more than 10.0 mm was conducive to a poor result and was the result of a loose fit.

The results in the present paper support the view that a solid femoral fit is essential to the success of all arthroplasties, be they cemented or noncemented, and that the presence of a calcar-bearing collar probably inhibits such an outcome and prevents distal migration of the prosthesis and successful weight transmission. The view that there is no essential difference between so-called cemented arthroplasty and uncemented arthroplasty, and that all are subject to the same complications to a greater or lesser degree, is supported.

Debate in the meeting

The meeting noted that a rocking or shearing motion of the collar of the prosthesis on the flat, cut surface of the calcar is implicated in bone resorption, but that this particular movement is not commonly measured. Elaborating on the mechanisms of migration and micromovement, the author (TFS) noted that there seemed to be a certain amount of 'adaptive change' while the bone gets used to the new stresses placed on it, with whatever prosthesis, whether it be a hip or a knee. If those stresses are within an acceptable physiological limit the bone stabilizes, but if those stresses are outside that physiological limit then migration continues. Migration is really a form of slow breakage.

The author (TFS) disliked the distinction between cemented and uncemented prostheses. The cement creates a one-off, made-to-measure prosthesis. The disadvantage of uncemented prostheses is that it is necessary to machine the patient to fit the prosthesis which means that sophisticated tools are needed. If it were possible to achieve the same degree of fit with an uncemented prosthesis as with a cemented prosthesis the author would expect the result to be exactly the same. Cemented prostheses are two-phase prostheses built from a metal core and a cement mantle that is all—and the meeting agreed.

9 Stabilities of standard and custom hip stems under axial and torsional loading

J. Hua and P. S. Walker

Introduction

The stability of an uncemented femoral stem within the medullary canal has important implications for the interface response and the clinical success of the prosthesis in total hip replacement. Earlier studies showed that cemented stems had less relative motion between the stem and the bone under cyclic loading ('micromovement') than cementless stems (Nunn *et al.* 1989; Schneider *et al.* 1989*a*; Phillips *et al.* 1991), and short-term clinical results were excellent. However, laboratory tests showed evidence of microfracture of the cement mantle under torsional loading (Sugiyama *et al.* 1989) and major stress shielding of the cemented stems (Zhou *et al.* 1990). Such factors are likely to contribute to the deterioration of results with time (Harris *et al.* 1991). Therefore cemented stems are generally not considered adequate for the young and active patients.

Clinical and animal studies have shown that uncemented stems with hydroxyapatite or porous coating can induce bone tissue ingrowth into the stem (Oonishi *et al.* 1989; Bloebaum *et al.* 1991), and are expected to achieve longer survival of the femoral prosthesis. However, the achievement of bone ingrowth will rely on the stability of the femoral stem in the canal. Schneider *et al.* (1989), Burke *et al.* (1991), and McKellop *et al.* (1991) compared the initial stabilities of cementless hip prostheses and showed that micromovements varied considerably between different designs. Anatomically shaped stems showed less rotational micromovement than straight stems. Recently, micromovement testing has been recommended as an important criterion, either in the laboratory to evaluate a design of a femoral stem or intra-operatively to assess the immediate fixation of a femoral prosthesis (Harris *et al.* 1991).

It is believed that stem–canal fit and cross-sectional fill are of paramount importance in controlling micromovements and in relieving thigh pain (Whiteside 1989). Owing to wide variations in the geometry of femoral canals, off-the-shelf femoral stems cannot produce consistent optimal fit-and-fill, but for custom stems, the optimal fit in the proximal regions and the maximum fill in the distal regions can be consistently achieved (Hua *et al.* 1991). The hypothesis of the present study is that asymmetrical ('anatomic') stems will produce less micromovement than symmetrical ('straight') stems, especially in rotation, but that custom stems will have the least micromovement of all. To test this, a comparative study of symmetrical, asymmetrical, and custom prostheses was conducted.

Materials and methods

Eight adult cadaver femurs free of abnormality were radiographed and formalin-fixed prior to test. Three femurs were right sides and five were left. Femoral stems were computer-aid-designed and computer-aid-manufactured by the Hip Design Workstation within the authors' department. The symmetrical stems and asymmetrical stems were designed from an average femoral canal whose geometry was obtained by studying 26 American adult femurs. The only difference between these two types of stem was that the asymmetrical had an anterior flare (Fig. 9.1). Both the symmetrical and the asymmetrical stems were selected from a seven size system. The custom stem was designed individually for each femoral canal by digitizing the internal cortical outline of the antero-posterior and lateral views of the radiographs. The Hip Design Workstation then predicted the three-dimensional canal shape. The design algorithms for symmetrical, asymmetrical, and custom stems were designed to achieve a close fit in the proximal medial part and maximum fill in the distal part, with clearance in the middle third. In order to make these three types of prostheses comparable, they were given the same stem length and neck cutting angle, and were all made of titanium-6 aluminium-4 vanadium alloy (Fig. 9.1). They were also collarless, so that the stability of the prostheses were entirely dependent on the stem shape.

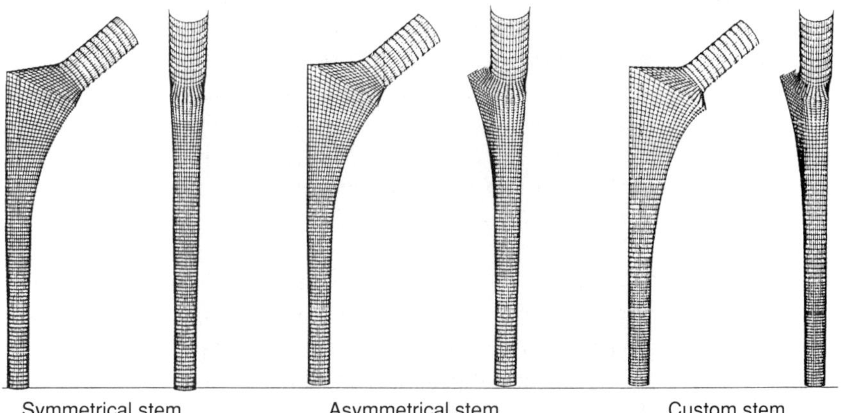

Fig. 9.1 Stems for testing (for femoral bone specimen 6).

The femurs were resected just above the femoral condyles and potted in a metal tube at 12° laterally and 10° posteriorly, so that the force applied vertically on the head of the prosthesis would produce bending moments in the femur resembling a single-leg stance. To measure the vertical micromovement of the implants, a hole was drilled 20 mm below the lesser trochanter in the antero-posterior plane. A smaller hole was drilled at the same place through the stem of the prosthesis. After the implant was inserted into the femur, a metal rod was tightly fitted into the stem with enough clearance with the adjacent bone so as not to impede any motion during cyclic loading conditions. Two 15-mm sided plastic cube targets were screwed and glued on to each end of the rod. A plastic apparatus holding two transducers vertically was fixed to the proximal femur with four pointed screws (Fig. 9.2). The average signal from the anterior and posterior transducer, was used as the measure of vertical micromovement. The micromovements in the antero-posterior and medio-lateral planes were measured at the tip of the implant. One hole was drilled laterally and another posteriorly. A plastic apparatus holding the transducers horizontally was fixed to the femur by three pointed

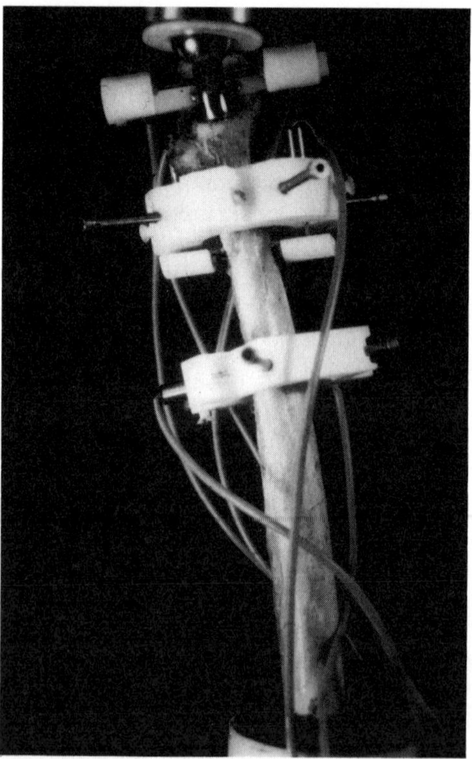

Fig. 9.2 Devices for measuring vertical, lateral, and posterior micromovements.

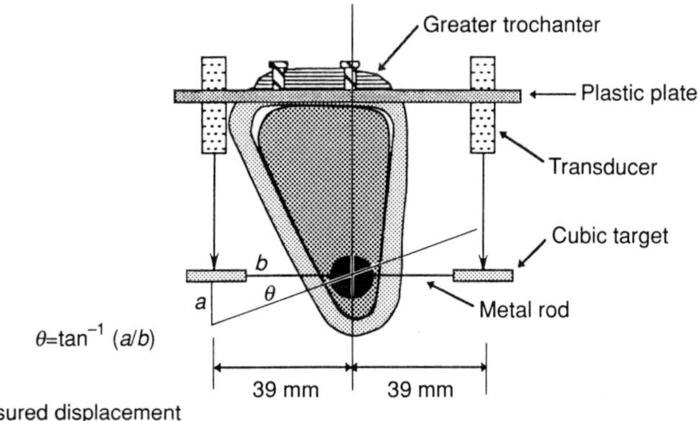

Fig. 9.3 Devices for measuring rotational micromovement.

screws, level with the tip of the implant (Fig. 9.3). Rotational micromovement was detected by two transducers placed parallel in the anterior and posterior plane at neck level. These two transducers were held by a plastic plate which was screwed tightly onto the cutting face of the greater trochanter. A small hole was drilled at the bottom of the neck of the implant, and a metal rod, pointed anteriorly and posteriorly, was passed through the hole and fixed tightly (Fig. 9.3). Plastic cubed targets with a smooth surface were screwed and glued onto the ends of the rod. Rotational micromovement was defined as the average of the anterior and posterior readings.

All the targets were perpendicular to the pointer of the transducers. Six linearly variable differential transducers (LVDT from Sangamo, a division of Solartron Transducers) with a resolution of 0.001 mm through a range of ± 2.5 mm were used. The femoral prostheses were inserted and tested in the sequence: symmetrical, asymmetrical, and custom. A force of 1000 N was applied vertically to the prosthetic head. After preconditioning with three loading cycles to seat the implant into the canal, the micromovements of the stem relative to the femoral bone were measured simultaneously through six channels of LVDT signals by a linked computerized data acquisition system (Labview, National Instruments). A cyclic load with an amplitude of 500 N and a frequency of 1 Hz was then applied. The readings were repeated at the end of every 500 cycles, up to 2500 cycles. The difference in the displacement with and without load applied was defined as micromovement, and the difference between the initial position of the stem and the position after cyclic loading was defined as migration or sinkage.

After each stem was tested in the alignment of the single leg stance, the femur was rotated to 70° from the vertical, faced up anteriorly. A plastic pillar was used posteriorly to support the proximal part of the femur. A single force of

Fig. 9.4 Devices for measuring rotational micromovement under torsional loading.

150 N was applied to the prosthetic head, which produced torsional loading through the prosthesis applied to the prosthetic head, to simulate stair climbing and chair rising situations (Fig. 9.4).

Results

The micromovement data from eight femurs for the symmetrical, asymmetrical, and custom stems were averaged and compared. ANOVA was used for statistical analysis. When significant differences occurred at $p<0.05$, the paired t test was used to determine which stem types were different.

Vertical micromovement

In the initial loading stage (0 cyclic loading), there were only small differences between the three types of prostheses with a range of 80–100 μm (Fig. 9.5). After cyclic loading the micromovement of all the prostheses were again similar, but the range was now reduced to 20–30 μm. The micromovement continued to reduce up to 2500 cycles when the values were 6 ± 6 μm for symmetrical stems, 19 ± 13 μm for asymmetrical stems, and 11 ± 12 μm for custom stems. The values for the asymmetrical stems were statistically higher than for the other two stems. There were no significant differences between the symmetrical and the custom stems.

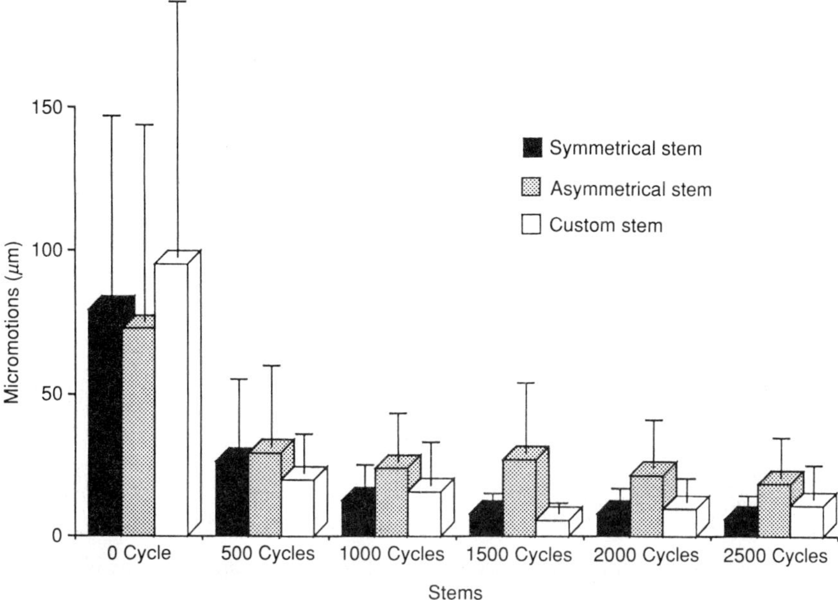

Fig. 9.5 Comparison of vertical micromovement after cyclic loading among symmetrical, asymmetrical, and custom stems.

Rotational micromovement

The asymmetrical and the custom stems were much more resistant to the torsional force than the symmetrical stem (Fig. 9.6). The differences between the asymmetrical, the custom, and the symmetrical stems became significant after 500 cyclic loading (43 ± 38 μm for the symmetrical stem, 12 ± 15 μm for the asymmetrical stem, and 6 ± 9 μm for the custom stem). The custom stem was the best through each stage of cyclic loading, but there were no statistical differences between the asymmetrical and the custom stems.

Lateral micromotion at the tip

The lateral micromovement was monitored by a transducer pointed onto the lateral side of the tip of the prosthesis. Usually, the stem tip moved laterally during cyclic loading, but a few stems moved medially at certain stages of cyclic loading. In total, the lateral micromovements for all the categories were very small, and the differences were that the asymmetrical and custom stems had less lateral micromovement than the symmetrical stem (Fig. 9.7).

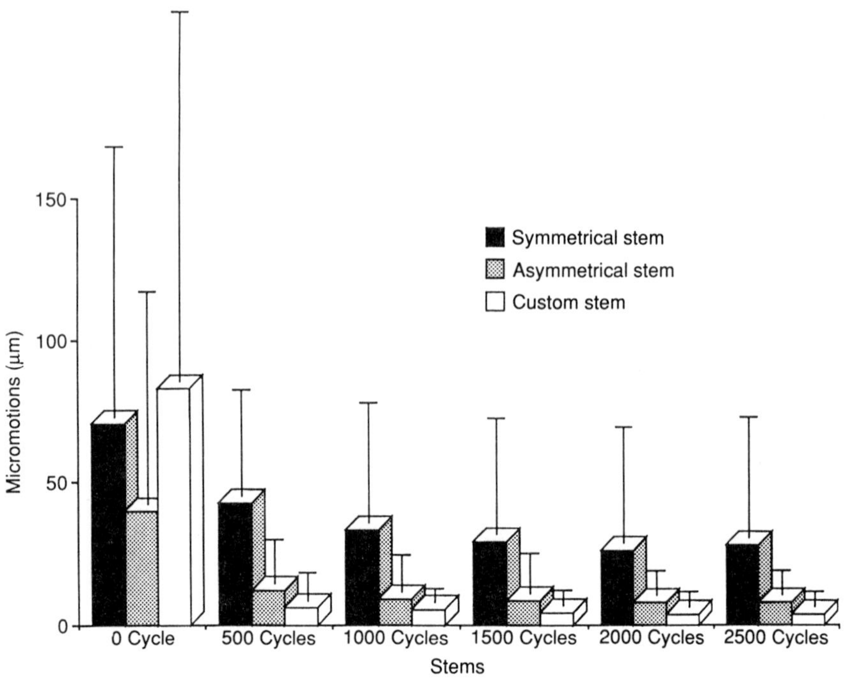

Fig. 9.6 Comparison of rotational micromovement after cyclic loading among symmetrical, asymmetrical, and custom stems.

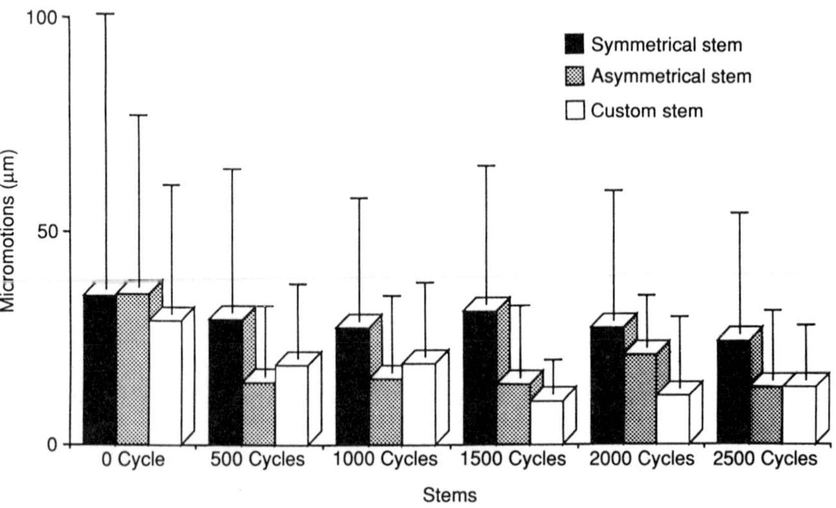

Fig. 9.7 Comparison of lateral micromovement after cyclic loading among symmetrical, asymmetrical, and custom stems.

Posterior micromovement at the tip

The posterior micromovement was measured by a transducer mounted on the posterior side of the tip of the prosthesis. On average, the stem tip moved posteriorly for the symmetrical and the custom stems, while for most of the asymmetrical stems, the stem tip moved anteriorly (Fig. 9.8). Except for the initial stage (0 cyclic loading), the custom stem was significantly better than the asymmetrical stem ($p<0.05$). For the symmetrical and asymmetrical stems, the results were similar overall: only at the initial stage (0 cyclic loading), was the symmetrical stem more stable than the asymmetrical stem ($p<0.05$). Although there were no statistically significant differences between the custom and the symmetrical stems, the custom stem showed substantially less average micromovement than the other two types of stem.

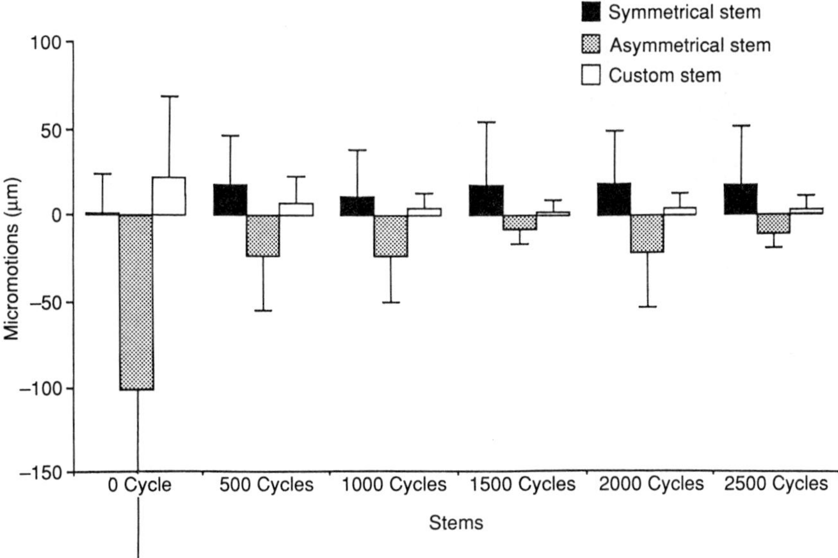

Fig. 9.8 Comparison of posterior micromovement after cyclic loading among symmetrical, asymmetrical, and custom stems (negative values indicate that the tip of the stem moved anteriorly).

Torsion

The prostheses were more likely to rotate when they were subject to torsional force, even though the loading force was very small (150 N). Compared with the asymmetrical and the custom stems, the symmetrical stems had a much greater tendency to rotate (Table 9.1), and the differences were significant

Table 9.1 Rotational micromotion under torsional loading (μm)

Bone no.	Symmetrical	Asymmetrical	Custom
1	124.5	16	20
2	80	15	7.5
3	34.5	11	20
4	131	29.5	8
5	84	2.5	15
6	46.5	6	1
7	20.5	5	5
8	386.5	21	12
Mean	113.4	13.3	11.1
SD	117.3	9.1	6.9

($p < 0.05$). The custom stem showed slightly better results than the asymmetrical stem, but there were no significant differences between them.

Migration (sinkage)

The migration at the end of each 500 cycles was measured by comparing each cyclic loading stage with the initial stage (0 cyclic loading). Figure 9.9 shows that the total migration for all three types of stem was around 150 μm after 2500 cycles of loading. There were no significant differences. Most of the migration had occurred by 1000 cycles of loading.

Discussion

Femurs vary in shape and bone quality. Ideally, each stem of same type has to be tested in various femoral canals, but the comparison between different groups of stems should be made in the same geometry of femoral canals. McKellop *et al.* (1991) used a synthetic composite model femur to compare three different press-fit hip prostheses with a cemented stem in micromovement. The advantage of this method is that the different hip prostheses were compared in the same geometry of the femoral canal, while the disadvantage is that all the implants were only tested through one type of femoral canal, which may introduce bias because one particular prosthesis may just agree with this femoral canal. Moreover, the elastic modulus of the epoxy reinforced with fibreglass powder (7–10 GPa) is lower than the human cortical bone (15–20 GPa), which may affect the interface micromovement between the implant and cortical wall of the femur. Phillips *et al.* (1990) used 13 pairs of femurs for

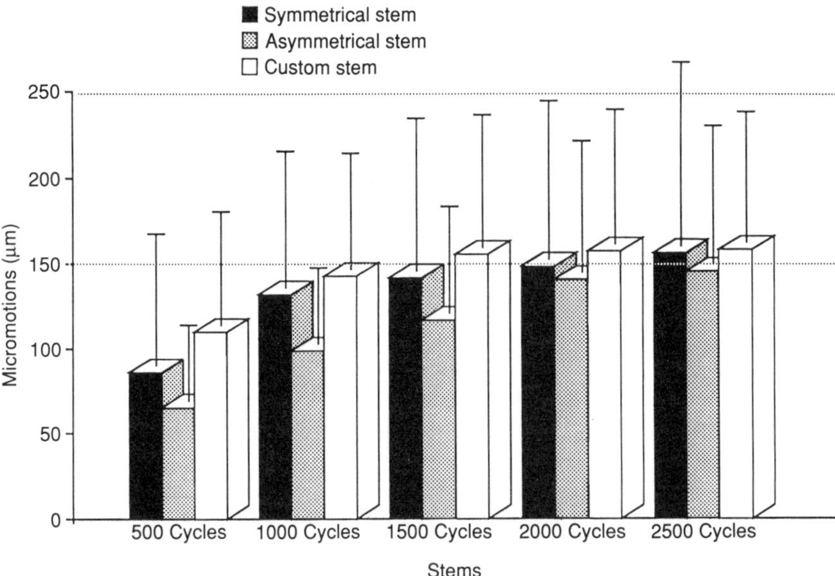

Fig. 9.9 Comparison of migrations of symmetrical, asymmetrical, and custom stems after cyclic loading.

comparison between 2 types of press-fit stem and another 5 pairs for comparison between press-fit and cemented stems. This seems a valid method but such large numbers of paired cadaveric femurs are difficult to obtain. Many earlier investigators tested micromovement by reinserting different types of implant into the same femoral canal (Nunn *et al.* 1989; Schneider *et al.* 1989*b*; Burke *et al.* 1991). The advantage of this method is that the prostheses can be compared in the same geometry of the femoral canal but challenged by the different types of geometry of each femoral canal. The disadvantage of this method is that the previous reaming of the canal may affect the fit of the following stem, and also the cyclic loading may alter the quality of the femoral bone, if the force is beyond the limits of the femoral bone. The fatigue properties of the femoral cortical bone have been reported to be 1.3×10^4 cycles at a stress amplitude of 84.1 MN/m^2 and 4.6×10^6 cycles at 46.6 MN/m^2 (Swanson and Freeman 1971). In this study, loading forces and cycles were far below these fatigue limits.

The asymmetrical stem was implanted after the symmetrical stem by simply reaming off the anterior part of the cancellous bone to accommodate the anterior flare of the asymmetrical stem. For the custom stem which had maximum contact with the internal cortical wall of the femoral canal, the femoral canal was prepared by reaming off the remaining cancellous bone. Three types of the implant had the same neck cutting angle and stem length, which can eliminate progressive neck resection and make the results comparable in the horizontal displacements at the stem tip.

The formalin fixation and post-mortem changes before formalin fixation have been shown to be associated with very slight reductions in the modulus of elasticity and hardness of the bone (McElhaney *et al.* 1964; Ashman *et al.* 1982). To summarize the literature regarding the effects of embalming on the mechanical testing of bone, Finley *et al.* (1989) concluded that embalmed specimens provide a suitable model for a parametric evaluation of prostheses.

To detect rotational movement of the femoral prosthesis, most previous investigators have mounted one transducer in the proximal medial part of the prosthesis. Although the femoral prosthesis often rotates posteriorly, the actual rotatory centre of the cross-section of the prosthesis cannot be predetermined. Mounting the transducer in different positions can produce different values for stem rotation. In this experiment two transducers were used in order to detect rotational motion wherever the rotary centre was.

Attention has recently been drawn to the importance of the torsional stability of femoral components. Wroblewski (1979) and Harris *et al.* (1991) pointed out that torsional loading and rotational instability of the stem were the critical factors in failure for both cemented and cementless prostheses in total hip arthroplasty. Harris *et al.* (1991) suggested the use of a torque wrench micrometer for assessment of intra-operative stability of femoral component. However, even in the common cases of stair-climbing and chair-rising, a variety of values have been reported for the position and force applied to the femoral head (Rydell 1966; Williams and Svensson 1968; Paul and McGrouther 1975; English and Kilvington 1978; Berme and Paul 1979; Crowninshield and Brand 1981). A report by Kotzar *et al.* (1991) of telemeterized data *in vivo* of hip joint force in two patients after total hip replacement revealed that, during the early post-operative period, the forces generated by patients while walking were well below those often predicted. Davy *et al.* (1988) used a telemeterized total hip prosthesis implanted in one patient and measured at 31 days post-operatively. They found that for the single-limb stance the joint-contact force was 1160 N (2.1 times body weight) with the resultant force located on the antero-superior portion of the ball. For stair-climbing, the force was 1450 N (2.6 times body weight) with the maximum at 70° anteriorly. The present study used these parameters to simulate physiological loading with the one exception that 150 N was used as a torsional force. In laboratory tests, if the femoral bones were mounted 70° from vertical with 1450 N load, excess motion would occur and the bone would be likely to fracture, hence it seems impossible exactly to simulate the physiological situation in a laboratory test. However, as a comparative study of rotational micromovement of femoral components under torsional force, 150 N was found to be sufficient.

Under axial loading, the stabilities were achieved by wedging the femoral stem into the canal, mostly in the medio-lateral plane. The symmetrical stem with straight antero-posterior plane showed less vertical displacement after 500 cycles, but more rotational micromovement, especially under the torsional force. In contrast the asymmetrical stem was much more resistant to rotational micromovement, but sometimes its anterior curvature did not fit well with the femoral canal, which resulted in less stability in vertical and posterior displacement. This study indicates that there are two factors affecting rotational stability. Firstly the radius of the proximal medial part of the stem has to fit with the canal. If it is too large the stem cannot fit in contact with the medial cortex and will be positioned in varus; if the radius is too small the stem will be unstable in rotation. Secondly, the anterior flare of the stem can resist rotational micromovement. Custom stems produced optimal fit with the femoral canals in the proximal medial and proximal anterior regions, and therefore were very stable in rotational and vertical micromovements. Regarding the lateral micromovement at the tip of the stem, there were no significant differences in these three types of stems. However, in the sagittal plane, some of the asymmetrical stem tips moved anteriorly instead of posteriorly. This phenomenon suggests that if the anterior curve of the stem contacts the inner cortical wall before the stem fits properly within the canal, the stem might toggle and the micromovement pattern might be altered.

Laboratory experiments can never perfectly simulate the true physiological situation and no biological boundary and fixation will occur in the cadaveric specimen. However, from this study, one conclusion can be drawn—immediate vertical and rotational stabilities of a femoral stem within a canal can be enhanced by optimal fit in the proximal anterior and proximal medial regions. To achieve all these design goals, the custom stem is probably the best choice.

Acknowledgements

The first author (J.H.) was supported by the Sino-British Fellowship. We gratefully acknowledge the funding and equipment provided for this work by the Arthritis and Rheumatism Council, by Zimmer Inc. (USA) and by Howmedica, Pfizer Health Care Products (UK).

Debate in the Meeting

It is well known that an uncemented, symmetrical stem is not rotationally stable, but it is easy to manufacture and easier to insert than an anatomical stem. Every stem has its own design features and advantages. This paper addressed a single difference between a symmetric and an anatomic stem—an anterior curvature. The authors have shown that this single feature is sufficient to ensure rotational stability.

Extra rotational stability is bought at the price of added stress over some part of the bone/prosthesis interface. The meeting was concerned that the tip of a curved stem might over-stress the shaft of the femur. The larger the area used to shed the rotational stress, the smaller the strain will be.

This debate continued after the paper by Callaghan *et al.* which follows.

10 The effect of femoral stem geometry on micromovement in uncemented porous coated total hip arthroplasty—comparison of a straight and curved stem design

J. J Callaghan, C. S. Fulghum, R. R. Glisson, and S. K. Stranne

Introduction

The interface micromovement between prosthesis and bone during axial and torsional loading in paired, fresh-frozen, cadaveric femora that had been implanted with (i) uncemented, collarless, isthmus-filling straight-stem prostheses, and (ii) uncemented, collarless, proximal-filling curved-stem prostheses were compared to determine the effect of stem geometry on interface micromovement.

Method

Seven matched pairs of fresh femora were implanted with a straight Harris–Galante and a curved anatomic uncemented femoral component. Axial loads and combined axial and torsional loads were applied to the implanted femoral prostheses using a jig which stimulated a hemipelvis with trochanteric strap. Extensometers were used to measure prosthesis bone interface micromovement. The prostheses were then removed and reinserted with cement applied to the proximal porous coating to simulate bony ingrowth. The axial and combined axial and torsional loads were reapplied and interface micromovement recorded.

Results

No statistical differences in micromovement were recorded between the straight and curved stem implanted femora during axial loading. When large torsional moments (22 N m) were applied, significantly less micromovement occurred at the interface of femora implanted with curved stems than at the interface of femora implanted with straight stems both proximally ($p=0.019$) and distally ($p=0.0013$). When cement was applied proximally both proximal and distal micromovement was decreased with axial and torsional loading in both the straight stem and curved stem implanted femora.

Discussion

Avoidance of interface micromovement between prostheses and bone in the uncemented condition has been demonstrated to be important for optimal bony ingrowth into porous surfaces. In addition, the clinical results including the potential for thigh pain may be dependent on the avoidance of bone prosthesis interface motion. The findings of the present study support the use of stem geometry which is curved and proximally fills the canal in the anterior–posterior dimension rather than a stem geometry which is straight and rectangular in cross section proximally with the intent to fill the isthmus distally, if avoidance of interface micromovement is an important criteria for optimal clinical outcome when uncemented porous coated total hip arthroplasty is utilized.

Debate in the meeting

In this and the paper by Hua and Walker, Chapter 9, pp. 64 ff. there has been simulation of the contact forces applied to the system. These contact forces are large because they are balancing muscle forces, but models have not been described, either theoretically or experimentally, which try to simulate the forces applied in the vicinity of the prosthesis by muscles through their insertions. These forces must have an enormous effect on the distribution of stress in the bone round the prosthesis and must cast doubt on the significance of any modelling to the *in-vivo* situation.

This view was contested in that abductor muscles, at least, have been modelled both mathematically and by cadaver experiments.

Again the issue was raised that, in this study, torsional stability was being achieved by, in effect, forcing a bent nail into a straight hole, which would seem to generate excessive stresses at the tip as it was swept round by twisting of the nail. The author (J.J.C.) conceded that this concern has not yet been resolved. There is not yet a good way of measuring contact stresses experimentally.

11 Stability of uncemented, femoral long-stems in revision total hip arthroplasty—a Roentgen stereophotogrammetric study

I. Önsten, Å. Carlson, and L. Sanzén

Introduction

Since 1987 the authors have used the BIAS (Zimmer) femoral prosthesis in revision cases with marked femoral bone loss. This titanium alloy stem has a proximal triangular body aiming at press fit, a long antecurvated intramedullary stem, and is supplied with anterior and posterior pads of pure titanium fibre mesh.

The aim was to prospectively evaluate the *in-vivo* stability of this prosthesis using Roentgen stereophotogrammetric analysis (RSA).

Patients and methods

Twelve patients with a mean age of 67 years had 13 revision total hip arthroplasties using the BIAS femoral stem. Spongious auto-grafts were routinely used. Five to nine tantalum ball markers with a diameter of 0.8 mm (Industrial Tectonics) were inserted into the proximal femur at the time of surgery. RSA examinations were performed with the patient in the supine position, using the uniplanar technique (Selvik 1974). Prosthetic movement was calculated as the displacement of the centre of the prosthetic head in relation to the femoral bone markers. The precision of RSA in this application was determined from one double exposure of each hip. The standard deviations in millimetres from the expected zero were 0.15, 0.08, and 0.25 for the transverse, longitudinal, and sagittal axes respectively. Correspondingly, the limits for significant translations with a 99 per cent confidence interval were ± 0.5, ± 0.3, and ± 0.8 mm for the transverse, longitudinal, and sagittal axes respectively. RSA examinations were performed post-operatively and at 3, 12, and 24 months.

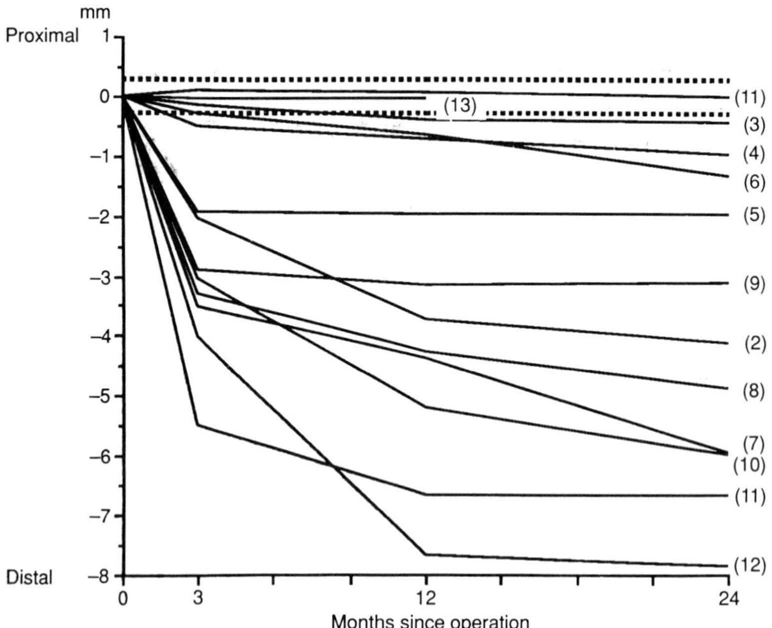

Fig. 11.1 Translation along the longitudinal axis over time.

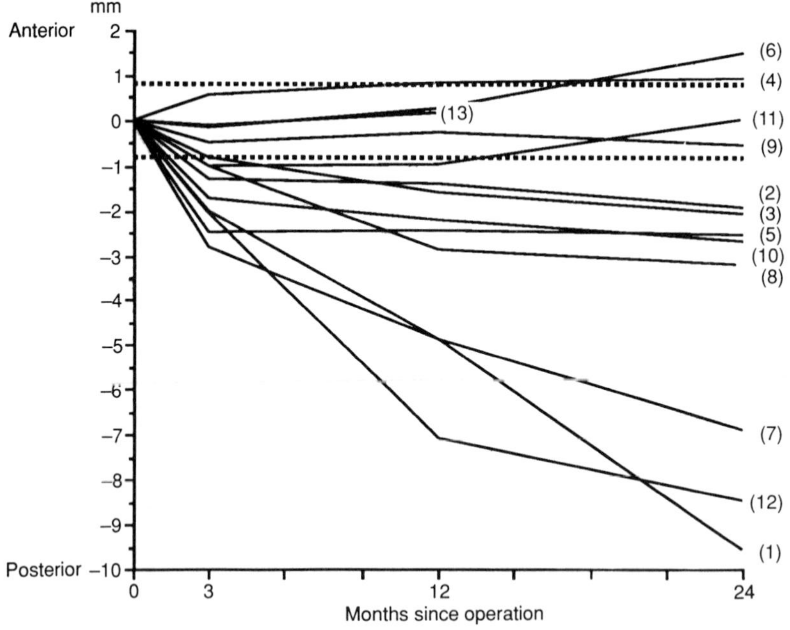

Fig. 11.2 Translation along the sagittal axis over time.

Results

The micromovement of the stems in the longitudinal and sagittal axes is illustrated in Figs 11.1 and 11.2. One stem was stable at 12 months, none at 24 months. At 24 months, 11 of 12 stems had subsided 3.9 (0.4–7.9) mm and 8 had migrated posteriorly 4.6 (1.9–9.5) mm.

Discussion

Posterior translation of the femoral head in cemented revisions has been found earlier by Snorrason and Kärrholm (1990). In a study of the uncemented LMPCH femoral prosthesis, Nistor *et al.* (1991) found large internal rotations. It is assumed that the posterior translations found in the present study could be attributed to internal rotations.

Gross instability in combination with poor clinical results have led the authors to abandon this concept in revision hip arthroplasty.

12 Fretting as a source of particulate debris in total joint arthroplasty

A. J. C. Lee, R. M. Hooper, R. S. M. Ling, R. Brooks, G. A. Gie, and R. Hale

Introduction

Polymethylmethacrylate bone cement has been in use in orthopaedic surgery for over two decades and has proved to be an effective, long-term biocompatible material for the fixation of total joint arthroplasties. However, aseptic loosening of the femoral stem and concomitant lysis of the femoral medullary cavity remains a cause for concern among many surgeons and was the major reason for the development of modern uncemented prostheses. Long-term loosening of the femoral stem by breakdown of the implant–bone interface may be the result of the short- or long-term properties of the implant materials *per se*; may be caused directly by the methods or techniques used to implant the components; or may be due to other long term effects such as the production of wear debris in the region of the implant.

In a paper published many years ago, Willert and Semlitsch (1977) were the first authors to draw attention to the potential effects of wear debris on the long-term fixation of a prosthesis. They concluded that the response to foreign body particles, where the internal transport system to the perivascular lymph spaces is insufficient to handle the volume of particles, may extend to the whole environment surrounding the joint and may lead to loosening of the joint components. More recently, Witt and Swann (1991) have drawn attention to the tissue response to metal wear debris in the revised titanium alloy/polyethylene total hip replacements, remarking on the burnishing of the relatively rough, shot blasted surface of the removed femoral components. They suggested that the surface damage had been caused by repeated rubbing against a hard surface. In an abstract, Hale *et al.* (1991) have reported on surface polishing of 36 femoral stems removed for loosening. They hypothesized that the polishing on three stems for which partially intact cement mantles were available was due to abrasion of the matt stem surface against the inside of the cement mantle. The polishing resulted in removal of metal, producing fine metal and acrylic debris together with an increase in the internal dimensions of the cement mantle. Caravia *et al.* (1990) have reported on damage to the stainless steel counterfaces during sliding wear tests of UHMWPE pins on stainless steel discs with particles of bone and bone cement debris added. Both bone and bone cement with barium sulphate or zirconium additives scratched the steel counterface, cement particles with zirconium additive producing significantly greater damage. Damage was also dependent on particle size and geometry, surface roughness, and contact stress. It was concluded that particles were likely to cause surface roughening and increased wear rates in artificial joints. DiCarlo and Bullough (1992) in a review paper have described the biologic responses to orthopaedic implants and their wear debris. Both *in-vivo* and *in-vitro* experimental studies were presented with discussions regarding their clinical importance. The paper presented studies of clinically successful implants and the expected morphological features of incorporation and acceptance by the host, followed by description of the local and systemic effects which complicate the use of implanted materials, particularly when implant failure is observed.

This paper reports the results of a detailed analysis of the wear damage on 49 cemented total hip femoral components and some preliminary laboratory investigations of the wear process.

Materials and method

A total of 49 cemented total hip femoral components obtained at revision operations have been examined (Table 12.1).

Table 12.1 Femoral components examined

Exeter matt surface	21
Exeter polished surface	10
Charnley	3
McKee Farrar	3
Others	12

In every case the location and extent of wear was recorded using manual surface mapping, optical microscopy, and surface profilometry. In addition, samples from the worn areas of three stems were obtained by spark machining for examination using scanning electron microscopy (SEM) and energy dispersive X-ray microanalysis (EDS).

Preliminary laboratory tests have been carried out to investigate the mechanism of wear between cement and stem. Reciprocating sliding wear tests of 1200 cones machined from cast PMMA bone cement against stainless steel samples cut from an unused stem have been performed under the following conditions: compressive load 4 N, stroke 10 mm at 2 Hz, temperature 20 °C. The tests were carried out in the presence or absence of slurry containing barium sulphate. After 60 000 reciprocating cycles both the worn PMMA cones and the steel counterfaces were examined using optical microscopy, SEM, and EDS. Preliminary experiments to investigate fretting wear have also been carried out. A metal specimen cut from a femoral stem was reciprocated under a PMMA bone cement cone. The load was 5 N, the stroke was between 0.0045 and 0.07 mm at 10 Hz for 10^6 cycles at 20 °C. The worn bone cement cones and the steel counterfaces were examined using optical microscopy, SEM, and EDS. Following the fretting tests, the extent of work hardening below the fretted surface was measured using a Leitz Miniload hardness tester with a Knoop indenter.

Results

Visual observation of the burnished areas on the matt finished stems revealed that the average distribution of wear on the stems was as follows: 43 per cent postero-medial 30 per cent antero-lateral, 20 per cent medial, and 7 per cent lateral. The polished Exeter stems did not exhibit burnishing or wear that was immediately visible to the naked eye. The predominance of postero-medial and antero-lateral wear (Fig. 12.1) with some medial and little lateral wear corresponds exactly with the areas of high torsional (x-axis) stresses predicted by Berme and Paul (1979) in their analysis of load actions transmitted by joints.

A profile of a 10 mm length of the anterior surface of a Charnley Cobra implant, immediately above the line marked on the surface of the implant and shown in Fig. 12.1, was obtained using a Rank Taylor Hobson Form Talysurf®. The raw profile obtained is shown in Fig. 12.2 and clearly demonstrates the difference in surface roughness between the polished region 'b' and the unpolished region 'a'. Further profiles were obtained to show surface features in more detail. Figure 12.3 shows the surface profile along a 2.5 mm length of unpolished surface and Fig. 12.4 shows a 2.5 mm surface profile from the polished area. The surface profiles indicate that significant polishing has taken place—surface roughness R_a is 1.1570 μm in the unpolished area and 0.8046 μm in the polished area, a change of 30 per cent.

Surface profilometry from the other implants examined indicated that an average reduction of 40 per cent in surface roughness had occurred and, in consequence, it was calculated that, in a typical case, some 0.5 μm of metal had been removed from the surface in the polished regions of the stem. This implies that a significant amount of fine metal and bone cement debris had been generated within the femoral medullary cavity. The shape of the asperities before and after polishing also indicates that wear is produced by some form of abrasive cutting rather than by a form of plastic deformation. The implant shown in Fig. 12.5 was removed at a revision operation which was carried out

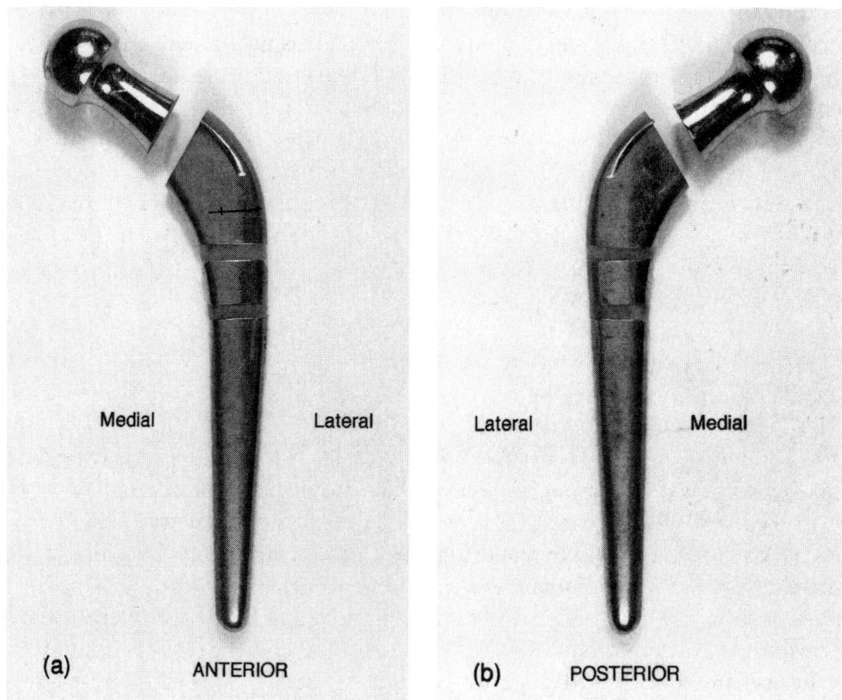

Fig. 12.1 Charnley 'Cobra' stem showing (a) antero-lateral and (b) postero-medial polished areas and position of surface profile on the anterior lateral surface.

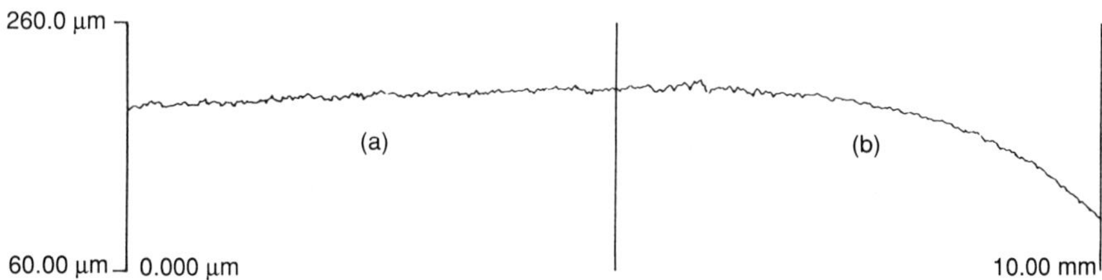

Fig. 12.2 Form Talysurf® profile of Charnley 'Cobra' implant stem, showing (a) unpolished surface and (b) polished surface.

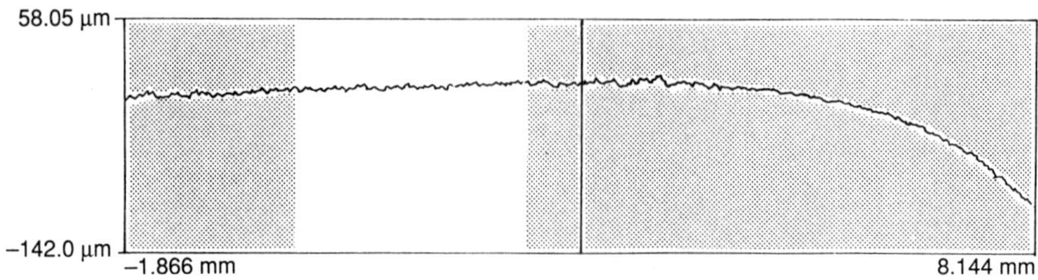

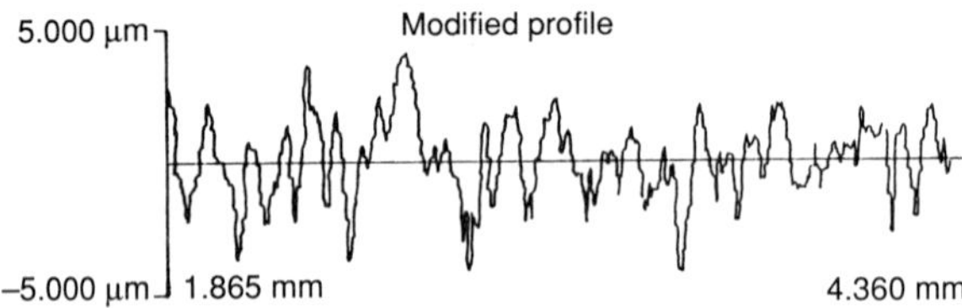

Fig. 12.3 Detail of surface profile in indicated region of unpolished surface—average surface roughness $R_a = 1.1570$ µm over a 2.514 mm length.

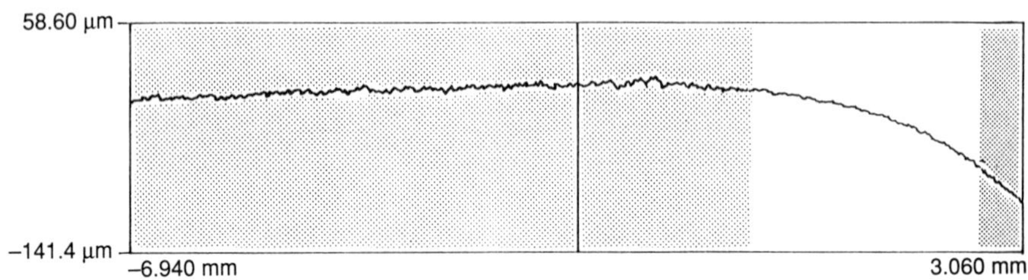

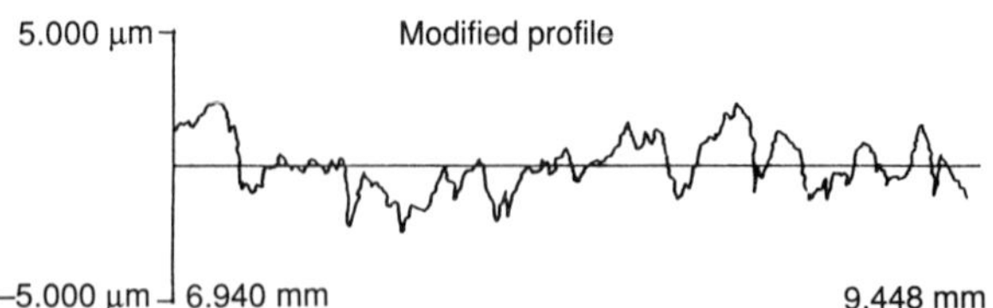

Fig. 12.4 Detail of surface profile in indicated area of polished surface—average surface roughness $R_a = 0.8046$ µm over a 2.513 mm length.

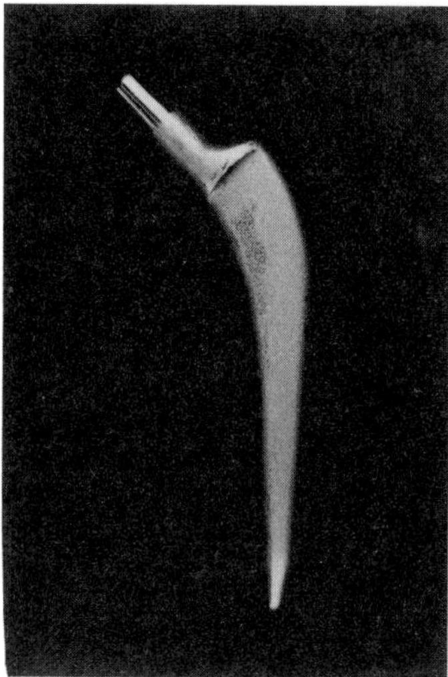

Fig. 12.5 Vives implant showing polished area on postero-medial surface.

for reasons other than stem loosening—the implant was absolutely fixed at the time of removal. Polishing wear is again evident and a surface profile 5-mm long extending from the unpolished into the polished region on the antero-lateral surface, one-third of the distance down from the collar towards the tip, was taken. The profile is shown in Fig. 12.6, with details of the unpolished surface at 'a' and the polished surface at 'b'. The two detail profiles have different scales, however, the difference in general profile shape can be seen. The polished surface shows removal of asperities from the valleys as well as from the peaks, indicating that the amplitude of polishing is finer than that of the surface undulations and must, therefore, arise from a fretting type process based on a combination of transverse and normal movements.

The final clinical example is a Charnley implant (Fig. 12.7), revised following acetabular complications and confirmed as firmly fixed by the surgeon who carried out the revision. The polishing pattern is typical of the polishing wear observed on all the implants and is present despite the complete absence of stem loosening.

In three cases the mantle of cement was recovered with the stem at revision. Patterns of wear corresponding to those on the metal stem were noted on the internal surface of the cement, which was smooth ($R_a = 0.7$ μm), and it was calculated that approximately 20 mm^3 of cement had been worn away. In only one case was there any visible evidence of corrosive action or chemical wear, as reported by Waterhouse and Lamb (1980), who interpreted their observation of polishing wear as the formation and removal of oxide layers.

A number of sliding wear tests were carried out as noted above. Although

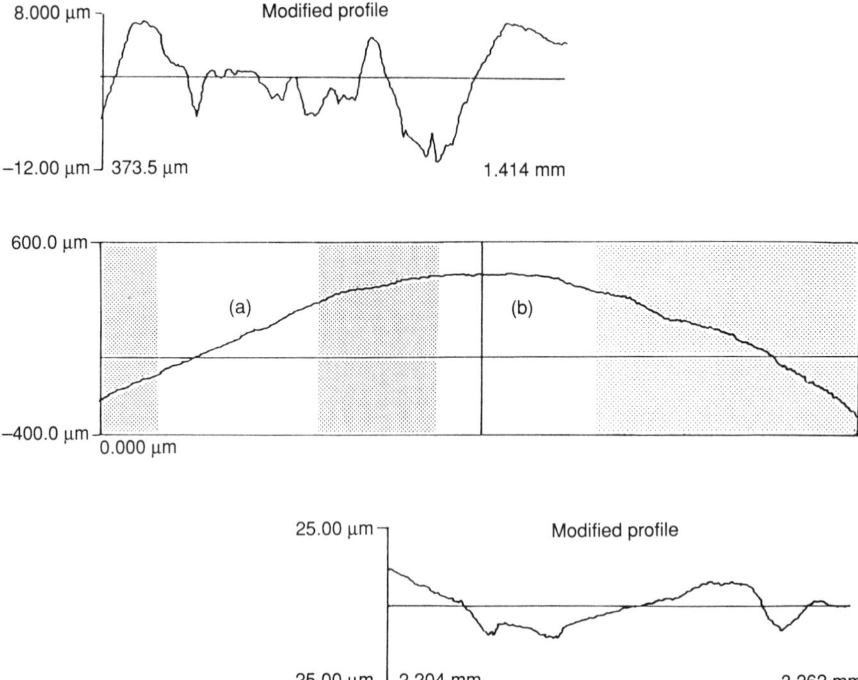

Fig. 12.6 Form Talysurf® profile of antero-lateral surface of Vives implant showing details of profile in (a) unpolished area and (b) polished area.

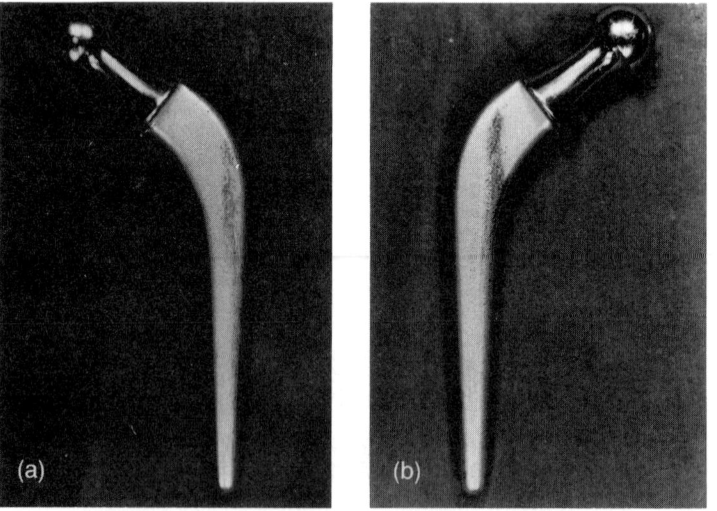

Fig. 12.7 Charnley implant—(a) anterior surface (b) posterior surface.

the size of the relative movement between stem and cement was many times greater than that expected *in vivo*, a number of important observations can be made. Figure 12.8 shows an electron micrograph of the tracks and debris produced when a cone of PMMA with barium sulphate slides on a sample of polished stainless steel implant material. The deformed surface of the cone (Fig. 12.9) which was in contact with the steel shows similar grooving, EDS analysis showed that barium sulphate was concentrated in small voids in the cement. Similar observations were made of the fragments of cement mantle which were recovered during revision operations. Tests with cones made from PMMA cement with no radiopacifier present also produced tracks on the stainless steel implant material (Fig. 12.10), showing that the hard radiopacifying material is not necessary to produce abrasion of the implant stem.

Preliminary tests in a simple fretting test machine demonstrated that polishing of metallic stems by PMMA bone cement can be produced by micromovement between the two materials. Microhardness tests on the material below the wear scar indicated that an increase in hardness from 230 to 290 kg/mm^2 had been produced by the test, suggesting that fretting induces some work hardening of the implant stem material.

Fig. 12.8 Electron micrograph of pre-polished stem material after sliding test number 1 with barium sulphate present in the bone cement.

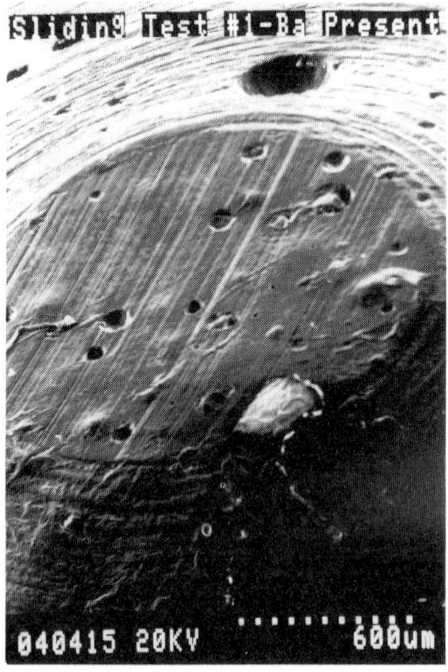

Fig. 12.9 Electron micrograph of the counterface of the bone cement cone in sliding test number 1.

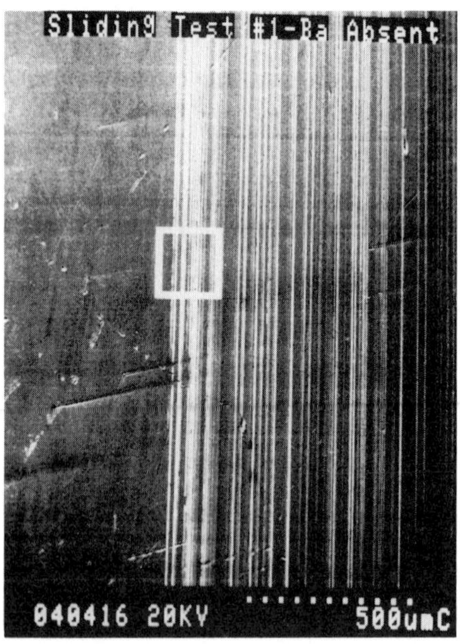

Fig. 12.10 Electron micrograph of pre-polished stem material after sliding tes number 1 with plain bone cement.

Discussion

The observation of polishing patterns on total hip femoral stems removed at revision procedures for both loosening of the stem and for reasons not connected with the stem (where the stems were firmly fixed with no sign of any loosening), strongly suggests that a wear mechanism similar to low amplitude fretting is operating. The pattern of wear, on the postero-medial and antero-lateral surfaces with some medial and a little lateral polishing, corresponds to the pattern of forces predicted by Berme and Paul (1979) to give torsional reciprocating rotation around the vertical axis of the midline of the stem. The presence of a collar on the stem leads to significant polishing under the collar with some indication of the so-called 'windscreen wiper' pattern of polishing with the collar acting as fulcrum.

Profilometry using the Form Talysurf™ surface measuring equipment gave an extended profile of the surface which could also be examined in detail. The profilometry and the SEM observations strongly suggest that fretting in the form of microadhesion or -abrasion is the major factor in the production of the polishing wear. Surface asperities are removed rather than deformed during the polishing process, producing fine particles of metallic and PMMA debris.

The observation that pitting corrosion and cracking is not generally present on the surface of the stems excludes chemical wear as a major factor. In the preliminary fretting tests in the laboratory the metal under the observed surface wear was work hardened. It is also likely that the metallic debris will be work hardened during and after removal from the surface by polishing wear, leading to increased rates of wear by further abrasion of PMMA and metal. The wear process will continue to build upon itself, enlarging the internal surface of the cement mantle. Debris may escape from the stem–cement site of production through localized defects in the cement mantle (if there are any) which may lead to localized bone lysis (Anthony *et al.* 1990). If no such defects are present, then debris may track back up the interface between cement and stem into the articulation of the joint where it is likely to remain within the boundaries of the pseudocapsule, giving an increased possibility of aseptic loosening of the prosthesis.

Conclusion

Observations of cemented total hip femoral stems removed at revision operations, using optical and electron microscopy together with surface profilometry, have identified an important long term mechanism for the production of fine particulate metallic and PMMA debris. The debris so formed may contribute significantly to the long term loosening rate of total hip replacements.

Debate in the meeting

The meeting questioned the assumption that wear particles cause loosening through activation of macrophages. The mechanical motion in shear may be the most important component causing aseptic loosening. There does seem to be increasing evidence that, where the volume of wear debris becomes significant and is not carried away, it does indeed lead to loosening, whatever the mechanism. Debris comprising metal, polyethylene, and acrylic particles is always found in the cystic lesions well down the femur which are explored histologically.

13 Workshop—prosthetic joints

Chaired by M. A. R. Freeman

There are four regimes of micromovement in relation to the development and use of prostheses: (i) micromovement predicted by finite element and other theoretical studies, (ii) strain-induced micromovement at the bone/implant interface in cadavers, (iii) strain-induced micromovement at the bone/implant interface in patients, measured inter-operatively, post-operatively, or both, and (iv) micromovement observed in patients with the passage of time—migration

Micromovement with the passage of time—migration

There is increasing evidence that, if it is known accurately enough how an individual implant migrates during years 1 and 2, it will be possible to say how much it is likely to move in the following years. For example, recent analysis of data from Lund on healing in the tibia indicates that micromovement between 0 and 1 years corresponds to healing, whereas micromovement between 1 and 2 years indicates ultimate stability. Little movement took place after 2 years in implants which subsequently survived successfully for 10 years. In her unpublished contribution to this meeting, Grewel (London) stated that, with the exception of wear, a low rate of migration between 1 and 2 years was associated with 100 per cent survival at 5 years or more, whereas survivorship falls with a higher migration between 1 and 2 years.

If this measurement on individuals generally holds true, it is even more true for characterization of a group. The Exeter hip, in which migration continues beyond year 1, may not represent an exception to the rule in this respect because the implant may be considered as a composite of cement and metal. Its characteristic sinkage between cement and femoral component represents an internal adjustment, micromovement with respect to the bone/implant interface does not necessarily occur.

It is not only the magnitude of movement which is important. The early-warning signs of possible premature failure may be associated with a change in the pattern of movement, perhaps from one direction to another, as much as the magnitude of movement. This is particularly so for acetabular sockets.

Outside this workshop, others at the meeting recognized the importance of micromovement in differentiating the performance due to two or three design variations within a relatively short period of time. But there were reservations in that some felt that it would be premature to condemn a prosthesis on the basis of such 'tenuous evidence' when so many other parameters of success or failure might have importance. Only longer-term, prospective randomized trials can unequivocally establish prosthetic joint performance. So at the moment, when a claim has been made that a better way of reconstructing the top of the tibia, or a better design of prosthesis, or a better surgical approach has been found, it seems to be held that the better method should not be advocated without 5 or 10 years' experience of survivorship, pain, or disability. In practical terms these reservations mean that, not only are benefits denied to a wider population for 10 years, but also that perhaps a thousand people are placed at unnecessary risk (one hundred hips per year). And even then the end-points may not be clear.

The workshop accepted the general hypothesis that abnormal early micromovement, particularly between 6 months and 2 years post-operatively, is a prognostic indicator of premature prosthetic joint failure. It is incumbent upon the orthopaedic community, then, to establish some criterion that can be used to compare a new design or technique against a current standard during this early period.

Stress-induced micromovement

It is not slow migration rather than dynamic, load-induced micromovement which inhibits bone ingrowth into an implant. Stress-induced micromovement may be measured in cadavers, but migration can only be measured clinically. Both measurements are valuable and important to perform. It would be most valuable to be able to follow up the clinical measurements on an individual during surgery, along the lines of Yoshida's method (Chapter 16, p. 124), with RSA studies. Stress-induced RSA studies conducted at 6 weeks post-surgery have been conducted at Lund. That stage may be sufficiently late that the micromovement observed may be the cause of later loosening rather than the result of poor initial fixation.

Theoretical studies

The order of magnitude of stability in cadaver studies is the same as the order of magnitude of stability in a patient. Finite element studies of the same configurations have also demonstrated the same order of magnitude of micromovement. It seems that a relationship may hold through from mathematical modelling to cadaver studies, to early micromovement, and to late micromovement in which the earlier measures are able to predict the later outcomes.

Convincing our colleagues

The orthopaedic establishment remains to be convinced of the relationship between earlier theoretical calculations, cadaver studies, early clinical measurements, and later clinical measurements. It is possible that links are not universal, and that every prosthesis may act differently. To prove our hypothesis, long-term studies will have to be conducted to demonstrate that these links exist.

In the case of the link between theoretical analysis and laboratory models, McTague had presented a paper at the meeting on strain gauge measurements (unpublished) which illustrated the importance of establishing links between theory and practice. Strain-gauge reinforcement appears to be dependent on many factors, for example the strain-gauge backing, the substrate material, the effect of solder tabs, the thickness of the section under investigation, the modulus of the section, the level of excitation of the strain gauge and so on. It is not possible to apply an empirical formula which will correct for all these factors, so it is necessary to conduct a fundamental test using laboratory conditions with actual strain-gauges, actual excitation levels, and on similar materials of similar thickness.

The meeting had also been concerned about the great difficulty of modelling the interface between prosthesis and bone by laboratory analysis or mathematical finite element analysis. McTague had considered that this was a reason not to rely on a mathematical analysis alone but to to conduct fundamental laboratory tests as well. Laboratory tests will often reveal over-simplifications in finite element analysis which can be corrected without difficulty to produce a mathematical model which accurately reflects the real world situation. It is essential to validate fundamental theory and finite element analysis by experiment.

Concerning the link between early *in-vivo* measurements and later outcome, the ability and will to measure micromovement *in vivo* exists only in a very few places in the world, and those chiefly in Europe. There are even fewer groups who have measured migration accurately for more than 5 years and whose data may corroborate the hypothesis. RSA is a research tool, invaluable for validating any other technique, but as an everyday clinical tool it is too expensive, too complicated, and too time consuming. Can an alternative, simpler method be found which does not make the great experimental demands of RSA, particularly in the case of the hip × The technique would probably have to use a standard X-ray and might possibly involve special markers on the prosthetic components. Although some believed that such a technique might be developed, the workshop was very concerned that the important rotational movements which happen in the femoral component might be missed. RSA remains the definitive method for measuring micromovement *in vivo*.

References to Section 2

Amstutz, C., Harlan, M. D., Nasser, S., More, R. C., and Kabo, J. M. (1989). The anthropometric total femoral hip prosthesis. *Clinical Orthopaedics and Related Research*, **242**, 104–22.

Anthony, P. P., Gie, G. A., Ling, R. S. M., and Howie, C. R. (1990). Localized endosteal bone lysis in relation to the femoral components of cemented total hip arthroplasties. *Journal of Bone and Joint Surgery*, **72B**, 971–9.

Ashman, R. R., Donofrio, M., Cowin, S. C., and von Buskirk, W. C. (1982). Postmortem changes in the elastic properties of bone. *Orthopaedic Transactions*, **6**, 223–4.

Berme, N. and Paul, J. P. (1979). Load actions transmitted by implants. *Journal of Biomedical Engineering*, **1**, 268–72.

Bloebaum, R. D., Merrell, M., Gustke, K., and Simmons, M. (1991). Retrieval analysis of a hydroxyapatite-coated hip prosthesis. *Clinical Orthopaedics and Related Research*, **267**, 97–102.

Burke, D. W., O'Connor, D., Zalenski, E. S., Jasty, M., and Harris, W. H. (1991). Micromotion of cemented and uncemented femoral components. *Journal of Bone and Joint Surgery*, **73B**, 33–7.

Caravia, L., Dowson, D., Fisher, J., and Jobbins, B. (1990). The influence of bone and bone cement debris on counterface roughness in sliding wear tests of ultra-high molecular weight polyethylene on stainless steel. *Proceedings of the Institution of Mechanical Engineers H, Journal of Engineering in Medicine*, **204**, 65–70.

Chafetz, N., Sheldon, B., Murray, W. R., Genant, H. K., and Kohr, E. L. (1985). Subsidence of the femoral prosthesis. *Clinical Orthopaedics and Related Research*, **201**, 60–7.

Crowninshield, R. D. and Brand, R. A. (1981). The prediction of forces in joint structures: distribution of intersegmental resultants. *Exercise and Sports Science Reviews*, **9**, 159–81.

Davy, D. T., Kotzar, G. M, Brown, R. H., Heiple, K. G., Goldberg, V. M., Heiple, K. G. Jr *et al.* (1988). Telemetric force measurements across the hip after total arthroplasty. *Journal of Bone and Joint Surgery*, **70A**, 45–50.

DiCarlo, E. F. and Bullough, P. G. (1992). The biologic responses to orthopaedic implants and their wear debris. *Clinical Materials*, **9**, 235–60.

Dorr, L. D. (1983). Total hip arthroplasties in patients less than 45 years old. *Journal of Bone and Joint Surgery*, **65A**, 474–9.

Engh, C. A. (1983). Hip arthroplasty with a Moore prosthesis with porous coating. A five year study. *Clinical Orthopaedics*, **176**, 52–66.

English, T. A. and Kilvington, M. (1978). *In vivo* records of hip loads using a femoral implant with telemetric output (a preliminary report). *Journal of Biomedical Engineering*, **100**, 72–8.

Finlay, J. B., Rorabeck, C. H., Bourne, R. B., and Tew, W. M. (1989). *In vitro* analysis of femoral strains using PCA femoral implants and a hip-abductor muscle simulator. *Journal of Arthroplasty*, **4**, 335–45.

Gruen, T. A., McNeice, G. M., and Amstutz, H. C. (1979). 'Modes of Failure' of cemented stem-type femoral components. *Clinical Orthopaedics and Related Research*, **141**, 17–27.

Hale, D. G., Lee, A. J. C., Ling, R. S. M., and Hooper, R. M. (1991). Debris production by the femoral component in cemented total hip replacement. *Journal of Bone and Joint Surgery*, **73B** (*Suppl.* 1) 18.

Harris, W. H., Mulroy, R. D., Maloney, W. J., Burke, D. W., Chandler, H. P., and Zalenski, E. B. (1991). Intraoperative measurement of rotational stability of femoral components of total hip arthroplasty. *Clinical Orthopaedics and Related Research*, **266**, 119–26.

Hua, J., Iguchi, H., and Walker, P. S. (1991). Comparison of fit and fill between custom stems and standard stems. *Proceedings of the ISSCP*, San Francisco, p. 65.

Kattapuram, S. V., Lodwick, G. S., Chandler, H., Khurana, J. S., Ehara, S., and Rosenthal, D. I. (1990). Porous-coated anatomic total hip prostheses—radiographic analysis and clinical correlation. *Radiology*, **174**, 861–4.

Kotzar, G. M., Davy, D. T., Goldberg, V, M., Heiple, K. G., Berilla, J., Heiple, K. G. Jr *et al.* (1991). Telemeterized *in vivo* hip joint force data: a report on two patients after total hip surgery. *Journal of Orthopaedic Research*, **9**, 621–33.

Ling, R. (1986). Observations on the fixation of implants to the bony skeleton. *Clinical Orthopaedics*, **210**, 80–96.

Loudon, J. R. (1986). Femoral prosthetic subsidence after low-friction arthroplasty. *Clinical Orthopaedics and Related Research*, **211**, 134–9.

London, J. R. and Charnley, J. (1980). Subsidence of the femoral prosthesis in total hip replacement in relation to the design of the stem. *Journal of Bone and Joint Surgery*, **62B**, 450–3.

McElhaney, J., Fogle, J., Byars, E., and Weaver, G. (1964). Effect of embalming on the mechanical properties of beef bone. *Journal of Applied Physiology*, **19**, 1234–6.

McKellop, H., Ebramzadeh, E., Niederer, P. G., and Sarmiento, A. (1991). Comparison of the stability of press-fit hip prosthesis femoral stems using a synthetic model femur. *Journal of Orthopaedic Research*, **9**, 297–305.

Nistor, L., Blaha, D., Kjellström, U., and Selvik, G. (1991). *In vivo* measurements of relative motion between an uncemented femoral total hip component and the femur by Roentgen stereophotogrammetric analysis. *Clinical Orthopaedics and Related Research*, **269**, 220–227.

Nunn, D., Freeman, M. A. R., Tanner, K. E., and Bonfield, W. (1989). Torsional stability of the femoral component of hip arthroplasty. Response to an anteriorly applied load. *Journal of Bone and Joint Surgery*, **71B**, 452–5.

Oonishi, H., Yamamoto, M., Ishimaru, H., Tsuji, E., Kushitani, S., Aono, M. *et al.* (1989). The effect of hydroxyapatite coating on bone growth into porous titanium alloy implants. *Journal of Bone and Joint Surgery*, **71B**, 213–16.

Paul, J. P. and McGrouther, D. A. (1975). Forces transmitted at the hip and knee joint of normal and disabled persons during a range of activities. *Acta Orthopaedica Belgica*, **41** (*Suppl.* 1), 78–88.

Phillips, T. V. V., Nguyen, L. T., and Munro, S. D. (1991). Loosening of cementless femoral stems. A biomechanical analysis of immediate fixation with loading vertical, femur horizontal. *Journal of Biomechanics*, **24**, 37–48.

Phillips, T. W. Messieh, S. S., and McDonald, P. D. (1990). Femoral stem fixation in a biomechanical comparison of cementless and cemented prostheses. *Journal of Bone and Joint Surgery*, **72B**, 431–4.

Rothman, R. H. and Cohn, J. C. (1990). Cemented versus cementless total hip arthroplasty. *Clinical Orthopaedics and Related Research*, **254**, 153–69.

Rydell, N. W. (1966). Forces acting on the femoral head-prosthesis: A study on strain gauge supplied prostheses in living persons. *Acta Orthopaedica Scandinavica*, **37** (*Suppl.* 88).

Schneider, E., Eulenberger, J., Steiner, W., Wyder, D., Friedman, R. J., and Perren, S. M. (1989*a*). Experimental method for the *in vitro* testing of the initial stability of cementless hip prostheses. *Journal of Biomechanics*, **22**, 735–44.

Schneider, E., Kinast, C., Eulenberger, J., Wyder, D., Eskilsson, G., and Perren, S. M. (1989*b*). A comparative study of the initial stability of cementless hip prostheses. *Clinical Orthopaedics and Related Research*, **248**, 200–9.

Selvik, G. (1989). A Roentgen stereophotogrammetric method for the study of kinematics of the skeletal system. *Acta Orthopaedica Scandinavica*, **60** (*Suppl.* 232).

Selvik, G. (1974) A Roentgen stereophotogrammetric system for the study of the kinematics of the skeletal systems. Thesis. University of Lund, Sweden. Reprint in: (1989). *Acta Orthopaedica Scandinavica*, **60** (*Suppl.* 232).

Snorrason, F. and Kärrholm, J. (1990). Early loosening of revision hip arthroplasty. A Roentgen stereophotototogrammetric analysis. *Journal of Arthroplasty*, **5**, 217–29.

Sugiyama, H., Whiteside, L. A., and Kaiser, A. D. (1989). Examination of rotational fixation of the femoral component in total hip arthroplasty: A mechanical study of micromovement and acoustic emission. *Clinical Orthopaedics and Related Research*, **249**, 122–8.

Swanson, S. A. V. and Freeman, M. A. R. (1971). The fatigue properties of human cortical bone. *Medical and Biological Engineering*, **9**, 23–32.

Walker, P. S., Schneeweis, D., Murphy, S., and Nelson, P. (1987). Strains and micromotions of press-fit femoral stem prostheses. *Journal of Biomechanics*, **20**, 693–702.

Waterhouse, R. B. and Lamb, M. (1980). Fretting corrosion of orthopaedic implant materials by bone cement. *Wear*, **60**, 357–68.

Whiteside, L. A. (1989). The effect of stem fit on bone hypertrophy and pain relief in cementless total hip arthroplasty. *Clinical Orthopaedics and Related Research*, **247**, 138–47.

Willert, H. G. and Semlitsch, M. (1977). Reactions of the articular capsule to wear products of artificial joint prostheses. *Journal of Biomedical Materials Research*, **11**, 157–64.

Williams, J. F. and Svensson, N. L. (1968). A force analysis of the hip joint. *Biomedical Engineering*, **3**, 365–70.

Witt, J. D. and Swann, M. (1991). Metal wear and tissue response in failed titanium alloy total hip replacements. *Journal of Bone and Joint Surgery*, **73B**, 559–63.

Wroblewski, B. M. (1979). The mechanism of failure of the femoral prosthesis in total hip replacement. *International Orthopaedics*, **3**, 137–9.

Wykman, A. and Goldie, I. (1984). A prospective randomized study comparing results of cemented and non-cemented endoprosthetic fixation in the surgical treatment of osteoarthrosis of the hip. *Engineering in Medicine*, **13**, 181–4.

Zhou, X. M., Walker, P. S., and Robertson, D. D. (1990). Effect of press-fit femoral stems on strains in the femur—a photoelastic coating study. *Journal of Arthroplasty*, **5**, 71–82.

Section 3—Measurement Techniques

14 A method for the measurement of the migration of prostheses from single standard radiographs

G. Liossis, A. K. Forrest, and M. A. R. Freeman

Introduction

The standard radiograph remains the routine diagnostic method for the assessment of the condition of joint implants. For the majority of surgeons it only provides qualitative information about the fixation of prosthetic components such as the appearance of radiolucent zones in cemented systems. Numerical measurement of the migration of prostheses is rare, and usually inaccurate. A method developed and tested at Imperial College and the Royal London Hospital aims to provide a usefully accurate quantitative tool. Numerical knowledge of the amount of migration is not only important for diagnostic purposes but also for recording the condition of a prosthesis over a long period, something of principal importance for the evaluation and improvement of prosthetic designs.

Measurement of the migration of bone implants has been a research topic for several years. A number of methods have been developed with varying degrees of accuracy. They can be broadly divided into two main categories: methods using single standard radiographs and stereo methods using pairs of radiographs taken simultaneously at an angle. Stereo methods provide a full description of the three-dimensional motion of the prosthesis. They have reached very high accuracies in the order of 0.1 mm, but they have not found wide clinical use because they are laborious and require specialized equipment and staff. Their high cost of operation means that only a very small number of research centres can afford them. Thus something simpler and cheaper is needed for the majority of hospitals (Turner-Smith 1990). Single radiograph methods use the standard radiographs taken under normal hospital conditions and attempt to extract the maximum amount of information possible. The constraint of using a single radiograph means that these methods cannot provide a full description of the three-dimensional motion. However, loosening modes are usually limited to predictable patterns and the majority of migration is frequently contained in a plane.

None of the currently available single radiograph methods has found wide clinical use either because they are insufficiently accurate or because their accuracy is not well established, i.e. it is unclear under what radiographic conditions they were measured, what parameters influence them and how. Exceptions to this are some recent methods (e.g. Russe 1988), where effort has been expended on the detailed examination of errors.

Background

Single radiograph methods

The main advantage of single radiograph methods is that they utilize the conventional radiographs included in patient follow-ups. Implementation is inexpensive since most of them only require a pencil and a ruler, or at most a digitizer and a personal computer. Operational costs are low too, because measurements can be carried out fast and no specialized staff are necessary. As a result, these methods can be used for a large number of patients in a large number of hospitals.

Sutherland method

Sutherland *et al.* (1982) developed a pencil and ruler method for the measurement of migration of the hip joint components. Figure 14.1 shows the landmarks used for measurement. The position of the acetabular cup is defined by the point which bisects the long axis of the ellipse projected on the radiograph by the metal ring at the rim of the cup, and which is referred to as the hip centre (HC). The landmarks defining the position of the pelvis are the inferior aspect of the teardrop (TD), the Kohler line (K) and the sacroiliac line

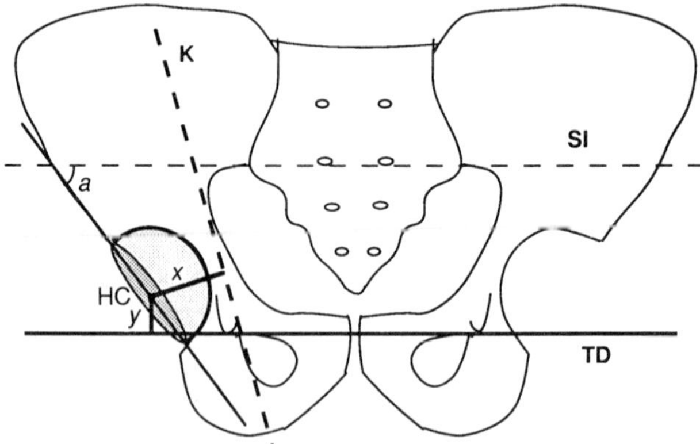

Fig. 14.1 The landmarks used by Sutherland *et al.* (1982) for the measurement of acetabular migration are the Kohler line K, the inferior aspect of the teardrop TD, and the sacroiliac line SI.

(SI). Transverse (x) and vertical (y) acetabular migrations are measured as the change in distance between the hip centre and the Kohler and teardrop lines respectively. The angle of inclination of the acetabular component (a) is measured as the angle formed by the sacroiliac line and the long axis of the acetabular ring. A variation of the method is used for the measurement of the migration of the femoral component. Sutherland estimates the error of the method for any of the migration components to be no more than 5 mm. The advantages of the method are that it is inexpensive, simple and fast, and can be used both in prospective and retrospective studies. Its main disadvantage is the low accuracy of the results.

Wetherell method

In his study Wetherell *et al.* (1989) examines the accuracy of a number of methods and proposes new lines to be used as landmarks in the skeletal bed. For the determination of vertical migration (y) he proposes a line linking the acetabular margins, the acetabular line (AM in Fig. 14.2). For the transverse migration he proposes a line tangential to the pelvic brim and passing through the midpoint of the obturator foramen at its greatest horizontal span (obturator/brim—OB in Fig. 14.2). Using a dry bone model Wetherell proved that these lines are the most reliable landmarks. In a set of measurements where the pelvis was rotated from -10 to $+10°$ (relative to a neutral position), first around a transverse and then around a vertical axis, he measured the mean errors of his technique to be no more than 0.5 mm and 1 mm for vertical and transverse migration respectively. The method is inexpensive and simple

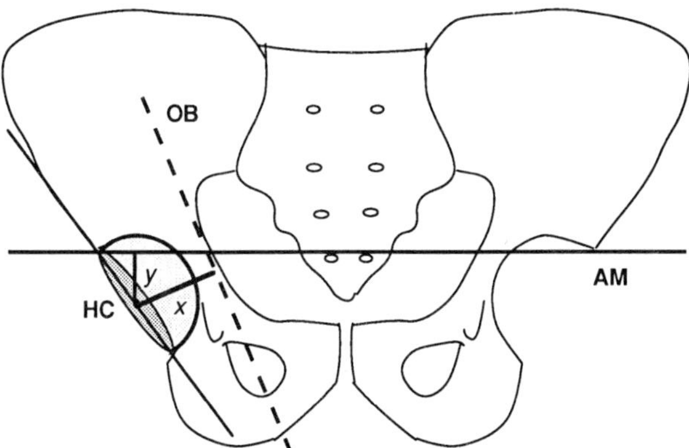

Fig. 14.2 The landmarks used by Wetherell *et al.* (1989) for the measurement of acetabular migration are the line linking the acetabular margins AM and a line tangential to the pelvic brim and passing through the midpoint of the obturator foramen OB.

and theoretically it provides good accuracy. In practice it may be difficult to locate the bone landmarks on patients accurately.

The Maxima technique

This is an image analysis method developed at the University of Manchester (Hardinge *et al.* 1991; Jones *et al.* 1992). The system consists of a personal computer, a light box, a video camera, and a high resolution monochrome television camera. The image of the backlit radiograph is captured by the video camera whose output is fed into a frame grabber in the computer which converts it into a digital image. The digital image is stored in the computer and is subsequently processed by the image analysis software. Measurements are made either automatically by the computer, or interactively with the aid of the user. The image analysis software has two main functions: to improve the quality of the image and to quantify predefined parameters. A poor radiograph can be manipulated to enhance its intensity, contrast, or consistency of exposure by filtering the image mathematically. The measurements that can be obtained by the technique include stem subsidence, cup migration, cup wear, and stem loosening. No markers are necessary for the definition of landmarks on the bone and therefore the method can be used both in prospective and retrospective studies. The authors estimate the precision of the technique to be ± 0.5 mm. This estimate is based on the variation of subsidence values for the femoral component in a series of radiographs of the same patient and not on a detailed theoretical or experimental investigation of the errors.

Ebra method

This method was developed at the University of Innsbruck for the measurement of hip prosthesis migration. It is computerized, requiring a digitizer for the measurement of points on radiographs connected to a personal computer for processing. A total of 18 points are necessary for the determination of the migration of a polyethylene acetabular cup or 21 for a metal backed cup (Russe 1988). The first 8 points are natural landmarks on the pelvis, the rest are points on the periphery of the prosthesis. The computer software incorporates a quality control system which determines which of the radiographs are suitable for processing. Depending on an accuracy criterion whose value is set by the user, a number of unsuitable radiographs may be rejected. The method uses a technique for the elimination of systematic errors. An increased number of radiographs for a patient leads to more effective elimination of systematic errors. As a result the accuracy of the method can increase when more radiographs of the same patient become available with time. The method demands a minimum of four radiographs per patient and therefore the first results will come later than in other methods. Using mathematical calculations the authors estimate the accuracy to be better than ± 1 mm, with most of the errors being below ± 0.7 mm.

University of Washington method

Lippert *et al.* (1982*b*) developed a method at the University of Washington in the early eighties. It requires the bone to be marked with ball markers and the prosthesis to have landmarks which are visible from various angles. Additionally it requires a control frame containing steel balls which fits around the patient's hip or knee joint. The control frame balls are all contained in an imaginary plane which passes through the joint and which is parallel to the X-ray film. Using the positions of the points on the frame and their respective positions in the image, the transformation from the one plane to the other is established. Using this transformation, all the positions of prosthesis landmarks and ball markers on the film are transformed to the control frame plane. The distance between the prosthesis and the bone landmarks is computed for various loading conditions. Loosening is indicated by a change in the distance between the landmarks in the bone and those on the prosthesis. The accuracy reported is ± 0.2 mm. This figure was obtained using a model of the knee joint. It is not clear how the experiment was conducted and it appears that no proper consideration was given to changes in patient orientation or changes in radiographic settings.

MIRA method

MIRA was developed in Sweden for the measurement of the migration of the tibial component of the knee (Carlsson *et al.* 1992). The prosthesis contains four landmarks, two visible on antero–posterior and two visible in lateral radiographs. Four radiopaque ball markers are implanted in the bone in specified positions below the prosthesis landmarks so that four distances can be measured on standard radiographs. The aim of the method is to measure subsidence of the tibial component. In a study of 21 knee arthroplasties in patients, the accuracy of the method was established by comparison to Roentgen stereophotogrammetric analysis (RSA, a stereo method of proven high accuracy). The standard deviation of the error was measured to be 0.5 mm. The method requires a digitizer and a personal computer.

Other single radiograph methods

Kirkpatrick *et al.* (1982) uses a chariot as a patient support system in order to obtain consistent radiographic visualization of the hip. Amstutz *et al.* (1986) standardizes the radiographic conditions further by adding a grid. The grid is engraved on a plexiglass sheet which is supported by four uprights and centred over the centre of the pubic symphysis by the radiographer. No measurement accuracies were reported. Collet *et al.* (1985) describes another method without giving accuracy details. Nunn (1989) developed a pencil and ruler method with an accuracy of ± 3 mm.

In a study by Ilchman *et al.* (1992), four of the single radiograph methods measuring acetabular migration are compared with RSA. The study examines the accuracy of the Sutherland method, the Wetherell method, a method developed by Sultzer, and EBRA. While the first three produce results of similar accuracy (mean error slightly less than 1 mm), EBRA produces more accurate results with a mean error of around 0.4 mm. The mean error is defined as the mean of the absolute differences between the values obtained with the compared method and the corresponding RSA values. A factor increasing the accuracy of EBRA is its control system which rejects some of the most unsuitable radiographs thus avoiding rather than correcting big errors.

Stereo methods

Since the 1970s a number of groups have developed X-ray stereophotogrammetric systems. They have now reached accuracies in the order of 0.1 mm. Stereo systems need to be calibrated, which means that the exact position of the X-ray source relative to the film has to be determined. Calibration objects are used during radiography which are typically plates of plexiglass or mirror glass rigidly fixed together and marked with small metal markers (Selvik 1989; Takamoto 1976). Once a radiographic system has been calibrated, given the position of an object point on the film, it is possible to obtain the mathematical expression for the ray which emanates from the source, passes though the object point and hits the film. The three dimensional position of the point can be obtained as the intersection of two different rays whose mathematical expressions are known. Stereo systems therefore require simultaneous exposure from two X-ray sources. Typically the beams of the X-ray sources are either perpendicular to each other (biplanar system) or at a smaller angle, e.g. 40° (convergent system) (Veress 1989). The choice depends mainly on the anatomy of the joint, e.g. convergent methods are usually preferable for the hip and biplanar methods for the knee.

Radio stereophotogrammetric analysis (RSA)

Selvik developed his system in 1974 at the University of Lund (Selvik 1989). RSA is currently the best established stereo method and a large number of research studies have been completed for a wide range of measurements. Accurate marking of both the bone and the implants is critical. The bone is marked with small tantalum balls. Mjöberg *et al.* (1986) reports the measurement of migration for both the acetabular and the femoral component of the hip. The standard deviation of the error was measured to be around 0.1 mm, with small variations for the different components of migration along the transverse, vertical and longitudinal axes. Ryd (1986) reports the application of the method to the knee with accuracies better than 0.1 mm.

Other research groups have developed their own stereo systems in the USA (Brown *et al.* 1976; Green *et al.* 1983; Veress *et al.* 1977) and Germany (Probst 1980).

Method

Radiographs of the same patient cannot be compared directly for two main reasons:

1. The position and orientation of the patient (relative to the film) varies. It is practically impossible to fix the patient in the same position every time that a radiograph is taken.
2. The setting of the radiographic equipment does not remain constant (i.e. the distance between the X-ray source and the film is not fixed).

As a result the relative positions in the films of various landmarks on the skeletal bed change, and it therefore becomes very difficult to determine accurately whether a component has migrated. The majority of single radiograph methods aim to find landmarks whose position relative to the prosthesis does not change significantly. The method developed at Imperial College and the Royal London Hospital is based upon a different approach: the use of the prosthesis as a calibration object in which the most important parameters of the radiographic process are estimated and used to mathematically transform all radiographs of the same patient in the same coordinate system.

In order to measure the migration of a prosthesis, identifiable reference points in the skeletal bed are necessary. Natural references are extremely difficult to locate accurately and we consider marking of the bone with small ball-bearings as necessary for accurate results. The technique of implanting ball markers is now standard and does not present any particular complications. In the authors' method, two implanted ball bearings act as a reference against which the migration of the implant is measured. Three identifiable points on the prosthesis define it as a rigid body. A total of five points are measured in every radiograph using a digitizer. The five positions are stored in files in a computer and subsequently processed to produce a result in the form of translational migration, consisting of linear components along a vertical (y) and a transverse (x) axis, and a rotational component around an axis perpendicular to the film. Figure 14.3 shows an acetabular cup

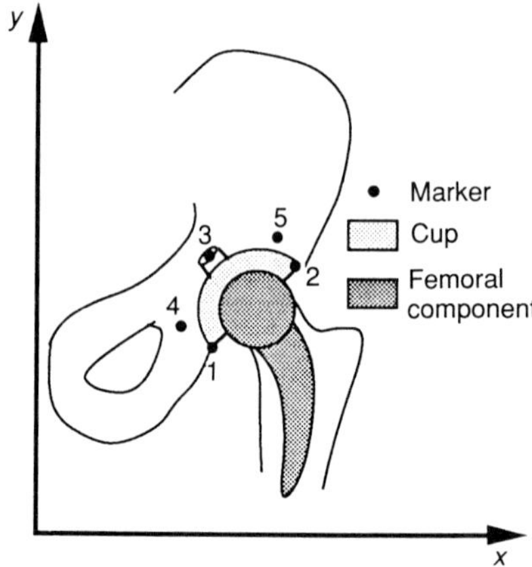

Fig. 14.3 Reference points on the acetabular cup.

with three reference points on it (marked 1–3) and two ball markers implanted in the bone (marked 4 and 5).

The method can be applied to measurement of migration of any prosthetic component provided that it can be marked appropriately (i.e. the prosthesis contains three clearly identifiable points, and two ball markers are implanted in the bone). For good accuracy all five reference points should lie in the same plane. The authors have already successfully tested the method in the acetabular component of the hip and the tibial component of the knee. The femoral component of the hip is another area where the method can be applied but the accuracy for that component has not yet been established.

Results

The theoretical foundations of the authors' method were tested using a mathematical model and the results validated by practical work.

Mathematical simulation

A program was devized to simulate the radiographic process whereby three-dimensional objects are transformed into two-dimensional images. The transformation depends on a number of parameters such as the distance between the film and the source, the six degrees of freedom for the movement of the patient, the three-dimensional positions of the reference points on the prosthesis, and the positions of the ball markers. Two simulated radiographs were created, one of which contained a migration transformation combining rotational and translational components. A total of 30 parameters were necessary for the complete description of the transformations in the two radiographs. The error of the method was obtained as the difference between the known migration used as a parameter in one of the radiographs, and the calculated value obtained by processing the simulated radiographs with the authors' method. The values of the parameters could be varied and the effect on the error observed. From statistical analysis of a large number of simulated radiographs the authors concluded that translations of the patient and the distance of the X-ray source from the film have only a small effect on the error. Different orientations of the patient have a much larger effect, while the position of the ball markers with respect to the reference points on the prosthesis is the most influential factor. Ideally all five reference points should lie in the same plane. The accuracy of the method reduces drastically if the ball bearings are placed at a distance of more than 1 cm from the plane defined by the prosthesis reference points. Using the symbolic computation package MACSYMA, we derived an expression for the error as a function of the radiographic parameters. Due to the large number of parameters, the equation is several hundred lines long and a simple expression for the error could not be obtained.

Tests on a plastic pelvis

The method was tested for the acetabular component of the hip in a series of experiments. A metal backed acetabular cup together with two ball markers was implanted in a plastic pelvis and a series of radiographs were taken where the pelvis was rotated from -10 to $+10°$ (relative to a central position) in steps of 2°, first around a vertical and then around a transverse axis. Any migration found by comparing two radiographs is an error since the prosthesis remained in the same position for the whole duration of the experiment. All errors were below 1 mm. The root mean square error for the linear and rotational components of migration are shown in Table 14.1.

In a second experiment, migrations were introduced in addition to the variations in orientation of the plastic pelvis. The migrations were created by moving the prosthesis by hand and their exact magnitude and direction was measured by taking stereo exposures and processing the films with RSA, a stereo method of proven accuracy. The prosthesis was moved by up to 5 mm and rotated by up to 6° relative to the plastic pelvis. The study consisted of only a small number of radiographs and the results are shown in the third row of Table 14.1. Migration errors were higher than for the previous experiments, but still within a reasonable range. Only one error was higher than 1 mm.

Test on knee-replacement X-rays (comparison with RSA)

Eighteen patients participated in an RSA study of the migration of the tibial component of the knee. The study started before the development of the

Table 14.1 Root mean square error of the method in the practical tests

Test	*x*-Migration (mm)	*y*-Migration (mm)	Total translational migration (mm)	Rotational migration (degrees)	Number of cases
Hip rotation around a vertical axis	0.27	0.09	0.29	0.33	10
Hip rotation around a transverse axis	0.08	0.16	0.17	0.34	10
Comparison with RSA (acetabular component)	0.45	0.75	0.88	1.17	6
Comparison with RSA (tibial component)	0.48	0.29	0.57	0.58	18

authors' method, but the ball markers were placed in a satisfactory position and the radiographs could be measured by the authors' method. The root mean square errors for the translational and rotational components of migration are shown in the fourth row of Table 14.1. A close correlation was obtained between the measurements made by the authors' method and those made by RSA, especially for the measurement of vertical migration (i.e. subsidence) where the root mean square error falls below 0.3 mm.

Discussion

The mathematical investigation of the method now proposed showed that from the large number of parameters influencing the error, the two most critical are the orientation of the patient and the positioning of the ball markers in the bone. Provided that the pelvis of the patient is not turned excessively (i.e. more than $\pm 15°$) and that the ball makers are implanted near specified positions (i.e. placed in or near the same plane), the root mean square error of the method should be less than 0.5 mm for x and y translations and 0.5° for in-plane rotation.

The aim of the practical tests was to verify these conclusions. The results from the first two experiments on the plastic pelvis were very satisfactory since all root mean square errors were below 0.3 mm. In the comparative study with

RSA they deteriorated, but this can be attributed to three factors. First, the accuracy of RSA is not as good for the prosthesis used in this study, as it is for other prostheses. The error can occasionally reach values of 1 mm or more and from the observation of certain parameters it appears that the RSA measurements in the authors' tests were not as reliable as expected. Second, the changes in orientation of the pelvis were not controlled and measured as in the first experiments and it is possible that the plastic pelvis might have been rotated excessively (i.e. more than 15°). Thirdly, the magnitudes of migration used were very high (up to 5 mm). In the comparative study in the knee, where the migrations were smaller and where RSA results have been demonstrated to be very accurate (Ryd 1986), good results were obtained, with root mean square errors below 0.5 mm for both vertical and transverse migration. The determination of rotation was also very accurate with a root mean square error of 0.58°.

The authors believe that the new method offers a quantitative tool which facilitates the acquisition of more precise information from standard radiographs.

Its main advantages are:

1. Ease of use. No specialized staff or radiographic equipment is necessary.
2. High speed of processing radiographs. Only a small number of points on the radiograph need to be recorded and measurement can be completed very quickly.
3. Low cost of implementation. Only a digitizer and a personal computer are necessary.
4. Flexibility. The method is not restricted to a specific component. It has been successfully applied to the acetabular cup and the tibial component of the knee. The femoral component of the hip as well as the wear of the acetabular cup have been shown theoretically to be applicable areas but no practical tests have been completed yet.

The method is not restricted to the measurement of antero-posterior radiographs. It can be equally well be applied to the measurement of lateral radiographs, provided that the prosthesis is appropriately marked. In this way the components of the three dimensional motion (z-migration) that are impossible to measure on antero-posterior radiographs can be measured.

Acknowledgement

All the work related to the RSA measurements was carried out with the help of Soren Toksvig Larsen from the University of Lund Hospital.

Debate in the meeting

The meeting wondered how it was possible to place the balls exactly where they should be, as they were required to lie in a plane. The author admitted that in the hip this was very difficult but in the knee the bone markers could be placed fairly accurately. The method is reasonably accurate if the balls can be placed within ± 5.0 mm.

Was there any way that the accuracy of placing of the markers could be assessed? Certain parameters in the transformation do indicate whether something has gone wrong in a particular radiograph, but do not indicate whether the error is due to ball misplacement or some other problem.

15 Stereoradiogrammetry for the prediction of hip replacement survival

A. R. TURNER-SMITH AND C. J. BULSTRODE

Introduction

Settling of the components of a prosthetic hip implant by a millimetre or so during the first year is normal, but continued migration at a rate greater than 0.5 mm/yr appears to be associated with premature failure of a total hip replacement (THR) (Mjöberg 1986). Measurement of sinkage, therefore, provides a means of assessing new implant materials, designs and techniques which can indicate the likely longevity of an implant long before conventional prospective randomized trials. A useful clinical measurement system must be able to resolve movements of the order of 0.25 mm. An implementation of stereoradiogrammetry is described which meets this demand and incorporates some novel features to ease the process of digitizing. In this article the term 'stereoradiogrammetry' is used to stand for 'X-ray stereophotogrammetry', 'Roentgen stereophotogrammetric analysis', and other similar terms.

Stereoradiogrammetry

Direct measurement of migration from single radiographs is valid only if care is taken to re-align the patient and equipment accurately on successive examinations (Amstutz *et al.* 1986). Difficulty in identification of bony landmarks typically limits such measurements to ± 1 mm, although image-processing algorithms can improve measurement consistency.

Single film measurements, however, can only record directly two translations and one rotation. Translation in and out of the measurement plane, and rotation about the axes contained by the plane, can only be inferred by changes in magnification. One of these rotations is that of the femoral component within the femur, which is suspected to be a major contributor to the failure of THRs. Three-dimensional measurements are necessary to reveal all six of the possible movements of each component.

Fundamental to all stereo measurements is the identification of the same feature in two views (Fig. 15.1). Bone, however, provides few such unambiguous landmarks. A satisfactory solution was described by Aronson *et al.* (1974) who implanted small, radiopaque balls within the bone during surgery.

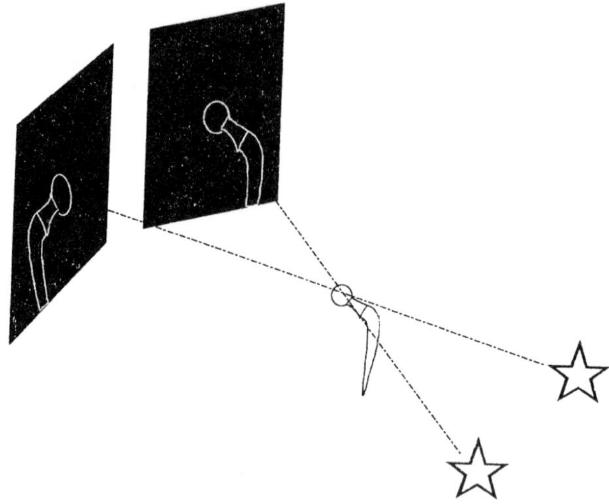

Fig. 15.1 Principle of stereoradiogrammetry.

Assuming that the bone acts as a rigid body, the matrix formed by a minimum of three marker balls defines the position of the bone unambiguously (Selvik *et al.* 1983; Lippert *et al.* 1982*a*), see Fig. 15.2. The use of more balls adds useful redundancy, rescuing the situation in which some balls have been obscured by the THR, improving overall accuracy, and revealing those balls that may have migrated within the bone (though in practice this seldom happens).

A full review of radiographic and photogrammetric techniques for radiogrammetry has been presented by Turner-Smith (1990).

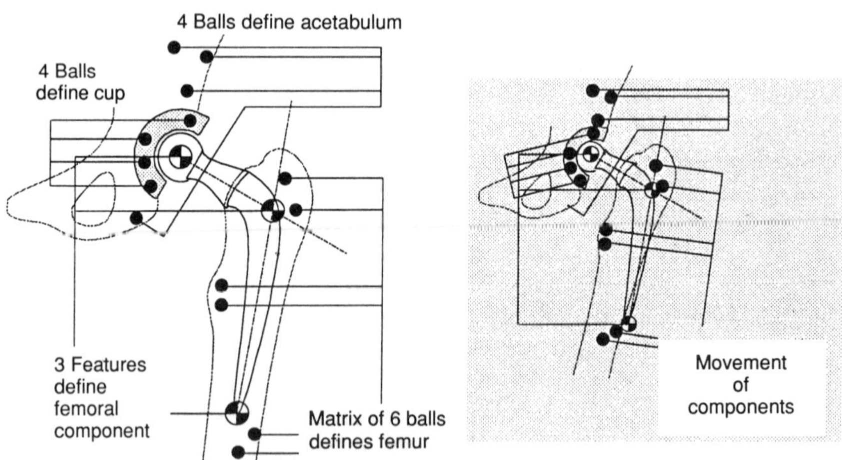

Fig. 15.2 The use of ball matrices to mark implant components.

Other implementations

Since the 1970s a number of groups world-wide have investigated the clinical use of stereoradiogrammetry. Brown *et al.* (1976) and Pearcy (1985) achieved a remarkable accuracy in identifying anatomical landmarks on the spine in their two biplanar views at 90° (error $< \pm 1.5$ mm). Following the work of Aronson *et al.* (1974), for the highest precision, steel or tantalum (high X-ray opacity) balls have been implanted to locate the bone, plastic components of prostheses, and even the edge of metal components. It is sometimes easy to mis-identify marker balls in the two views, but mis-identifications are revealed by large coplanarity errors or significant changes in the inter-marker distances.

Selvik completed his Roentgen stereophotogrammetric analysis (RSA) system in 1974, since when his clinical colleagues have published regularly on the performance of joint replacements. Baldursson *et al.* (1979) and Mjöberg (1986) have described the characteristic subsidence of both femoral and acetabular components of the THR. Lund's RSA system is a well-developed research tool that has been adopted by a number of clinical groups both in Sweden and beyond. The method of Brown *et al.* (1976) was adopted by Green *et al.* (1983) (at Cleveland) to measure hip and knee replacements. By measuring with and without weight-bearing through the joint, both reversible (loosening) and nonreversible (migration) movements were recorded. Cobalt–Chrome marker balls, 1.6–2.4 mm diameter, were used. These large balls were difficult to place in the bone and in the THR were positioned in the cement around the femoral component, precluding measurement of movement between implant and bone. In 1977 Veress *et al.* (at Seattle) developed an analytical method for tracking the motion of the human patella. The groups at Seattle and Cleveland compared their methods in routine clinical use (Lippert *et al.* 1982*a*). Chaftez *et al.* (1985) used a single X-ray source, moving it vertically between exposures. The limited resolution of their digitizing instrument and subject movement between exposures led to an overall system resolution of only about 0.4 mm, which was too large to establish statistically significant evidence of prosthetic loosening in their patients.

Findings in common

The findings of these groups confirm a common pattern of movement. After implantation in successful cases, during the first months both femoral and acetabular components generally undergo a period of rapid migration or settling, typically 1–4 mm, after which movement progressively reduces. But in some patients the movement continues, eventually leading to failure of the bone/implant interface. An accuracy of better than 0.25 mm is therefore desirable for reliable detection of THR loosening or migration.

Each of the implementations referred to above has its own advantages and disadvantages in terms of ease of use, method of calibration, or digitizing accuracy. The ultimate accuracy of all systems is primarily determined by the accuracy of the digitizing instrument. A successful design for a stereoradiogrammetric measurement system for clinical research will adopt an appropriate compromise between the finest accuracy on one hand and cost, ease of use, reliability, and adequate resolution on the other. For example, the identification of marker balls and stereo features of the prosthetic components can be immensely time-consuming.

The system described here attempts to find this compromise and automate the more taxing processes of digitization. It is primarily intended to measure THRs, but other joints and implants may also be measured.

Oxford system

Surgery

Pilot clinical trials are being conducted in Oxford on the Corin Medical Ltd, 'Hi-Nek' THR. The otherwise-invisible plastic acetabular socket has been modified to include a matrix of four balls for photogrammetric location. The pelvis and femur are marked with 1 mm diameter surgical-grade stainless steel balls. The balls have to be clearly visible on both films of the stereo pair and also be as widely separated as possible for accurate determination of orientation. Various locations have been tested in an attempt to discover easily identified anatomical landmarks which can be used to guide the surgeon.

Marker balls are placed as follows:

Acetabulum

1. The introducer needle is passed over the top of the acetabulum towards the greater sciatic notch and then withdrawn slightly and a marker ball placed in the bone as close to the greater sciatic notch as possible.
2. A second marker ball is placed in a similar way by blind location of the anterior–inferior iliac spine and then injection of a marker ball as close to this landmark as possible.
3. Two more balls are placed as high as possible on the wing of the ilium ensuring that they are not in the same horizontal plane.

Femur

4. Two marker balls are placed high on the greater trochanter, slightly anterior to the coronal plane, one above the other.
5. A second pair of marker balls are placed either in the neck of the femur or at

the base of the lesser trochanter, once again just anterior to the coronal plane. These two balls are separated vertically by 3 mm.

6. Two more balls are inserted in the distal femur 1 cm beyond the tip of the implant. These are separated vertically and rotationally by twisting the leg slightly.

Radiography

Stereoradiographs are taken at 10 days following surgery, shortly before the patient is discharged from hospital, and at annual intervals thereafter. In order to reduce artefacts due to patient movement during exposure, a support frame with arm rests and a lumbar pad is provided.

The anatomy of the hip favours an angle between beams of 60°. Films are placed perpendicular to the beams to minimize image blur, and to reduce the demand on flatness of the film. Patients are measured standing, with the X-ray beam horizontal. The tubes are set as far as possible from the film (3 m), which improves the sharpness of the image, reduces the effective source sizes, and improves image contrast. The layout of the equipment is illustrated in Fig. 15.3.

An electronic control is used to fire the X-ray tubes with a minimum, but controllable, delay in order to reduce artefacts due to subject movement. During the delay, two gravity-powered shutters are automatically released, which expose the second film and mask off the first from scatter. The total exposure time, including shutter time, is typically 0.8 s. Scatter, and

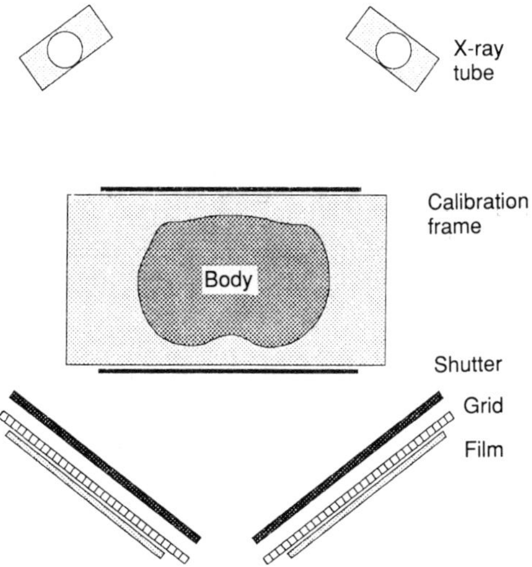

Fig. 15.3 Layout of equipment for stereoradiogrammetry.

unnecessary exposure to the patient is reduced by lead frames at the tubes, fitted with diagonal aluminium filters to reduce exposure of the thigh. The precise alignment of the patient is assisted by a low trolley on which he or she may be moved by small amounts, guided in two directions aligned with the beams.

Calibration

A calibration frame comprising a box, with a removable front panel located accurately on dowel pins, is placed around the patient. This is built of a bonded epoxy honeycomb material ('Fibrelam', manufactured by Ciba–Geigy) which is light, rigid, and transparent to X-rays. Glued to the surface of the front and back panels are sheets of glass-fibre printed circuit board material. A pattern of the calibration frame is machined on each panel, to a precision better than 25 μm. This generates a 'tartan' grid of light and dark lines on the radiograph which is easy to digitize. The panels themselves have proved stable and after a year's constant use remain flat to better than 125 μm over their entire surface of 600 × 400 mm.

Film measurement

A special digitizer system has been devized which is fast in use but which retains the nominal accuracy of a transparent electromagnetic digitizing tablet (GTCO Digi-pad, 400 mm × 600 mm working area, resolution 25 μm, calibrated to 75 μm). A CCD image sensor which views an area 6 mm × 8 mm on the film is connected to a computer. This sensor contains two coils from which both position and orientation of the camera can be determined. A feature is digitized by capturing its image in the computer, and moving a cursor on the computer screen to locate the selected object. The position of the cursor on the screen is mapped into digitizer coordinates to give the position of the feature point at the full resolution of the digitizing tablet.

Appropriately-shaped cursors may be selected on the screen image for the location of centres or edges of objects. As data is entered it is checked for accuracy, for example the linearity of coordinates belonging to a line, or the consistency of radius points which should belong to a circle. The crossing points of the grid lines are automatically identified by checking the position of one special diagonal line added to the grid for this purpose.

Analytical method

Three-dimensional photogrammetric reconstruction takes place in three stages, each of which has a simple physical interpretation: transformation of film coordinates to one plane, determination of the X-ray source positions with respect to the plane, and determination of the three-dimensional crossing point of rays forming the stereo images, see Fig. 15.4. Details are described elsewhere (Turner-Smith *et al.* 1990).

Performance

During the rectification procedure, a large number of measurements are used to find the crossing positions. The consistency may be estimated by noting how closely each ray passes its computed source position. Typical magnitudes

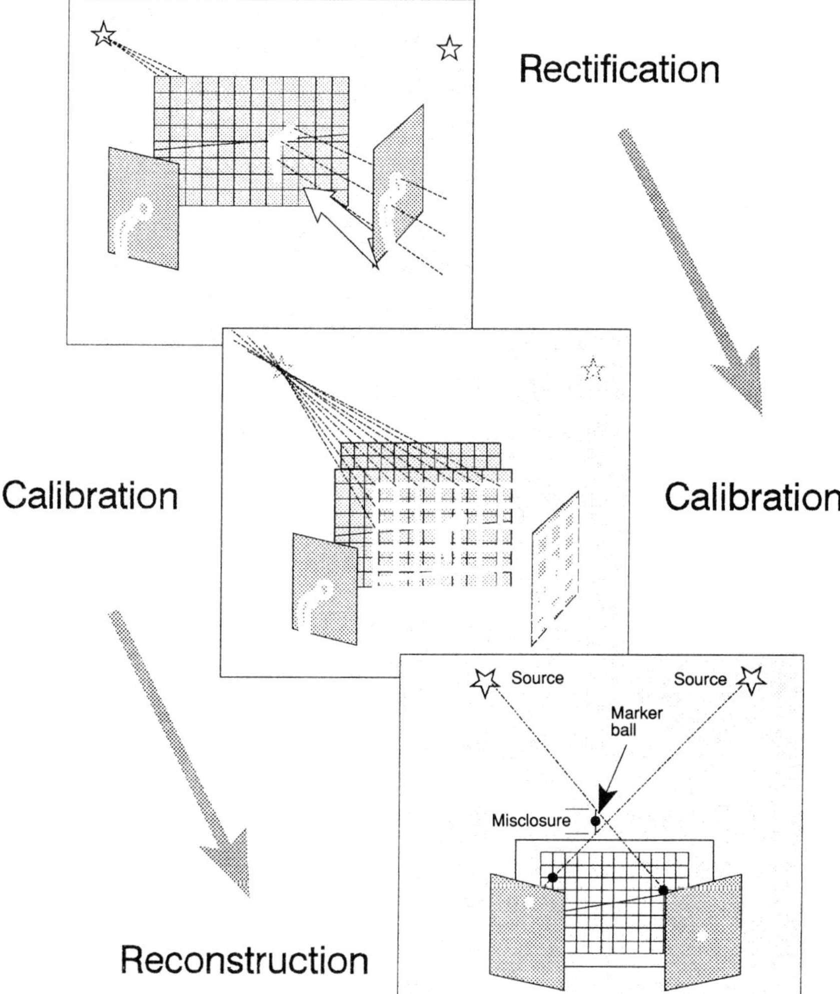

Fig. 15.4 Sequence of photogrammetric reconstruction.

of source position and standard deviation (not accuracy) are $X=1000$ mm± 0.15 mm, $Y=250$ mm± 0.15 mm, and $Z=2000$ mm± 1.5 mm.

The overall *absolute* accuracy of the system was estimated by measuring the standard deviation of the measured positions of an accurately constructed matrix of balls. Small relative movements can be measured with much higher precision. Accuracy of *relative* movement was found by repeated measurement of a random array of balls, in nine different orientations. Results are shown in Table 15.1.

Early results from repeated measurements of balls in patients show similar accuracies. The system can therefore confidently measure the relative position of bone and implant to better than 0.2 mm in three dimensions, which is sufficient for the purpose of measurement of micromovements of THRs within the bone.

Table 15.1 Absolute and relative accuracy of measurement

Number of balls	Dimensions	% Within tolerance	Tolerance	
			Absolute	Relative
1	X, Y, or Z	67 (1 SD)	± 0.10 mm	± 0.06 mm
1	3D	67 (1 SD)	± 0.17 mm	± 0.11 mm
1	3D	99.8 (3 SD)	± 0.52 mm	± 0.34 mm
3	3D	67 (1 SD)	± 0.10 mm	± 0.06 mm
3	3D	99.8 (3 SD)	± 0.26 mm	± 0.17 mm

Automatic location of balls and prosthesis

Pairing of stereo images

The two X-ray source positions and a ball centre in one film of a stereo pair define a plane. The intersection of this plane with the plane of the second film of a stereo pair is referred to as an epipolar line. The image of the ball on the second film must lie on the epipolar line, although in practice digitizing and calibration errors cause the fit to be imperfect.

Location of the prosthesis components

Acetabular cup, pelvis, and femur are located by matrices of at least four balls each. The femoral component is located by measuring three key geometrical features of its outline which, for the purposes of matching, can be treated as matrix points just like the other components:

1. The centre of the head is measured from the centres of the circles that it forms in the *film* planes, normal to the X-ray beams. The rectified images must not be used as the angle of rays passing through the reference plane distorts the circles.
2. The distal tip of the prosthesis is again defined on the film plane to ensure minimum distortion. On each film the tip is taken to lie at the intersection of the centre-line of the femoral shaft and a normal to it, drawn to form a tangent to the curve of the tip, see Fig. 15.5. This definition removes variability due to asymmetrical finishing of the component.
3. The shoulder of the prosthesis is defined as follows. The X-ray source and the centre-line of the femoral shaft on each film defines a plane. The intersection of the two planes defines a three-dimensional line which is the centre-line of the femoral component shaft. The shoulder is taken to lie at the foot of the three-dimensional perpendicular from the head to this line, see Fig. 15.6.

This algorithm is robust and can be applied to a variety of prosthesis designs. Repeated measurements of the 'Hi-Nek' in nine different orientations has

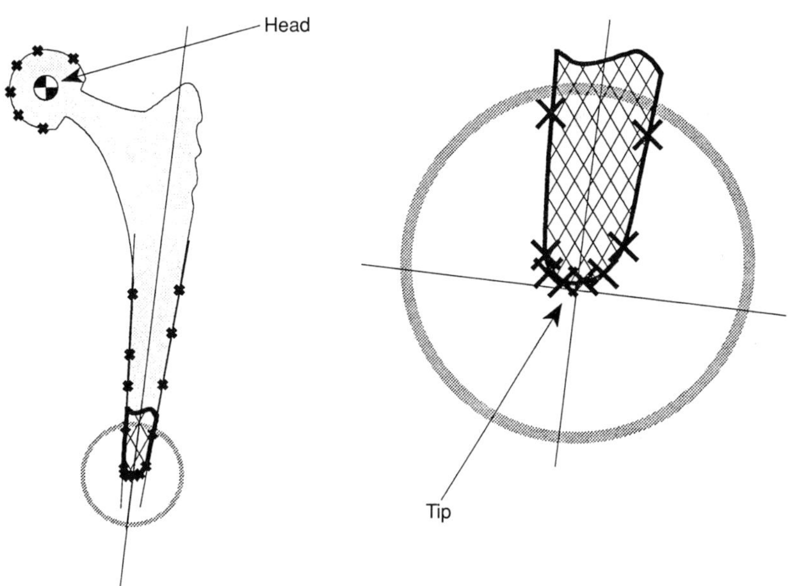

Fig. 15.5 Construction of femoral component tip.

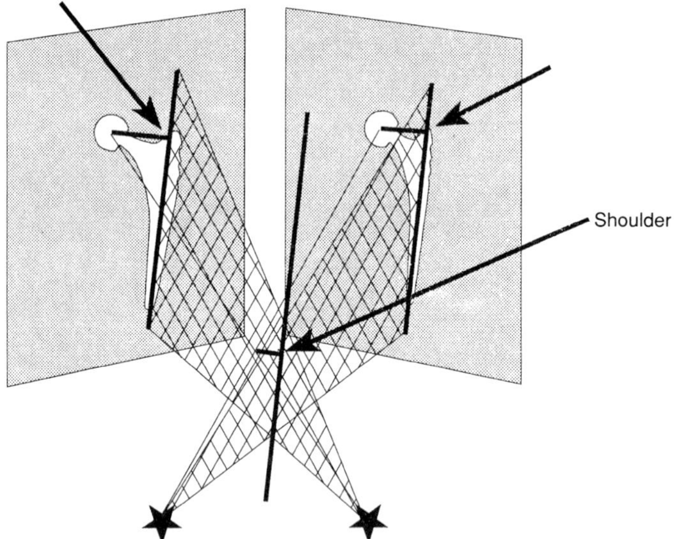

Fig. 15.6 Construction of femoral component shaft. Arrows indicate the shoulder.

shown that the head centre can be measured to ± 0.12 mm, the tip to ± 0.66 mm, and the shoulder to ± 0.14 mm, i.e. the three-dimensional position of the femoral component can be measured to ± 0.65 mm (± 1 SD), or ± 0.2 mm (± 3 SD).

Matching three-dimensional data sets

The ultimate aim of this work is to measure the movement, between prosthetic components and skeletal structures, from sequential film pairs. The positions of these objects are defined by three-dimensional ball and feature locations. An algorithm has been developed for matching two sets of locations which is independent of initial object orientations. The relative orientation and position of objects defined by sets of points is found by minimizing the sum of least squares interpoint distances by optimizing the three positions and three angles of an object at one time with respect to its position at a later time.

All the above algorithms have been implemented as a suite of FORTRAN subroutines on a Sun workstation.

Conclusion

Stereoradiogrammetry offers the only technique at present capable of predicting the lifetime of a new design of implant within a few years of implantation, by measuring component movements as small as 0.2 mm. However, its methods of radiography, calibration, digitizing, and image analysis have often appeared too complex and time-consuming for clinical research applications. A system has been described which is easy to use, and which requires a minimum of operator intervention.

Stereoradiogrammetry should prove of fundamental importance in future testing of new implants and materials, by avoiding large-scale studies on implants which are going to fail rapidly, and by accelerating testing so that new materials and designs can be brought into clinical practice without unnecessary delay.

Acknowledgements

The initial work on the system was supported by the Department of Health. Thanks are due to Agfa-Gevaert for the supply of high speed film, cassettes, and computing equipment.

16 Intra-operative measurement of tibial component micromovement in total knee arthroplasty

K. YOSHIDA, K. ASADA, AND H. SAKANE

Introduction

Is this component really stable or not? Tibial implant stability is critically dependent upon its design and the bone quality, and is also influenced by surgical technique. Quantificative and accurate measurement of micromovement immediately after operation would seem to be a reliable indicator of individual component stability. The authors have developed a measurement system for tibial component micromovement, which uses three noncontact displacement transducers. This information may be valuable for post-operative management and rehabilitation. In addition, there may be a possibility of selecting a fixation method based on the micromovement level, to control the quality of component stability during an operative procedure. Overall component stability immediately after fixation may be different from patient to patient, so we should measure the stability individually.

Method

Stability is defined as the amount of displacement when load is applied. A surgeon can push a component with a force of around 20–30 kg. If load is applied vertically to the tibial component, major displacements are vertical and horizontal displacements are minimal. Consequently, displacements of any points on the tibial component due to vertical forces can be calculated by three point measurements of vertical displacement.

Figure 16.1 shows the hardware mounted on a tibial component. Three steel plates (JIS SS41) are mounted on a tibial base plate to act as the opposing poles for three eddy current distance sensors. The sensors are mounted on another plate fixed by a Hoffmann external fixator to tibial bone with two screw pins 40 mm below the tibial plate. The fixation device was designed for three knee systems: PCA (modular total knee system, Howmedica Inc.), OCU (Kyocera Osaka City University model total knee system, Kyocera Inc.), and Whiteside (Whiteside Ortholock II knee system, Dow Corning Inc.). The sensors must be set a minimum of 50 mm apart from each other to avoid

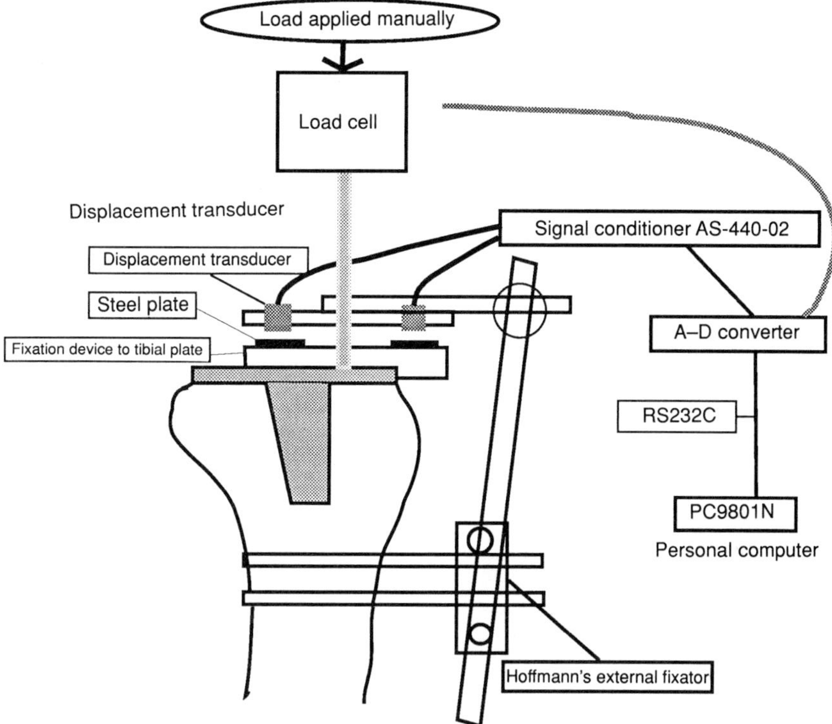

Fig. 16.1 Hardware for micromovement measurement.

magnetic interference between them. As a result, the two opposing steel plates must be set away from the tibial component. Loads were applied with a load cell (Kyowa Inc.) connected to an A–D converter. Position coordinates and load were calculated from corresponding sampled voltage data on a personal computer (NEC 9801) that displays the tibial component micromotion vs the load. The program was written in compiled BASIC and the micromovement was calculated and displayed on the screen in real time.

System accuracy

The accuracy of this system depends upon the tibial component deformity in response to the load, and the resolution of the sensor itself. The sensor used has 1 μm resolution over a 2 mm range. The system measured the tilt of the tibial component from three vertical displacements, but not other displacements such as rotation or horizontal translation, which could also affect the accuracy of the vertical displacement measurement. A comparison study was performed on a plastic bone model, between the output of this measurement system and values from a strain-gauge distance sensor mounted on the tibial component. The linearity of the measurement system seemed to be good enough, ranging between ±10 μm, that error mainly being generated by the noises from the cables connected between the sensors and the A–D converter (Fig. 16.2).

The measurement system determined the vertical translation of different parts of the component, consequently deformation of the tibial component itself was one of the major factors which affected the measurement accuracy. Bending tests on the tibial component showed that the deformation of the component was <9 μm per 100 N. The tibial component was considered to be floating on the tibial cancellous bone and the load applied on the component may not always act as a bending force.

Procedure

After the calibration procedure was carried out, the sensor unit and the fixation devices were sterilized with ethylene oxide gas pre-operatively. When the final setting of the tibial component was finished, the sensor unit was fixed to the tibial bone and the component fixation device with steel plates was set on the tibial component, which was connected temporarily to the sensor unit with a 1 mm plastic separator. The sensor unit was securely fixed with a Hoffman external fixation device to the tibial cortex, 40 mm away from the

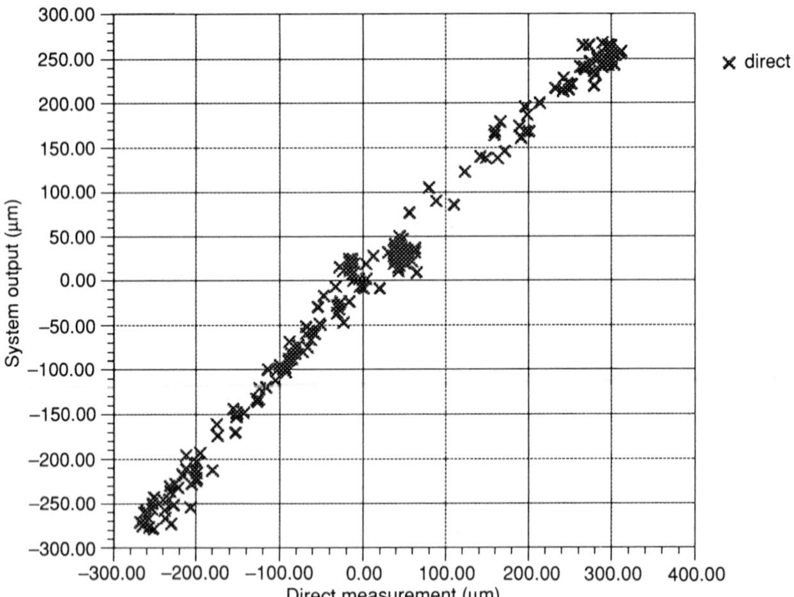

Fig. 16.2 Comparison of measurement system output with direct measurement on the component. The linearity of the measurement system is within 10 per cent.

tibial base plate to avoid jamming the component stem. When all cables from the sensors were connected to the microcomputer, the temporary separator was removed and the tibial component loaded manually with the load cell, as vertically as possible to the component. Maximal load on the component was usually 200–300 N, but was controlled by voice signal from the computer to around 200 N. Antero-medial, postero-medial, antero-lateral, and postero-lateral quadrants of the tibial component were pushed separately and the measured displacements of the component and the force were recorded automatically by the computer.

Experimental study

We tested our measurement device on a cadaveric specimen. As many researchers have reported, different designs or fixation methods have different qualities of stability. These may be evaluated by our micromovement measurement system.

Material and method

The proximal 15 cm part of a fresh cadaveric tibia was fixed in methylmethacrylate to mount on an autograph (Shimadzu autograph S500). The tibial bone was cut perpendicularly to the bony axis and the tibial base plate was fixed. Three different designs of tibial component were used: OCU, which has a 6-mm thick stem and two posterior screws; a 10-mm thick stem OCU II; and a Whiteside stem of 10 mm thickness but longer, with four screws (Fig. 16.3). If

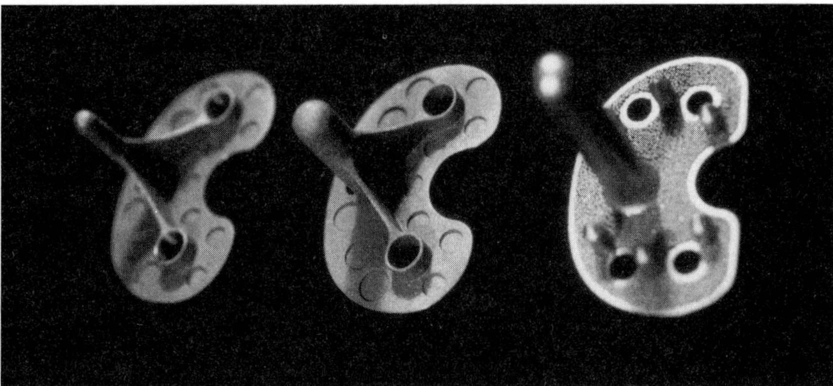

Fig. 16.3 Left: OCU (Osaka City University model, Kyocera Inc.)—6-mm stem, and two screws. Centre: OCU II—10-mm stem and two screws. Right: Whiteside (Whiteside Ortholock II, Dow Corning Inc.)—10-mm stem and four screws.

an implanted component and its fixation method did not affect the stability of the following procedure, and if there was minimum destruction of tibial bone, the same bone was used to compare fixation methods. These three components have different stem shapes but the same tibial plate dimensions. The posterior screws are at the same location. Consequently, these three models can be fixed sequentially without affecting next procedure. During the operative procedure, the component or fixation method may be selected in the same way.

We tested five cadaveric tibias to compare component stability. Load was applied at a speed of 10 mm/min up to 200 N on four positions on the base plate: antero-medial, postero-lateral, antero-medial, and antero-lateral.

Results

Using different components sequentially, each design of the tibial tray could be fixed securely on the bone. The stem was inserted into the tibial cancellous surface securely, and the screws were not loose even when the last Whiteside was inserted. In every case without exception subsidence was observed on the loaded side, and lift-up was observed on the contralateral side of load. When load was applied on the antero-medial quadrant, subsidence was observed on the antero-medial part and lift-up was observed on the postero-lateral part (Fig. 16.4). When the base plate was fixed on soft porotic bone, subsidence was observed on every part of the component. In this situation, the lift-up part was defined as the point of minimum subsidence. The load deformation curve was almost linear within 20 kg of load in all cases.

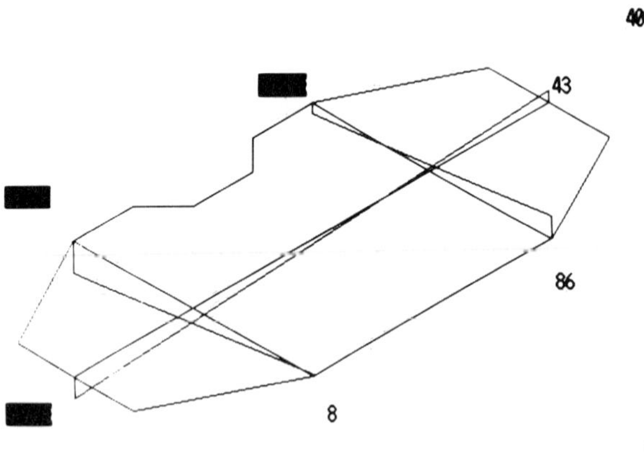

Fig. 16.4 Copy of CRT image. About 200 N applied on the postero-medial quadrant displaces the component on the same quadrant and causes lift-up on the contralateral side.

When load was applied for the first time, there was maximum movement of the base plate, but on loading for the second and subsequent load there was little change of deformation. Motion of the base plate caused by the first load may have very important implications for component stability, but it is influenced by the position before the load is applied. The deformation when the second load was applied was therefore used to compare the components.

With the exception of the OCU 6-mm stem, in Cases 1, 2, 3, and 4, subsidence was less than 150 μm when 20 kg load was applied, as shown in Fig. 16.5. In Case 5, subsidence exceeded 150 μm except in the Whiteside cases with screw. To evaluate the quality of the stability, each subsidence was compared with the value for the OCU 6-mm stem without screw. Screw fixation improved stability significantly, and there was a tendency for the stability of the small, thin stem of the OCU 6 type to be less than that of the OCU 10-mm stem or Whiteside. In the osteoporotic case, Case 5, there was no significant improvement using a thicker, long stem (Fig. 16.5).

Lift-up, which is defined as deviation of the contralateral side of maximum subsidence, is plotted on Fig. 16.6. In each case, screws significantly improved fixation, which was controlled to less than 100 μm of lift-up, except when the OCU 6 mm component was used. In Case 5, in which stability was controlled by stem fixation alone, movement exceeded 100 μm. Each lift-up was compared with the value for the minimum fixation model (OCU 6-mm without screw). Screws very significantly improved lift-up and there was minimum lift up using Whiteside with four screws. Stem size and thickness did not improve lift-up in every case, especially in Case 5 (Fig. 16.6).

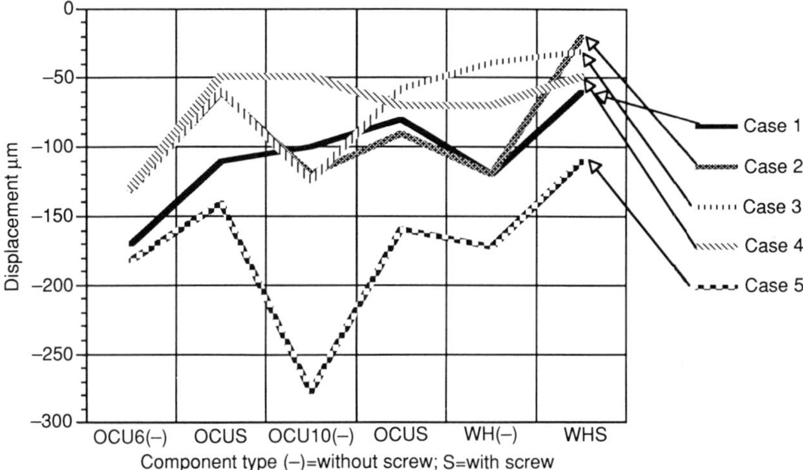

Fig. 16.5 Subsidence when 20 kg load was applied on the postero-medial quadrant. Subsidence was controlled to −150 μm, except in Case 5. The Whiteside with screw was the most stable of all.

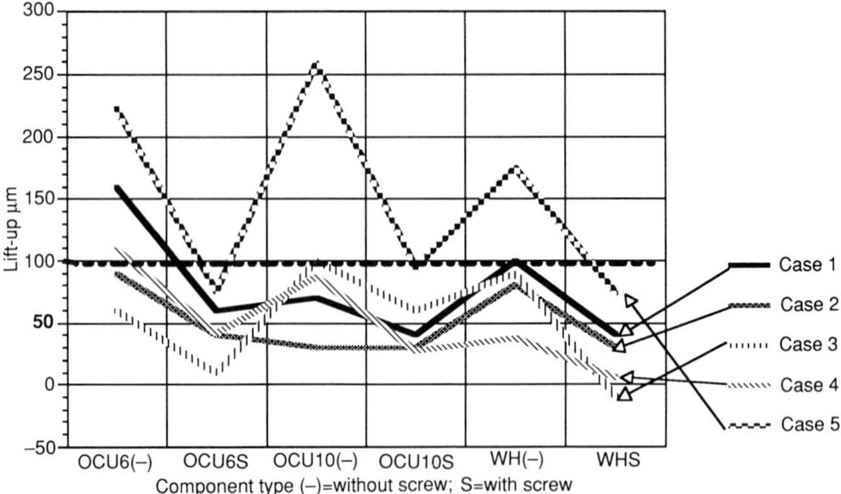

Fig. 16.6 Lift-up when 20 kg load was applied on the postero-medial quadrant. The screws significantly improve lift-up. There was minimum lift-up when using the Whiteside with four screws.

Clinical investigation

Material and methods

Trial clinical application of this measurement began in March 1990 when the trial protocol was applied, consistent with operation time and patient's status. Two factors were considered: firstly, the time necessary to perform the measurement; secondly, the amount of micromovement considered optimal or acceptable for the long-term stability of the component fixation. Fixation of the measurement apparatus usually took less than 10 min. After that, a further measurement for another fixation method takes less than 5 min. Consequently, for osteoporotic RA patients, it took more than 30 min to perform five or six measurements, changing several types of fixation or design. Figure 16.7 shows that stepwise application and stability measurements wasted a significant amount of time, so in clinical application it was decided to change to selecting the fixation method pre-operatively.

Lift-up to the extent of 150 μm is thought to disturb bone ingrowth into the porous surface of a component. It was decided to use 100 μm as a limit of lift-up. If the value of the measurement exceeded this value, application of cement or gentle post-operative care were indicated.

A preliminary flow chart used for the selection of the fixation method or component design since January 1991 is shown in Fig. 16.8. Indication of cementless or cemented fixation was usually decided roughly pre-operatively,

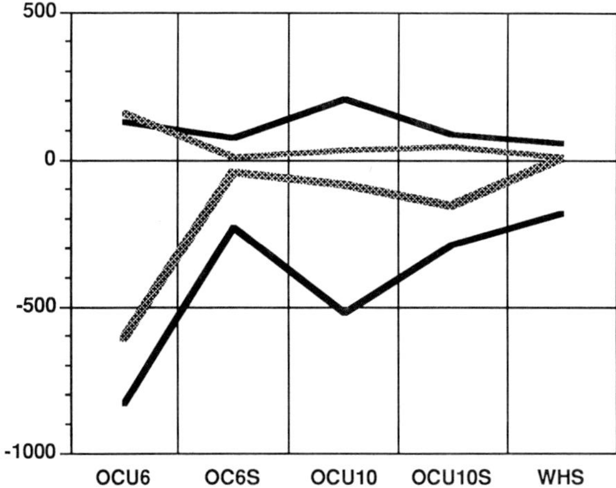

Fig. 16.7 The measurement was performed five times to select a secure fixation. The osteoporotic RA case was finally stabilized by a Whiteside implant. Stepwise application of the measurement wastes a significant amount of time.

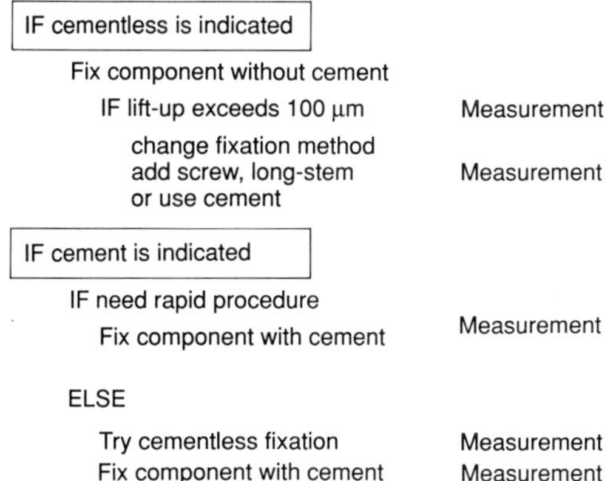

Fig. 16.8 Logic flow for choice of fixation. Need for cement fixation is indicted quantitatively.

and the more precise indication of the fixation was determined intra-operatively by reference to the measured value of stability.

Three types of total knee arthroplasty were used. Up to December 1990, the PCA was used for 10 cases (12 joints). The authors stopped using the PCA because it was hard to change the fixation method or the pre-operatively-indicated design to another choice intra-operatively. The OCU not only has two types of stem, but also has two screw holes. This model may be applied in a step-wise incremental manner of fixation, intra-operatively. It was used for 13 cases (16 joints) until the end of October 1991. The Whiteside model was used for osteoporotic cases, especially RA. In some OA cases this model was applied for cementless fixation as a result of excellent results from micromovement measurement. This model was applied for five Case 5 joints.

Assessment of the interface between the tibial component and underling bone was performed using fluoroscopic-controlled radiography to visualize the clear zone. The radiolucent line was categorized as shown in Table 16.1.

Table 16.1 Categorization of radiolucent lines into four grades

	Radiolucent line (RLL)
	Plane antero-posterior or lateral view under fluoroscopic control
−:	Invisible
+:	Under 1 mm thickness and 10 mm width
++:	Under 1 mm thickness and width not exceed half of component size
+++:	Under 1 mm thickness and width exceed half of component size
++++:	Over 1 mm thickness area is wide or over 2 mm in anywhere

Results

Six RA cases and four OA cases were measured for micromovement both before and after cement fixation. Micromovement of the component was significantly reduced by the application of cement. In RA cases, an average of −770 μm of subsidence was improved to −50 μm, and an average of 340 μm of lift-up was reduced to 40 μm. But the in the worst case, −2000 μm of subsidence had 140 μm of subsidence and lift-up, even after cement. In OA cases, −390 μm of subsidence and 280 μm of lift-up was reduced to −20μm of subsidence and 0 μm of lift-up on average (Fig. 16.9).

Average subsidence of cemented OA cases was −40 μm and lift-up was −10 μm. One case in which subsidence was −120 μm showed 60 μm of lift-up. Residual displacement when load was removed was within ±10 μm. In OA cases in which micromovement measurement indicated cementless fixation, average subsidence was −100 μm and lift-up was 50 μm. Two cases of PCA showed about 200 μm of subsidence and about 100 μm of lift-up, but the two Whiteside cases were measured to be well under 100 μm with residual micromovement almost at the 0 μm level (Fig. 16.10).

In RA cemented cases, micromovement was controlled to −50 μm of subsidence and 40 μm of lift-up. Residual displacement was within ±10 μm on average, but one PCA case showed −150 μm of subsidence and 140 μm of lift-up. Cementless fixation was applied to three RA cases. In one PCA case, subsidence was −600 μm and lift-up was 350 μm, but the other two were replaced by a Whiteside which showed under 100 μm of lift-up (Fig. 16.11).

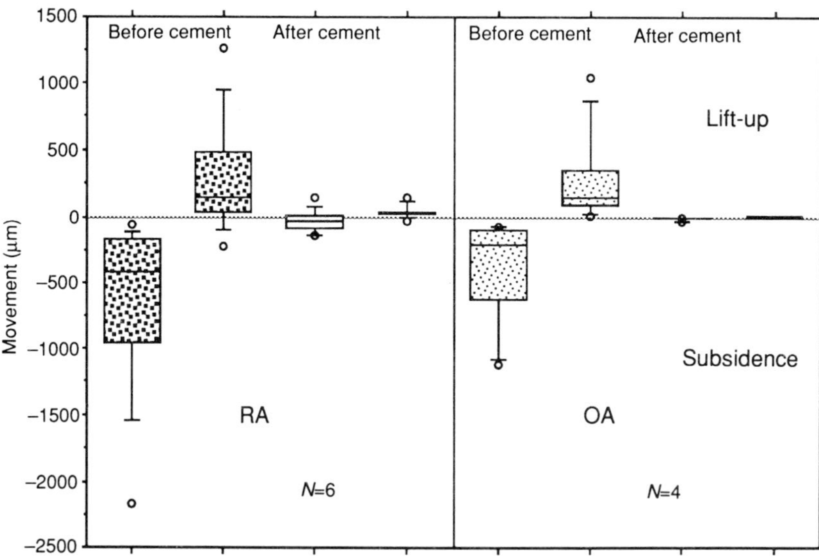

Fig. 16.9 Comparison of micromovement before and after use of cement. Both subsidence and lift-up were significantly reduced by cementing.

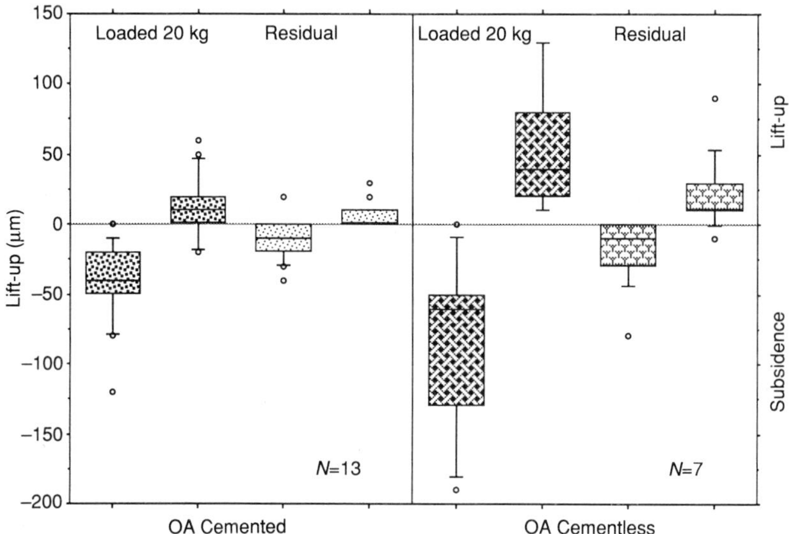

Fig. 16.10 With cement, OA cases were stabilized to well under 100 μm of lift-up. Without cement, lift-up exceeded 100 μm in two cases in the early trial period which were replaced by a PCA.

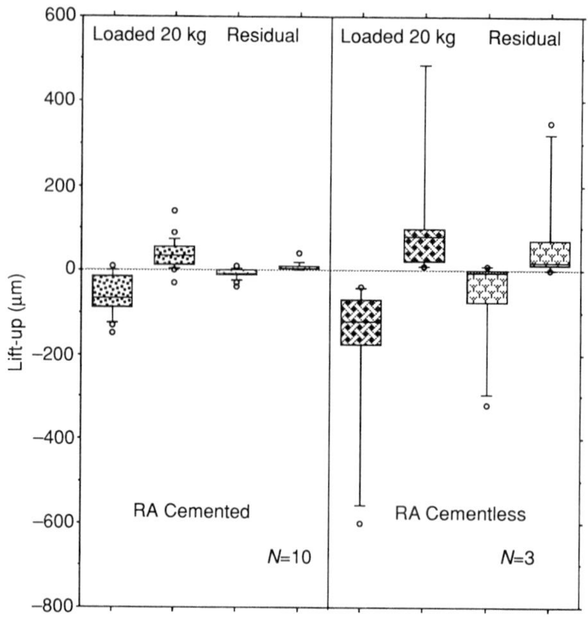

Fig. 16.11 In RA cemented cases, micromovement was controlled to −50 μm of subsidence and 40 μm of lift-up. During the trial period one cementless RA case showed −600 μm of subsidence and 350 μm of lift-up using a PCA, but lift-up of others was within 100 μm.

Radiographic observations of OA cementless cases revealed that radiolucent lines under the tibial component were restricted, with the exception of one OA case with OCU which had a significant amount of radiolucency. This patient had been complaining of pain in her knee when standing up. Subsidence and lift-up of two PCA cementless cases were higher than 100 μm, but their radiolucent line was minimal. There was only one case of an OCU cemented case which showed a restricted area of radiolucency, but follow-up time of cemented cases of this type are too short to evaluate stability (Table 16.2).

One RA cementless case using PCA showed a high level of micromovement exceeding 500 μm, but the radiolucent line was not severe in this 2-year follow up. Lift-up of the two Whiteside cases was under 100 μm, and there was no apparent radiolucent line. RA cemented cases replaced by PCA or OCU, in which lift-up was about 100 μm, had restricted areas of radiolucency, but it is hard to estimate whether more radiolucency exists in the cases showing greater micromovement, among the 11 joints of cemented RA cases (Table 16.3).

Table 16.2 Micromotion measurement and RLL of OA cases

	Displacement load+	Residual	RLL
OA cementless			
PCA1	−190∼130	−70∼90	+
PCA2	−180∼130	−80∼50	+
Kyocera1	−60∼20	−20∼10	+++
Kyocera2	−130∼40	−10∼20	++
Kyocera3	−130∼−60	−30∼20	+
Whiteside1	−60∼30	−10∼10	++
Whiteside2	−10∼20	0∼10	−
OA cement			
PCA1	−40∼10	−20∼10	+
PCA2	−20∼0	0∼10	+
PCA3	−70∼50	−20∼10	++
PCA3-1	−120∼60	−30∼30	++
PCA4	−60∼10	0∼10	−
Kyocera1	−30∼10	0∼10	−
Kyocera1-1	−10∼20	−10∼10	−
Kyocera2	−30∼0	−10∼0	−
Kyocera2-1	−40∼−20	20∼10	−
Kyocera3	−40∼20	−20∼20	−
Kyocera4	−80∼20	−30∼10	−
Kyocera5	−50∼20	−30∼10	++

Table 16.3 Micromotion measurement and RLL of RA cases

	Displacement load+	Residual	RLL
RA cementless			
PCA S1	−600∼530	−320∼350	++
Whiteside1	−180∼60	0∼10	−
Whiteside2	−70∼100	−10∼30	+
RA cement			
PCA1	−90∼60	10∼10	+
PCA2	−70∼10	−20∼10	++
PCA3	−150∼140	−40∼40	++
PCA4	−20∼30	0∼20	−
Kyocera1	−10∼60	0∼10	−
Kyocera1-1	−30∼40	0∼20	+
Kyocera2	−30∼10	−10∼10	−
Kyocera3	−100∼90	−30∼20	++
Kyocera4	−90∼40	−10∼10	++

Discussion

Various efforts to reduce migration of components, such as the addition to the standard polyethylene component of a metal back or a long stem, have significantly reduced both migration and inducible displacement (Albrektsson *et al.* 1990). Different fixation systems have different levels of fixation (Dempsey *et al.* 1989). Measurable migration and inducible displacement may be the rule rather than the exception in total knee arthroplasty. Accordingly, absolutely rigid fixation would not be necessary for successful function of a TKA (Ryd *et al.* 1986). However, how do we know whether or not the fixation was absolutely rigid and how do we control the stability? Hvid (1988) measured the trabecular bone strength during operation. The axial strength of trabecular bone at the knee is critical for the maintenance of support and fixation of the prosthetic components after total surface knee arthroplasty, but it is difficult to translate bone strength into component stability, which is influenced by component design or surgical technique individually. We are now able to measure accurately the micromovement of the tibial component itself immediately after the operative procedure, and to control fixation quality. A measure of micromovement immediately after the operative procedure must be the most reliable data for an individual case.

Friedman has claimed that the choice of prosthesis must be individualized according to the patient's goals and activity level (Friedman 1981). The perfect prosthesis is not yet available, but research in biomechanics, materials, and fixation may resolve some of the current limitations (Laskin 1988). This situation still holds true in total knee arthroplasty, because the mechanical characteristics and daily activities of the patients are very different from one patient to another. The authors could not find a perfectly designed artificial joint but have devized the best method of selection for individual patients.

The displacement of the tibial component relative to the tibial bone is three-dimensional, but the displacement caused by the vertical load is directed downward or upward. The authors' measurement system has only three vertical displacement sensors that may be used to detect major displacements of the component when loaded vertically. Theoretically, this measurement system can be modified to detect other directions, but the time required for this might extend and disturb the operative procedure.

The load applied for the measurement is restricted to that which is acceptable within the operative procedure. A manual method of application of the load was chosen which is only 200 N and is hard to control for loading rate, but nevertheless the reproducibility of the measurement was excellent. Simulation of the load of a walking cycle is extremely difficult with manual loading. The load was applied on the antero-medial, postero-medial, antero-lateral, and postero-lateral quadrants of the tibial component. Under this asymmetrical stress the tibial component was forced to lean down much more than it would under a balanced, normal load and so might have been more responsive to the quality of the underlying cancellous bone.

The program language used was BASIC with a compiler. The data sampling rate and computation was too slow to draw the load deformation curve accurately. If a detailed load deformation curve could be obtained, the measurement could be performed within the limit of elastic deformation as a nondestructive measurement.

Comparisons of the stability between different component designs applied on cadaveric bone were performed to evaluate the best artificial joint design (Shimagaki *et al* 1990), but comparison of all kinds of component now available is nearly impossible. If the authors' measurements could determine the most stable component, other factors such as minimizing the amount of resected bone might become the subject of controversy. There must be an optimum selection of the design of knee surface replacement for individual bone strength and activity. Bone strength could be evaluated with computer tomography to determine the severity of the osteoporosis in rheumatoid-arthritis cases, and pre-operative assessment for the selection of the design of the artificial bone might be possible. It is hard to evaluate the relationship between the surgical skill and the final result such as loosening of the component or deformation of the bone.

Quality of the component stability immediately after operation must be a most reliable indicator of surgical skill. Direct measurement of the micromovement of the component is the most reliable method for the evaluation of the strength of the prepared bone surface for replacement. The strength of the bone could be quantified as being the amount of component displacement under the applied load. According to the authors' cadaver study, it is possible to select and identify the fixation quality of different fixation methods, but the stepwise trial method used in this cadaver study may not be suitable for clinical application because the difference in stability between each step may not improve the absolute stability to an acceptable level. The influence of the parts and design, such as stem and screw, has been clarified by this experiment, and it may help us to select better designs to obtain better stability. The load applied in the measurement is an eccentric force applied on specified quadrants of the component. Screws are revealed to be effective against this eccentric load, especially against lift-up. Pitching and yawing of the tibial component may be well controlled by the long stem, as in the case of a yacht keel, but a round stem may not stabilize the component against rotational stress. Pegs, stems, and screws controlled the quality of the fixation that was selected during the measuring procedure with the authors' system.

The authors' measurement system works very nicely during operation, but it takes 10–15 min in a restricted operation time. The most time-consuming procedure of this measurement system is the fixation of the sensor unit to the tibial cortex with an external fixation device, in order to maintain a 2-mm gap between sensors and opposing steel plates fixed on the tibial component. The Hoffman external fixation device which the authors are currently using might be a better choice due to its easier setting. The measurement itself takes only 5 min with a real-time data acquisition and display system. The requirement for 15 min may delay the operative procedure completion by less than one pneumatic tourniquet period.

In clinical application of this measurement, stability of the tibial component may be controlled to <100 μm of lift-up, and a quantitative indication of cementless fixation obtained. The authors have successfully applied cement quantitatively, in cases where the micromovement of cementless fixation exceeds 100 μm. Cement fixation improved fixation quality dramatically and reproducibly, but there is still a possibility of instability after a long period, shown by a restricted area of radiolucency on short follow-up radiography. In spite of having controlled stability well with this measurement, one cementless OA case showed a significant radiolucent line and clinical problems. Lift-up of the porotic RA case was controlled fairly well to just 100 μm. These cases forced the use of the more stable component, or the addition of cement proximally. For long-term survival, it is necessary to be extra sure of the security of component fixation. The authors now use the Whiteside long-stem type for the cementless RA case to control stability and for early weight-bearing. Figure 16.12 is a chart for the selection of the fixation method now used to control stability.

The authors' measurement system detects micromovement of the tibial component precisely, but only gives a value immediately after fixation corresponding to 20 kg force applied manually. True long-term stability does not relate directly to the value of this measurement. The residual deformity is a result of the cancellous bone microfractures, accumulation of which might cause sinking or loosening of the component. It is possible to adapt post-operative management in the light of the stability measured immediately after fixation, as the RA cementless case with applied PCA. A radiolucent line on

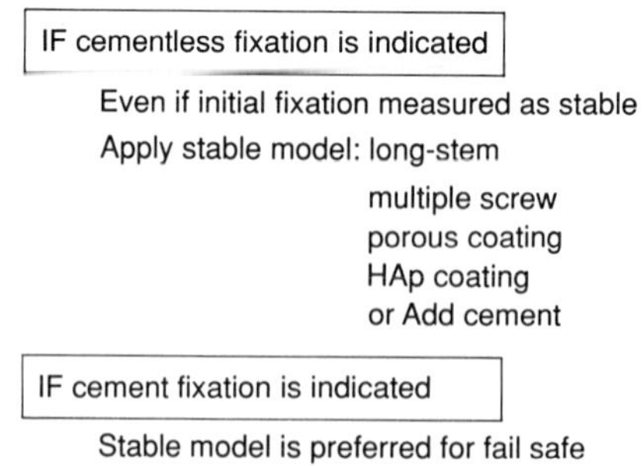

Fig. 16.12 How to select fixation method and design.

the plane radiograph may not be suitable for evaluation of stability of the component because the gap between the component and underlying bone tends to be compressed to the point of invisibility. But it is obvious that the standardization of the component stability made possible by this measurement method will clarify the factors which affect the longevity of the TKA, and should provide an initial evaluation of the wide variety of types and design philosophies of joint replacements before using more accurate and quantitative radiological evaluation methods such as Roentgen stereophotogrammetry (Albrektsson *et al.* 1990; Ryd *et al.* 1986).

Even though the measurement takes a little time, the quantitative information obtained from this system seems to be useful for the evaluation of the quality of the fixation of the component. Comparison of this data with the follow-up X-ray files may determine optimum fixation of the component.

Summary

1. The authors have developed a device to measure micromovement of the tibial component of a knee arthroplasty; it has a 10 μm accuracy and is easy to use during surgery.
2. The device enables selection of component designs or fixation methods to improve component stability according to the level of micromovement.
3. Tibial component stability is well-controlled to <100 μm of lift-up. The need for cement fixation is quantitatively indicated.
4. Even with micromovement measurement, unsatisfactory results may occur in cementless TKA.

Debate in the meeting

The meeting noted that the authors' technique was another example of micromovement being used to predict long term stability, in this case extremely early micromovement. The proof of this hypothesis will only be found in the years to come.

It was noteworthy that, with a load of only 20 kg, micromovement of as much as 100 μm was observed. In other cadaver studies, loads of the order of $1.5 \times$ body weight were applied to achieve the same micromovement. The authors emphasized that this method was intended to be a *non*destructive test.

17 Improvement in the monitoring of fracture repair by noninvasive mechanical techniques

A. F. G. Taktak, J. Edwards, P. Kay, and D. C. Laycock

Introduction

A variety of methods are used to assess the progress of fracture healing in long bones. Apart from radiography, orthopaedic surgeons rely on clinical observations such as mobility of the fracture, pain, tenderness, and time taken for a fracture to heal normally. Radiographs provide a picture of anatomical changes that occur during the healing process including formation of callus and changes in bone shape. The ability of clinicians to judge the stiffness of fractured tibiae from antero-posterior and lateral radiographs have been investigated by Nicholas (1979) and Panjabi *et al.* (1989). They found that a clinician can not detect early osseous union with reliability. Manual assessment of bone in bending is also widely used but gives only a rough estimate of union (Hammer *et al.* 1985). The bending moment applied and the amount of movement detected are only crude estimates and different surgeons have a tendency to reach different conclusions.

There is evidently a demand for more reliable methods for monitoring the progress of fracture repair by noninvasive techniques. Methods used in the past have varied from hanging weights on a supported leg and measuring deflection by means of a three-point bending test (Edwards 1981; Hammer *et al.* 1984), to the measurement of micromovement within an external fixation device (Cunningham and Kenwright 1989; Kay *et al.* 1989). Vibration studies have also been carried out in which oscillatory forces of variable frequency were applied to the surface of the leg overlying the bone. Then, the resulting amplitude, acceleration, and modal response of the tibia was monitored (Cornelissen *et al.* 1987). Resonant frequency was also used to determine bone stiffness (van der Perre and Christensen 1985; Nokes *et al.* 1984). However, all of these techniques suffer from measurement errors arising from distortions in the overlying soft tissue.

In order to devise a reliable technique whereby the changes of stiffness could be measured, two methods have been investigated in the present study. The first involved a static bending test in which measurements were made of bone deflections resulting from an externally applied four-point bending moment.

The second involved vibrational analysis of the impulse response of the leg due to sharp taps on the anterior surface of the tibial tuberosity.

Method

Static bending test

In order to carry out a static bending test of the human tibia *in vivo*, a frame was constructed which clamped around the leg as shown in Fig. 17.1. The patient was seated on a couch and the injured leg was inserted vertically downwards inside the frame. The leg was then clamped at the tibial tuberosity and the malleoli with the aid of four bars forming a rectangular shape.

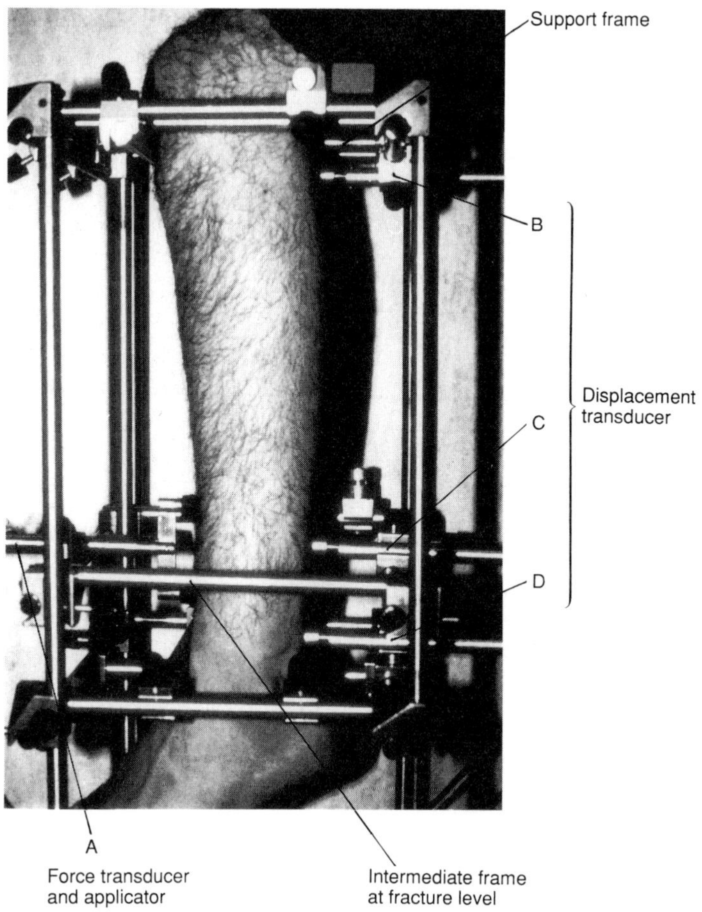

Fig. 17.1 Static bending test.

A further frame was positioned at the level of the fracture to carry an instrumented force applicator (A). This applied a four-point bending moment to the leg. Three in-line displacement transducers were used, as shown in the figure to measure displacements near the knee (B), at the fracture (C), and near the ankle (D). This arrangement compensated for the effect of gross deformations at the bone due to soft tissue compression at the upper and lower supports.

The outputs of the three displacement transducers and the force transducer were interfaced to a BBC microcomputer using its analogue-to-digital input. With simple geometrical theory, the computer calculated the applied bending moment in newton-metres and thc angle of bend in degrees. This angle was then plotted against the bending moment as shown in Fig. 17.2, and linear regression analysis was performed to determine the slope of the graph. Stiffness was then interpreted in terms of newton-metres per degree angle of bending.

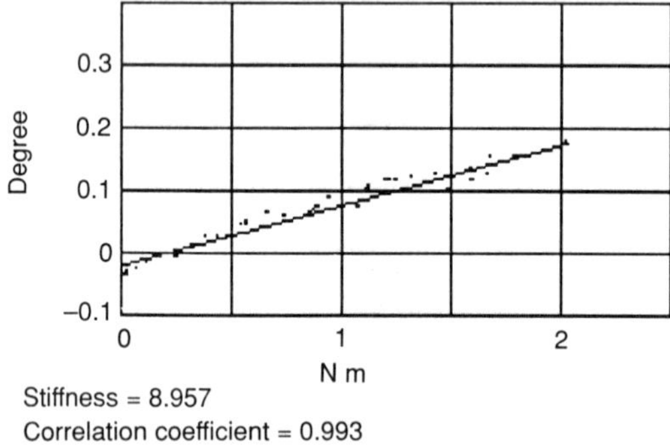

Fig. 17.2 Calculation of stiffness in the medial direction.

One noticeable omission from the literature concerns information on the variations in stiffness with direction around the fracture site. In particular there is no evidence concerning whether callus formation takes place symmetrically. To investigate this matter, the stiffness in different directions around the leg was measured. This was achieved by repositioning all four transducers around the frame in the horizontal plane in steps of 45° up to 360°. At each step, stiffness readings were recorded to give a total of eight measurements on each bone for each visit of the patient to the clinic throughout the process of healing.

Vibrational analysis

Two further tests were aimed at determining fracture stiffness by studying the dynamic behaviour of the tibia. One was based on the knowledge that the natural frequency of a linear mechanical system is known to be directly proportional to its stiffness (i.e. resonance). In the second, an impulse response signal was used to describe the mechanical properties of the bone.

Resonance measurements

Electronic equipment has been developed at the University of Salford which applies an oscillatory force onto the tibia *in vivo* with a range of frequencies scanning from 0 to 540 Hz. This force was applied to the tibial tuberosity using a modified loud speaker with its sound cone replaced by an impact shaft at different points along the anterior border of the leg. Resonant frequencies were noted when the rms value of the accelerometer's output reached a peak value (Fig. 17.3(a)). Response signals obtained proximally to the fracture were then averaged to obtain the resonant frequencies of the proximal segment of the tibia. Resonant frequencies were also determined in the same manner for the distal tibial segment.

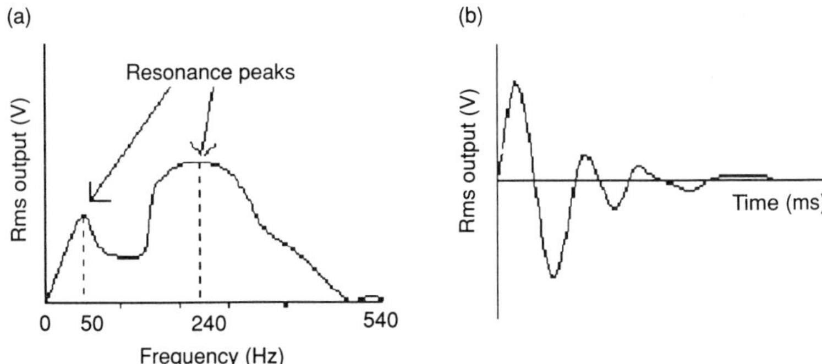

Fig. 17.3 (a) Resonance signal. (b) Impulse response signal.

Impulse response

By driving the speaker unit with a square wave electrical input, the same equipment was used to generate a sharp impulse applied at the tibial tuberosity. The accelerometer was placed at 1 cm intervals on the anterior border of the tibia, starting from the tuberosity and moving down to the medial malleolus. The output of the accelerometer was then converted by double integration. The resulting displacement waveform was then recorded by computer at each position and plotted against time (Fig. 17.3(b)). Finally, the recorded data was rearranged as a dynamic picture of the bone moving in space in response to the impulse, as shown in Fig. 17.4. The initial position of the bone prior to the impact is represented by a horizontal line (KA) where K represents the knee and A the ankle. In response to the impact of the external force on the tibial tuberosity near K, the bone moves downwards with time to a greater extent on the left than on the right. In fact, it pivots about a position P which is nearer to A so that the right hand end moves upwards. As the end K of an intact bone moves downwards with time, through positions 1, 2, 3, and 4 as shown in Fig. 17.4(a), there is only a slight curvature in the bone due to bending. However, a fractured bone was found to undergo a sharp bend at the fracture site as shown in Fig. 17.4(b). Software was developed to measure both the degree and period of this bending as an indicator of healing.

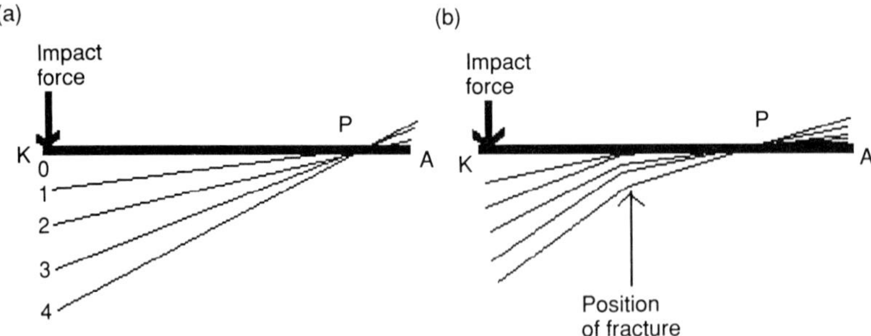

Fig. 17.4 Modal images. (a) Normal bone. (b) Fractured bone.

Results

Static bending results

Cadaveric experiments were carried out to ascertain the effects of soft tissue. It was found that the stiffness of a fresh tibia was around 45–55 N m/degree. Normal control tests were also conducted to determine the stiffness of normal tibiae. It was found that in normal legs, the stiffness was always 10 N m/degree or higher in all directions round the leg. Similar values were obtained when testing injured legs after complete union had occurred. Clinical healing was found to occur when the stiffness was around 20 per cent of that for a fresh tibia. Forty-six patients with tibial shaft injury were studied during healing. In a recently fractured bone, stiffness readings could be as low as 0.1 N m/degree.

Stiffness measurements were found to have a high correlation. Figure 17.5 shows the results of a patient who went through the normal stages of healing. The figure shows a consistent trend of measurements taken in three directions around the leg. Measurements in the other possible directions were ignored due to errors arising from the distortion of the surrounding soft tissue. A patient with a hypertrophic nonunion of the tibia was also monitored and his results were plotted in Fig. 17.6. In this case, there was a negligible increase in stiffness in all but the postero-lateral direction. This higher stiffness in one direction was found to be due to a fibula union which was radiologically confirmed. Clinical assessment of the patient also confirmed a delayed union

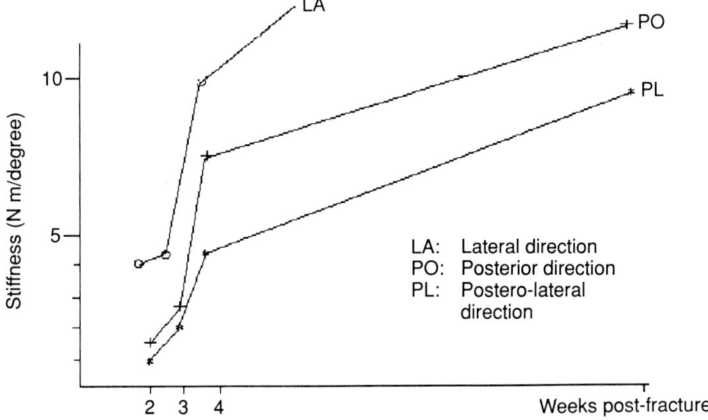

Fig. 17.5 Normal healing case.

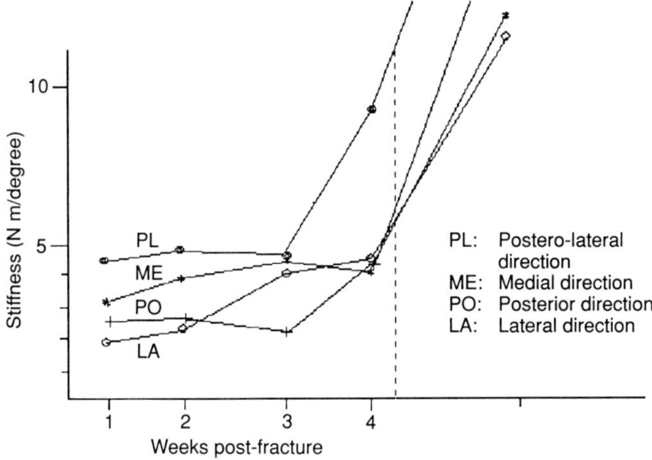

Fig. 17.6 Nonunion case.

and the patient had to undergo surgery. The dotted vertical line in Fig. 17.6 represents the date on which the patient received an intermedullary pin. Static bending measurements showed that the stiffness of the bone increased to a high value after the surgery.

Vibrational results

Over 60 patients were investigated using both resonance and impulse response techniques. Both fractured and normal legs were tested for comparison and standardization.

Resonance testing was carried out by two methods. The first used a frequency scanner to plot the frequency response signal, while the second analysed the Fourier components of an impulse response signal. Both of these tests were found to produce similar results. It was found that signals obtained proximally to the fracture site had a wide frequency band. However, at positions distal to the fracture, the higher frequencies disappeared resulting in a narrow band of low frequencies as shown in Fig. 17.7. Contrary to findings generally quoted in the literature (Edwards 1991), resonance testing did not show any correlation with healing. This probably arose because the measured frequencies were affected by other factors besides stiffness, such as bone weight and shape, callus weight and shape, and support conditions. All of these factors varied between tests carried out on different dates. Therefore, it was not possible to follow a trend of resonance with increase in stiffness during bone repair.

The signals from impulse response tests were processed by computer to produce a modal image of variations in bone shape with time. Results obtained were very promising and when standardized between patients, there was a good correlation with detection of delayed union.

When a modal image was plotted, the bone was noted to undergo an obtuse bending at the fracture site as illustrated in Fig. 17.8. During the initial downward movement of the bone following an impact at the left-hand end (K), the angle between the proximal bone segment (KF) and the distal segment (FA) decreases to a minimum value before opening out again. This minimum initial angle was called the primary bending angle (PBA). The PBA was found to be related directly to stiffness and was therefore thoroughly investigated. PBA was also a function of the length of the bone L and the amount of force applied. These determine the maximum downward displace-

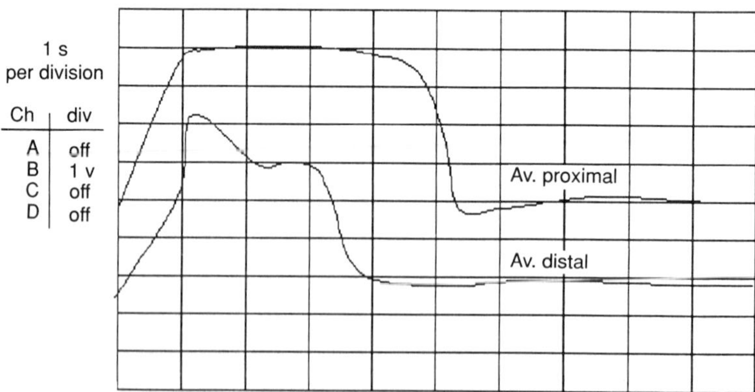

Fig. 17.7 Resonance signals in a clinical test.

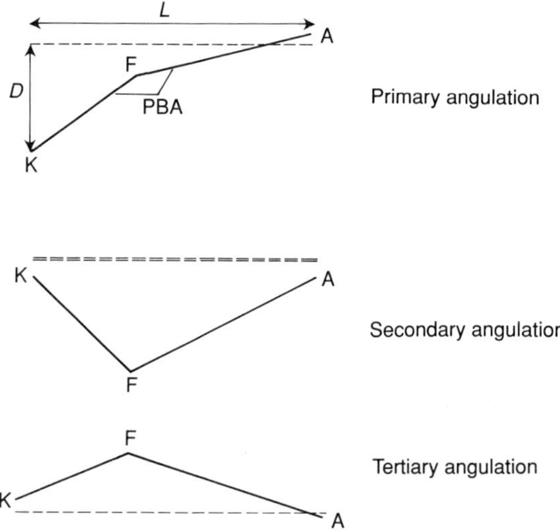

Fig. 17.8 Modal imaging technique.

ment D in Fig. 17.8. Both L and D were therefore rescaled and converted to fixed, constant values so that the PBAs could be compared between cases.

The time-period taken for the bone to reach minimum primary angulation was found to vary with stiffness. This period of the angulation was then measured by computer and used to represent stiffness. As the bone stiffness increased, the period of angulation decreased. Therefore, in order to observe an increasing trend with healing, the reciprocal of the angulation period (RAP) was plotted.

Two examples are illustrated in Figs 17.9 and 17.10 representing typical modal imaging results. Figure 17.9 shows an increase in both PBA and RAP with time for a patient who went through the normal stages of healing. The

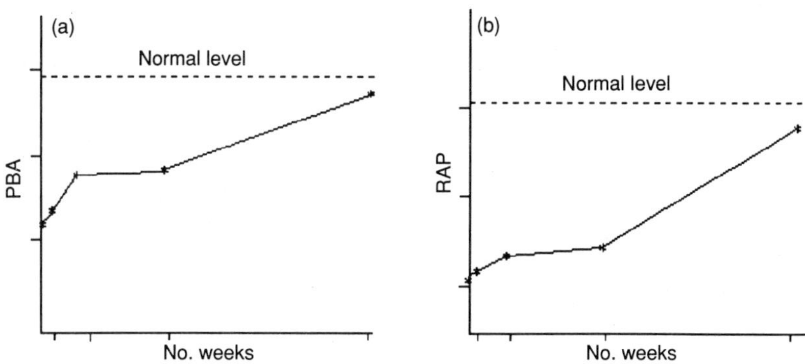

Fig. 17.9 Normal healing case.

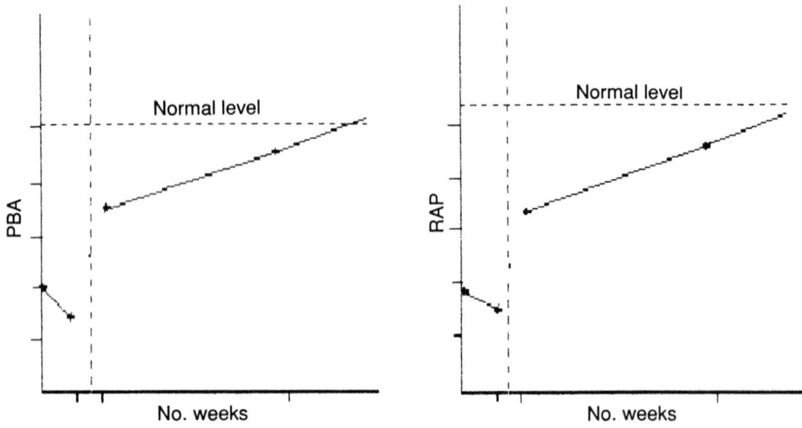

Fig. 17.10 Nonunion case.

normal levels were obtained by measuring PBA and RAP values at the mid-shaft position of the normal leg in the same patient. Figure 17.10 shows the comparative results for a case of nonunion. The dotted vertical line represents the date at which surgery was performed on the patient to promote union. The figure shows a significant increase of PBA and RAP values beyond that date.

This modal imaging technique was also used to observe a surgical procedure involving an Ilizarov external fixator (Fig. 17.11) where a section of the bone was infected and excized (defect). An osteotomy (transport segment) had been performed in the upper tibia. As the osteotomy attempted to heal (new bone), the transport segment was moved distally to elongate the callus (maturing callus). The impulse response test successfully identified the five zones of the bone as shown in Fig. 17.11.

Discussion

In the static bending test, when stiffness values reached 14 N m/degree or higher, measurement errors were found to be in excess of 18 per cent. These errors, which were unacceptably high, resulted from the difficulty of measuring small deflections through the skin. Fortunately, clinical healing was found to occur when stiffness values reached 10 N m/degree. Measurement errors in the region below this value were found to be around 2 per cent, which proved acceptable. Results from this static bending test showed good correlation with healing during the clinical trials and this method is now used on a regular clinical basis in the hospitals of Manchester and Salford.

In the vibrational analysis tests, resonance measurements were found to have little clinical significance whether measured *directly*, by driving the bone

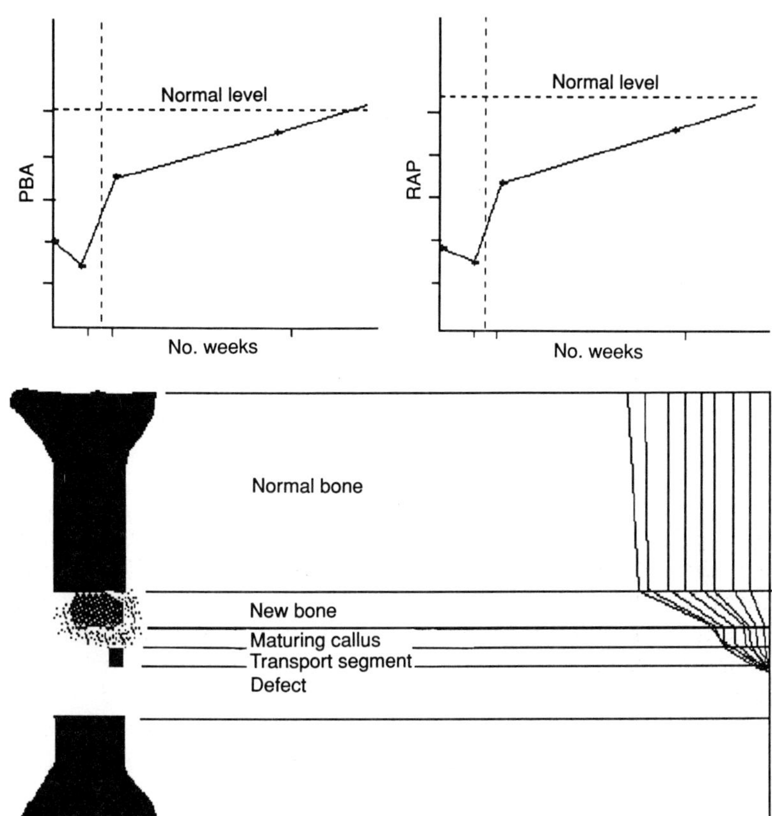

Fig. 17.11 Modal imaging of a bone treated by an Ilizarov fixator.

at different frequencies, or *indirectly*, by analysing the Fourier components of an impulse response signal. These two methods produced similar results. However, no general trend was found during healing, indicating that while measuring resonance was possible, relating it to stiffness was not credible.

The modal imaging technique, on the other hand, proved a good indicator of healing when the minimum angle of bend and the delay period resulting from a sharp impact were measured. Although these values were not absolute, they did correlate to stiffness and when standardized between cases, a threshold value for union was deduced.

Debate in the meeting

The meeting asked what clinically-verified value could truly be placed on bending and vibrational measures of healing. The authors replied that they had not attempted a study like that of Richardson who had demonstrated that, compared with traditional methods, less harm came to patients when an objective measurement of stiffness was used to decide when to remove an external fixator. However, the values from this stiffness measurement test do show a very high correlation with the best of existing measures, such as clinical impression and radiology. The stiffness value can be used as a useful numerical assessment for the degree of fracture repair which can be subjected to statistical analyses. This is a great improvement on 'time to healing' or 'fracture healed'. It would be interesting to compare what happens to patients who are released from support on a random basis of either this or a traditional method.

18 The use of externally applied mechanical vibrations to assess both fractures and hip prostheses

R. J. Collier, R. J. Donarski, A. J. Worley, and A. Lay

Manual sensing and X-rays

The current methods for assessing fractures involve in general two techniques. The first of these is the manual sensing technique in which a limb is stressed manually in order to see whether it feels rigid. The second is to use X-rays to investigate the structure of the fracture. The main assessment of the fracture occurs at around 12 weeks after the fracture and usually involves both techniques. The radiographic evidence is not always reliable, since the variation of X-ray absorption in the fracture region is not necessarily related to strength. There have been X-ray photographs which suggested a strong callus across the fracture which, by using manual sensing, was shown not to be united, and there have been X-ray photographs which suggested that a fracture was still present some years after the fracture had united. Radiographic evidence is exceedingly useful in confirming diagnosis and nowhere more so than in the problem of locating cracks in bone and other bone disorders. However, it is secondary to manual sensing for assessment of fracture healing.

Manual sensing is not totally reliable. There have been interdigitated fractures which have felt rigid in one direction but were actually flexible in another, and were missed by manual sensing. There is also the problem of what 'feels' rigid, which is made more difficult by the wide range of soft tissues which surround bones and the wide range of physical strengths of orthopaedic clinicians. Add to these the change in rigidity of a bone with age, occupation, and weight, and the sensing problem becomes formidable. It would be totally inappropriate for the orthopaedic clinician to stress all limbs with the same force applied in a similar position. Finally, the position and type of the fracture and the nature of any fixation also affect the 'feel of the bone'. The method of manual sensing is noninvasive, but can cause pain in some cases.

The first part of this article describes a new 'phase delay' technique which overcomes some of the problems of manual sensing described above. This should be distinguished from similar mechanical resonance methods which are less satisfactory. The phase delay technique can also be used to assess fractures while the limb is still in plaster, or has some form of fixation. So far, in all the patients measured, no pain has been experienced and the method has been used for a very wide range of fracture types, patient ages, and occupations.

Mechanical resonance

When a small sinusoidal mechanical vibration is introduced into a bone, typically a movement of around 10 μm, it will travel down the bone and reflect from the bone cartilage interface back up the bone. At the 'other end' of the bone it will again reflect at that cartilage interface and proceed back down the bone. If the delay in time for the round-trip is exactly equal to the time for one cycle of the mechanical vibration then the energy will build up and the bone is said to be in resonance. Higher resonances will occur when the time delay is equal to more than one cycle. If the losses in the bone are very small and the reflections are almost perfect, then the build-up of energy will be large at the resonant frequency. This is particularly true for *in vitro* dry bone. However, *in vivo*, the losses due to the damping effect of the soft tissues surrounding the bone and the absorption in the cartilage, result in a small build-up of energy. The net effect of this small build-up of energy is that the exact frequency of resonance is hard to measure. Also, a number of factors can change the frequency at which the resonance occurs. These include, for example in the case of the tibia, the position of the limb, the angle of the knee, the tension in the main leg muscles, the position of the various transducers used to introduce the vibration and detect the resultant movement, the thickness of the soft tissues surrounding the bone, and finally the condition of the cartilage at the ends of the bone. Various workers have measured these resonant frequencies using a variety of techniques, but because of the factors mentioned above, the errors in some of the measurements can be very high (van der Perre *et al.* 1989).

A considerable amount of work has been carried out to measure the change in resonant frequency that occurs as a fracture heals (van der Perre and Borgwardt-Christensen 1985). The resonant frequency of a less rigid fractured bone will usually be lower than the frequency associated with the unfractured bone. The resonant frequency of a fractured bone usually rises during the first 10 weeks after fracture and approaches the unfractured bone's frequency. In some cases, depending on use, there is a variation of the unfractured bone's frequency during this period. If the bone is not used, the frequency may reduce and if it is used more than normal for weight-bearing, then it may increase. Thus, ratios of the fractured to unfractured bone are not unambiguous.

The methods used by various workers to introduce the mechanical vibration involve a hammer to tap the malleolus and an electromechanical shaker. The hammer tap introduces frequencies which are in the upper audible range and thus higher resonant frequencies tend to be detected rather than the lower ones. These higher resonances sometimes have nodes at the fracture site and so may not change as the fracture heals. Also, these higher frequencies can occur quite close in frequency so there may be problems in correct identification. All these factors persuaded the authors to prefer time delay, as the vibrations travel down the bone, over resonance measurement. If a bone were perfectly damped there would be no resonance effect at all and the time delay or phase delay would be all that could be measured. A bone *in vivo* is heavily damped, so the time delay or phase delay method is easier than resonance measurement. Finally, the resonant frequency of a bone is a function of both the bone and its surroundings, whereas the phase delay is only a function of the fracture. Thus, phase delay is a more sensitive detector of the properties of the fracture.

Mechanical phase delay

A mechanical vibration imposed at one end of a bone travels at a certain velocity. The time to complete the journey to another point is called the time delay. If the time delay is expressed in terms of the fraction of one complete cycle of the vibration (360° or 2π rad) it is called the phase delay. The measurement of phase delay is straightforward if there is no reflected signal from the cartilage. In practice, there are small reflections but, because of the damping effects of the soft tissues, the reflected signal is relatively weak after a few centimetres. Phase delay can be measured at any frequency. There is not the problem of measuring frequency and identifying a resonance.

The velocity at which mechanical waves travel down the bone is approximately proportional to the square root of the frequency of the transverse vibrations. A typical set of phase measurements is shown in Fig. 18.1. The vibrations were introduced at points 1–6 and the phase delay measured at the point marked v. The amplitude curves are also shown in this case and the lowest resonant frequency—the so-called rigid body mode—is easily identified. The fracture healed normally and the two measurements were made 6 months after the fracture had healed. The phase measurements include the phase delay that occurs as the mechanical vibration passes through the skin and soft tissues and into the bone, and a similar phase delay as the vibration is detected further down the bone. In all these measurements it is assumed that this extra phase delay is a constant, independent of measurement position. This would clearly not be the case if the measurements were made through a considerable quantity of soft tissue, but the skin and tissue are reasonably uniform down the medial face of the tibia.

When a fracture is not rigid, the velocity in the fractured region is reduced,

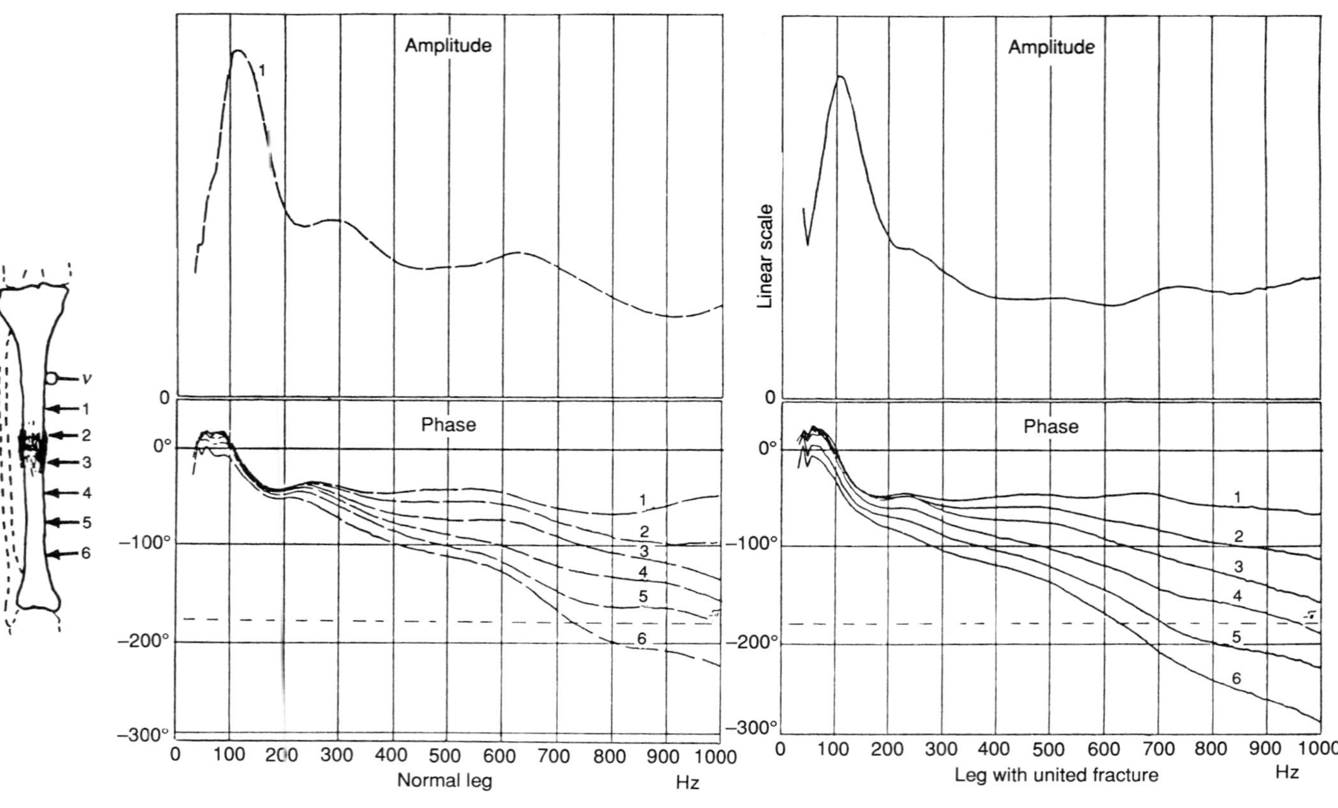

Fig. 18.1 Measurements of amplitude above and phase delay below on a patient with a united fracture and small callus.

thus the time delay or phase delay is longer. So a measurement of the increase of this delay gives a direct measurement of this velocity, which in turn is related to the rigidity of the fracture. Figure 18.2 shows a set of measurements on a normal leg and on a fractured leg. The fracture was measured when it had not united and then was measured again when it had united—as determined by manual sensing. The amplitude measurements are also shown in these cases. During the first measurement on the nonunited fracture, measurements were made at the seven positions shown. After the first two measurements, the patient complained of discomfort and changed his position, so position two was repeated. It can be seen from the amplitude curves that the frequency of the rigid body mode decreased after the patient movement. The large phase delay between positions 4 and 5 compared with the normal leg clearly show the lack of rigidity in the region of the fracture. The phase delays in the region of the fracture are over twice the value for normal bone. When the fracture was united, some 2 months later, it was possible to measure position 5 and the phase delay across the fracture had clearly decreased. In contrast, the amplitude curves do not reveal significant information. This demonstrates that the resonant effect, which arises from the properties of the entire bone as well as the fractured region, has not changed substantially. The phase delay measurement is more sensitive to the state of the fracture and has clearly shown the change that has taken place.

Figure 18.3 shows another feature of the phase delay method in that multiple fractures can be investigated. This is not possible with the manual sensing technique. In this case the patient had a non-rigid lower fracture which was investigated first. Later, both fractures were investigated and the curves show that between the measurements the phase delay across the lower fracture had decreased. In Fig. 18.3, this is the separation between curves 2 and 3 in the early measurement, and curves 3 and 4 in the later measurement.

Details of the measurement technique and a discussion of the problems with the resonant technique and the improvements possible with the phase delay technique are discussed in full by Donarski (1992).

Some conclusions on fracture assessment

It is unlikely that the phase delay method will replace manual sensing, despite the shortcomings listed in this article, as the latter is so well accepted in orthopaedic circles. However, the phase delay method does allow a quantitative assessment of fracture strength that could have a distinct part to play in future scientific research and in clinical measurements of problem fractures. Perhaps even the ancient problem of induced oesteogenesis might be reinvestigated using this technique.

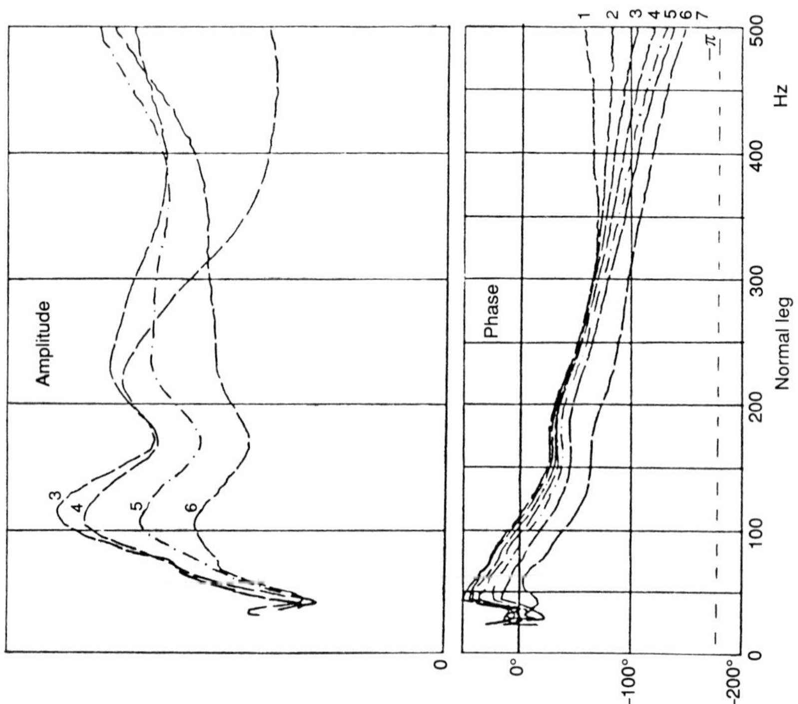

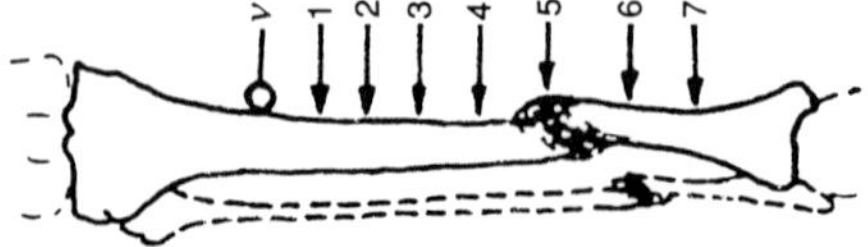

Fig. 18.2 Amplitude and phase responses of a normal leg and of a fractured leg before union and after union. The dotted lines were measured before the patient changed position.

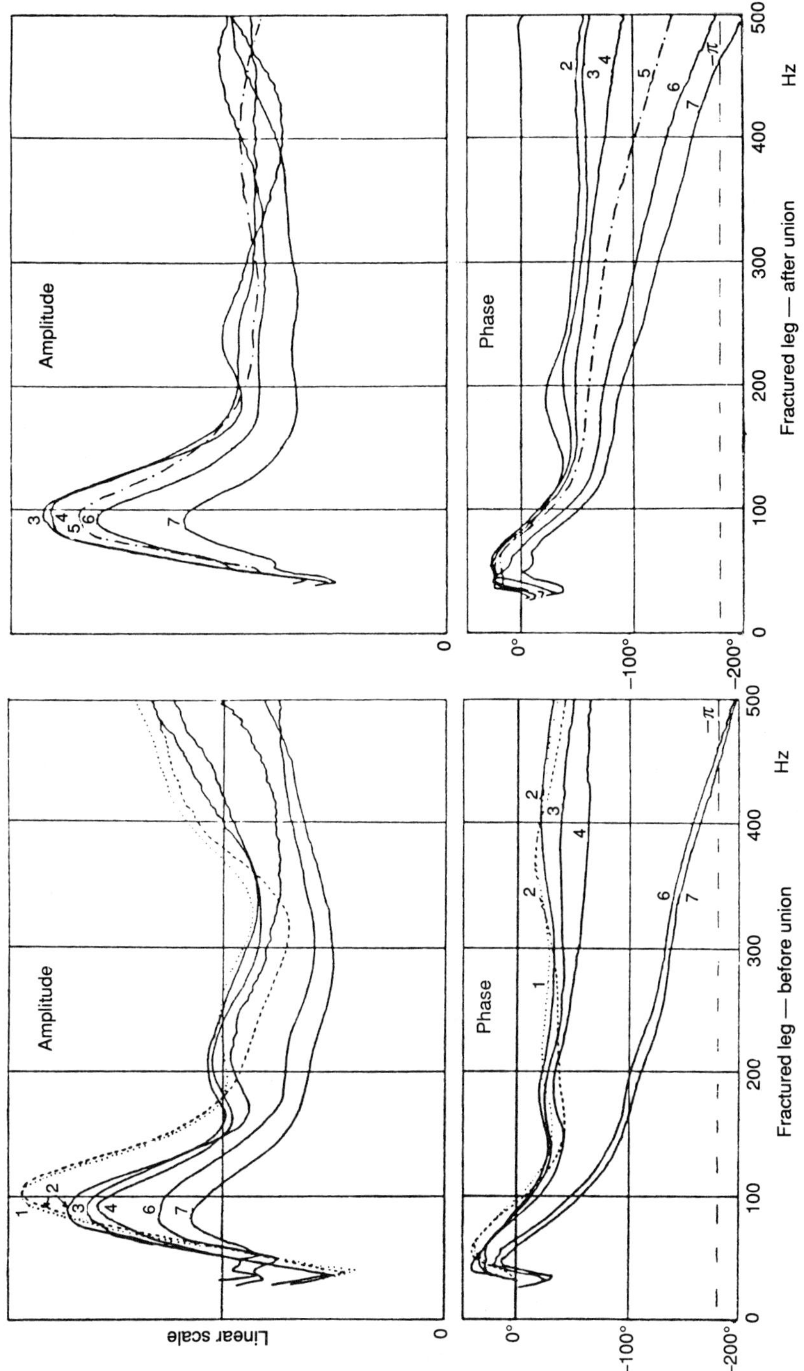
Amplitude
Linear scale
Phase
0
0°
-100°
-200°
$-\pi$
0
100
200
300
400
500
Hz
Fractured leg — before union
Fractured leg — after union

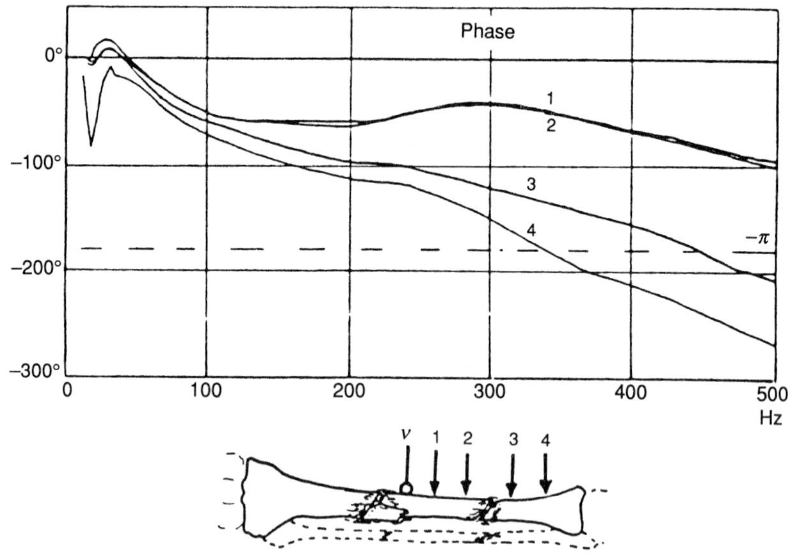

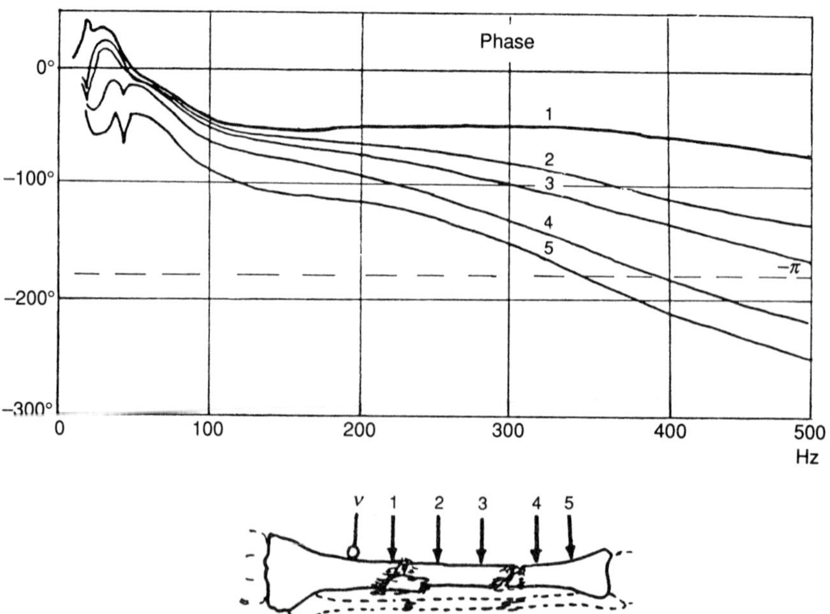

Fig. 18.3 Two measurements in a multiple fracture showing that the lower fracture had become more rigid between measurements.

A simple phase delay monitor for fracture assessment

A small hand-held device has been developed that gives a measurement of phase delay in a form which can be used to check fractures. The vibrations are introduced to the leg using a simple electromechanical shaker and the resultant signal is picked up further down the leg using an accelerometer. The simple phase delay monitor uses this changing phase delay as part of the feedback circuit of an electronic oscillator. The frequency of the oscillator is determined solely by the phase delay. The inverse of the frequency is the time for one cycle and this has to be the time for a signal to go round the loop shown in Fig. 18.4. By adjusting the electronic phase, the frequency can be set to any value. As the fracture heals, the frequency will rise.

In an initial trial of this device, which was carried out by Charles F. Thackray in Leeds, the frequency was set so that the oscillator ran at well above 100 Hz on healthy tibiae and well below on fractured tibia. The results of the clinical trial showed that a great variety of fractures could be monitored successfully by this device. Of more interest is that four cases of possible cracked tibiae were detected which were not detected by manual sensing. Figure 18.5 shows a typical set of results from the clinical trial. The device could be a useful adjunct to an orthopaedic clinic and used for research work where the more expensive equipment for the full phase delay technique is not justified. Since the initial frequency can be adjusted, it could also be used to compare the phase at one frequency with another. It can also be used along the shaft of a tibia to check the healthy bone either side of the fracture. Unlike the resonance method, this technique does not require a measurement of the unfractured bone.

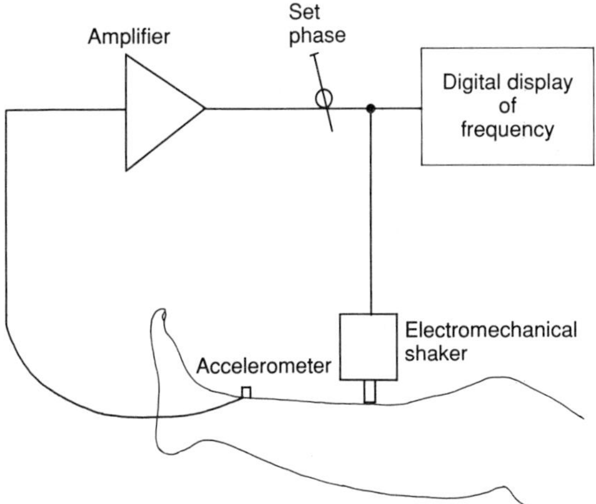

Fig. 18.4 A simple phase delay monitor for fracture assessment.

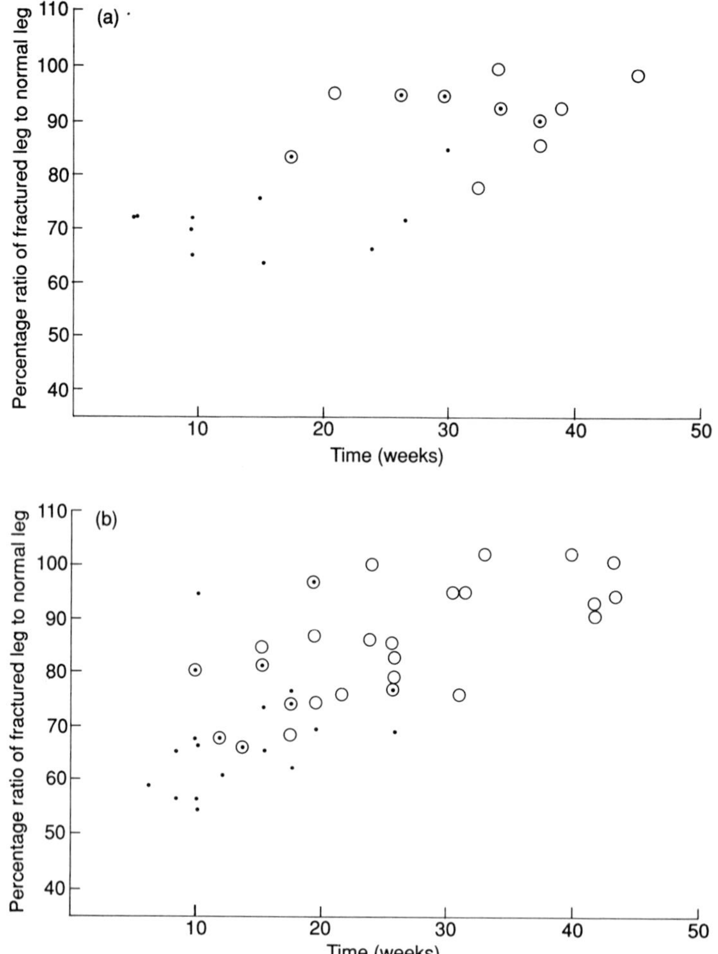

Fig. 18.5 Typical results from a clinical trial of the simple monitor for fracture assessment. Both show measurements are of transverse oblique closed fractures. (a) Severe displacement. (b) Moderate displacement. (Vertical axis = percentage ratio of the fractured leg to normal leg. Horizontal axis = time in weeks.)

Assessment of loose hip prostheses using mechanical vibrations

The assessment of loose hip prostheses is normally done by radiographic examination as manual sensing is not possible. Both false positive and false negative conclusions occur and these result in unnecessary revision surgery.

A new technique first suggested by the University of Kent and then taken up by an energetic group at Oxford (Rosenstein *et al.* 1989) uses the principle that

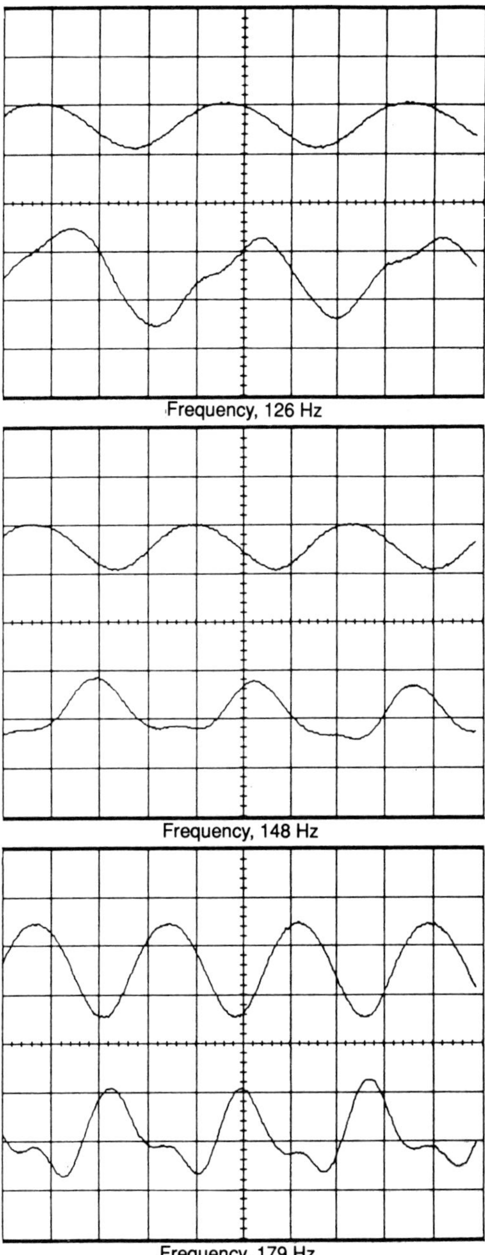

Fig. 18.6 Typical vibration measurement of a loose hip prosthesis. The upper trace is the input signal at the knee, the lower trace is the distorted sine wave detected at the greater trochanter.

a device that is loose will not transmit mechanical vibrations free from distortion. A mechanical vibration is introduced by a shaker into the femur, just above the knee cartilage and the output is measured in the region near the greater trochanter. A loose femoral implant might move in one direction more easily than in another, so that a sinusoidal mechanical vibration would be flattened in one direction. This flattening produces harmonics of the original sine wave since these harmonics when added to the original wave constitute the flattened wave, see Fig. 18.6. The harmonic content or distortion can be measured either subjectively by observing the distortion of the original sine wave, or objectively by using a spectrum analyser.

The sinusoidal shaker is not free from distortion itself, but this systematic error can be removed by measuring the vibration signal with an accelerometer on the shaker itself. Transmission of signals through the femur is, like the tibia, subject to resonance effects and so certain frequencies transmit with fewer losses than others. In order to eliminate this error, the transmission of a single frequency through the femur may be measured also to give a correction for the resonance effect. Thus, the harmonic content can be correctly established and the relationship between the looseness and the harmonic content investigated. Figure 18.6 shows a typical result obtained from a loose prosthesis.

This method has been applied to 30 patients. Since most of these underwent revision surgery soon afterwards, it is possible to say, even at this early stage, that there appears to be a good correlation between harmonic content and looseness. If no harmonic content is detected then there is a high probability that the femoral prosthesis is not loose. Similarly, if a high level of harmonic content is observed, then there is a high probability of looseness. However, the intermediate stage is not so clear and further research will be necessary before a quantitative measure of looseness can be established which would give clear guidance about the need for revision surgery.

Conclusions

The mechanical vibration techniques described in this article are useful to the orthopaedic clinician, but their true worth will be established only when they become widely available. Initial research has been carried out and now industrial development is needed.

Acknowledgements

The authors would like to acknowledge the support and encouragement of Mr J. B. King, London Hospital, Whitechapel, London, and Professor G. van der Perre, Katholieke Universiteit Leuven, Belgium. Without these two, much of this work would not have been possible.

Debate in the meeting

The meeting pointed out that a similar attempt at analysis of loose prostheses had been attempted some 15 years ago. Advances in data processing techniques since then should now enable a more reliable analysis. The very solid and the very loose prosthesis will always be identified clinically. This technique may clarify those in the grey area in-between. The phase method is certainly far less sensitive to muscle tone than the resonance methods.

19 A role for ultrasound in limb lengthening

A. H. R. W. Simpson, J. Derbyshire, and J. Kenwright

Introduction

An increasing number of patients now present to orthopaedic surgeons for the treatment of short stature or unequal leg lengths. The limb lengthening techniques in common use involve stabilizing the limb with an external fixation frame and dividing the bone. After a period of 5–14 days, controlled distraction of the bone is produced by the external frame, usually at a rate of 1 mm/day. The distraction is continued until the required length is achieved, and then a period of consolidation allows the new bone to strengthen.

Accurate imaging is required to monitor this process. Traditionally radiographs have been used to monitor new bone formation and measure the bone gap at the corticotomy site, however callus formation in the bone gap is typically not seen on radiographs within the first month and in some cases considerably longer. This limitation results in uncertainty during the initial stages of the procedure as to the success and appropriate rate of the osteogenesis. The clinician therefore cannot tell whether good quality bone is forming and he may stop, slow down, or even reverse the lengthening, or conversely proceed at too rapid a rate.

Dual energy X-ray absorptiometry (DEXA) scans and tension measurements have been suggested for monitoring new bone formation during limb lengthening. DEXA scans require special equipment which is not readily available in all hospitals. Tension measurements are still under evaluation, however most of the methods for measuring axial tension require adaptation of the external fixation device. The different external fixation systems, in particular the Ilizarov circular apparatus, require different adaptations for force measurement. Imaging ultrasound apparatus is available in the majority of general and orthopaedic hospitals and does not require any different adaptation for different fixator systems.

This study was designed to evaluate the role of ultrasound during limb lengthening with the aim of optimizing the procedure in terms of duration and outcome.

Materials and methods

A simple model for the bone gap was used (Fig. 19.1). The centre of the bone gap is considered to consist of immature callus which becomes increasingly more mature as it extends out towards the original bone ends. For this study the difference between 'mature' and 'immature' callus was taken to be the point at which the callus is strong enough to reflect a 10 MHz ultrasound signal.

Eight lengthening gaps were studied in six patients, four males and two females, with ages between 8 and 28 years who were undergoing limb lengthening by callatosis. The indications for limb lengthening procedures were post-traumatic growth arrest (1), Ollier's disease (1), infantile bone infection (1), congenital hemiatrophy (2), and adult bone infection (1). Ultrasound examinations of the distraction site using a 10 MHz transducer were performed at regular intervals as described by Young *et al.* (1989). The gap between the margins of maturing callus was measured. A note was made of the presence and size of any low reflectivity areas thought to be fluid-filled cysts. Radiographs were also obtained and the gap between the original bone ends measured by placing a radiopaque ruler in the same plane as the bone. The measurements obtained were displayed graphically, the bone gap and the ultrasound gap being plotted against time since the operation. The ultrasound gap was subtracted from the bone gap and this was judged to indicate the amount of new bone produced and this was also plotted against time. Ultrasound assessment was continued during the consolidation period after the distraction had ceased.

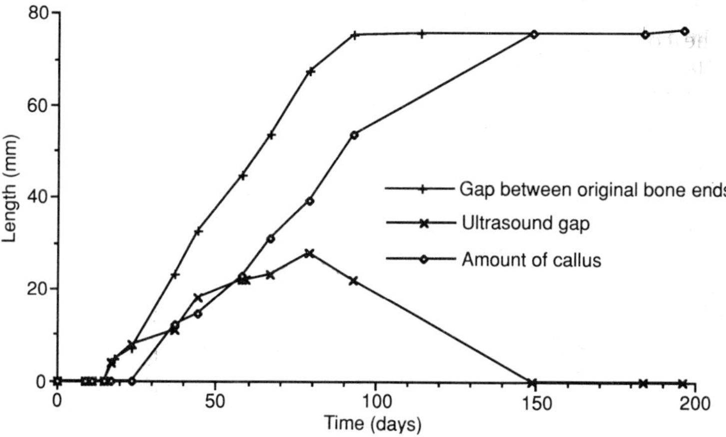

Fig. 19.1 Patient with good rate of callus formation.

Results

All the patients showed an initial lag phase before the production of measurable callus. During this phase the measurement of the gap on ultrasound and that between the original bone ends was the same. After this period it was not possible to detect the original bone ends as the callus had now matured sufficiently to be reflective.

Usually the lag phase was of the order of 10 days but in the three patients that had delayed healing, a prolonged lag period in excess of 25 days was found. After the initial lag period the measurements obtained on ultrasound differed from the gap between the original bone ends as seen on the radiograph, the difference being the amount of mature callus that had formed. The rates at which this callus formed fell into three groups:

1. Equal rates of callus formation and limb lengthening (Fig. 19.1). Four lengthening sites displayed this optimum pattern. All these sites healed well with the frames in place for a minimum length of time.
2. Rate of callus formation less than rate of limb lengthening (Fig. 19.2). Three sites fell into this category. For two of the three sites the rate of limb lengthening had to be reduced from 1 mm to 0.5 mm a day. The frame had to be on for a prolonged period of time during both the lengthening phase and the consolidation phase. Cysts were detected in two of these patients.
3. Rate of callus formation faster than limb lengthening (Fig. 19.3). The rate of limb lengthening of one patient had to be reduced and even stopped at one stage because of pain and stiffness of the soft tissues. The rate of callus formation for this patient exceeded the rate of limb lengthening. The

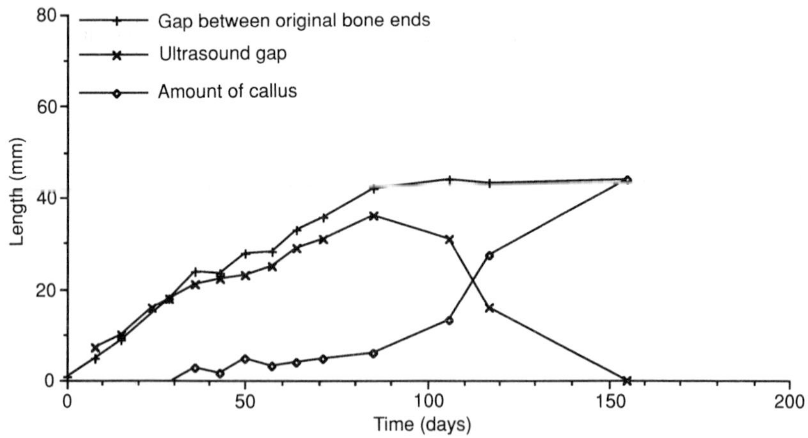

Fig. 19.2 Patient with delayed formation of callus.

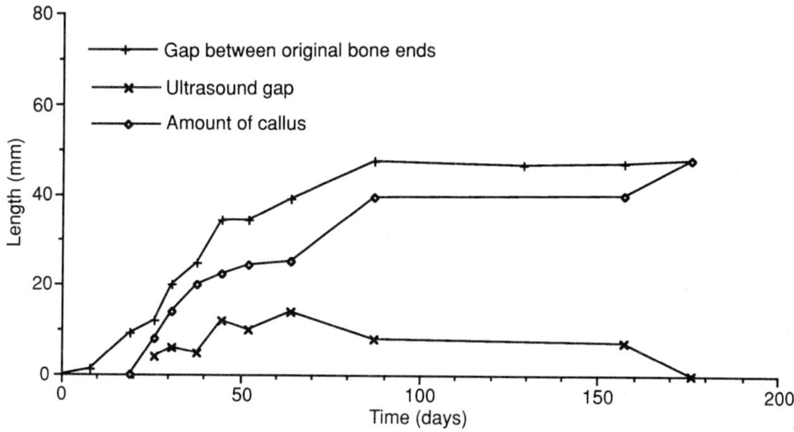

Fig. 19.3 Patient with rate of callus formation exceeding rate of lengthening.

patient appeared to be in danger of premature consolidation, however this did not occur.

Discussion

In this study, serial ultrasound examinations enabled the gap between the margins of the corticotomy or mature callus to be quickly and accurately measured. The quality of maturing callus was assessed and judged to be good if there was a uniform lamella pattern of echoes and no echo free regions. This concurs with the findings of Blane *et al.* (1991). As in this study, areas of decreased echogenicity in the ultrasound gap were found by Young *et al.* (1989). They aspirated these and found them to be fluid-filled cysts, the aspiration of which they felt improved osteogenesis. In this study, the smaller cysts were found to disappear by stopping or reversing distraction for 1 week.

To obtain a satisfactory result during the distraction osteogenesis, the rate of callus formation should parallel the rate of bone lengthening The four sites that achieved satisfactory bone lengthening had such curves. There were three sites in which the outcome was deemed to be unsatisfactory. In these cases there is a prolonged lag period and the curve of callus formation is seen to be shallow, i.e. there is too slow a rate of callus formation for the rate of bone lengthening. Three patients who had such curves were also the ones to develop cysts within the gap. When such results are seen to be developing, the rate of distraction needs to be slowed. It is hoped that this will enable the optimum rate of lengthening to be chosen before cysts have had a chance to develop and therefore pre-empt the problem of poor quality callus formation. Finally, in one patient the rate of callus formation was seen to be faster than the rate of distraction. Too slow a rate of distraction may result in premature consolidation and failure to achieve the target increase in length. However the patient in our study did not suffer this complication, and it may be that so long as an ultrasound gap of sufficient width is maintained, premature consolidation will not occur.

Radiographs may not show any evidence of callus formation in the first 4 weeks of lengthening, sometimes longer, whereas following a relatively short lag period we found that ultrasound can indicate whether callus is forming satisfactorily. Radiography is still required at intervals during distraction to ensure that there is no angulation at the lengthening site, however the number of films can be reduced.

Ultrasound thus provides a means of monitoring the quality and the rate of new bone production from the time of the initial corticotomy. The distraction can thus confidently proceed, the rate can be optimized, and the time that the patient has to wear the fixator can be minimized. It is a valuable technique which complements radiography in the follow-up of patients undergoing limb lengthening procedures.

20 Quantification of bone mineral density changes in the proximal femur following total hip replacement

J. A. K. Fenner, R. Wilkinson, I. G. Stother, and M. A. C. Craigen

Introduction

Failure of the femoral component following total hip replacement (THR) is a common clinical problem with many patients requiring revision of a replacement joint within a period of 10–15 years. It is well established that bone remodels following total hip replacement but the causes and specific details of the bone response are not clearly understood and remain to be quantified. In order to improve the success of THR, a more thorough understanding of the bone remodelling response *in vivo* is required.

Resorption of bone from the proximal femur is a significant factor contributing to the failure of total hip replacements. This has been attributed to stress shielding (Oh and Harris 1982), micromovement (Carlsson *et al.* 1983), thermal necrosis, and reaction to cement and wear particles amongst other causes. Several long-term studies have shown significant bone resorption in both cemented and noncemented THR (Gruen *et al.* 1979; Thomas *et al.* 1986; Engh *et al.* 1987).

Most methods of assessing the rate and degree of bone resorption have so far relied on serial radiography. Contrast in plain film X-rays arises because tissues of different density absorb the X-ray beam to different extents. However, quantification of tissue density from plain X-ray films is unreliable. Inherent errors make quantification of changes in the density of periprosthetic bone unreliable. Both accuracy and precision are poor in plain radiography. This is due to equipment related variables such as film lot, field variability, and exposure setting, and to operator/patient related variables such as leg positioning and analysis of regions. Even with careful control of these variables, precision error remains large, making detection of small changes in bone mineral density (BMD) difficult and imprecise.

Absorptiometry techniques for the measurement of bone mineral density such as single and dual photon absorptiometry (SPA and DPA), and dual energy X-ray absorptiometry (DEXA) have been developed which use collimated beams of known energy and intensity which can be scanned across the patient and the count rate monitored at each point (Tothill 1989). DEXA is the most recent development in absorptiometry. Several groups have used absorptiometry techniques to monitor changes in bone mineral density around orthopaedic implants (Bohr and Schaadt 1983; Richmond *et al.* 1990).

In the early post-operative period the expected changes in bone mineral density will be small and cannot reliably be detected by serial radiography for the reasons described above. If these small changes can be monitored in the operated hip after surgery by an alternative method such as absorptiometry, it will be possible to assess the early bone response and perhaps the patient prognosis. Changes in bone mineral density in specific regions around a prosthesis may be useful in predicting clinical outcome. DEXA is now well established as an accurate and precise method of measuring bone mineral density in the lumbar spine, femoral neck, and total body (Mazess *et al.* 1989; Sorenson *et al.* 1988). Specially-developed software has recently become available which allows this technique to be used to monitor changes in the density of bone adjacent to metallic orthopaedic implants (Lunar Product Information sheet 1990). The software provides automated analysis of bone density changes in regions or zones where bone remodelling has previously been observed (Gruen *et al.* 1979). Several groups have presented work using DEXA to evaluate bone remodelling around orthopaedic implants (Stulberg *et al.* 1990; Richmond *et al.* 1990). The study reported here evaluates the use of DEXA systems for monitoring changes in periprosthetic bone mineral density following cemented total hip replacement. Validation of the technique will be presented and both standard and alternative methods of analysing the basic scan data will be discussed.

Method

Data acquisition

The DEXA system used in this study was a Lunar DPX bone densitometer (Lunar Radiation Corporation, Madison, WI). This instrument (Fig. 20.1) uses a narrowly-collimated beam of X-rays which is passed through a cerium filter to allow two energies at 38 keV and 70 keV to be emitted. As each of the two energies is attenuated differently by soft tissue and bone, the effect of soft tissue can be eliminated leaving the final BMD result dependent only on the amount of bone mineral in the beam path.

The beam is passed through the subject in a rectilinear scan pattern to acquire an image of the entire bone–prosthesis complex. Bone mineral content (BMC) in the beam path is calculated from the recorded counts at the two energy levels. The BMC is divided by the area of the region of interest within the scan image to give the bone mineral density in g/cm^3 for that region. It

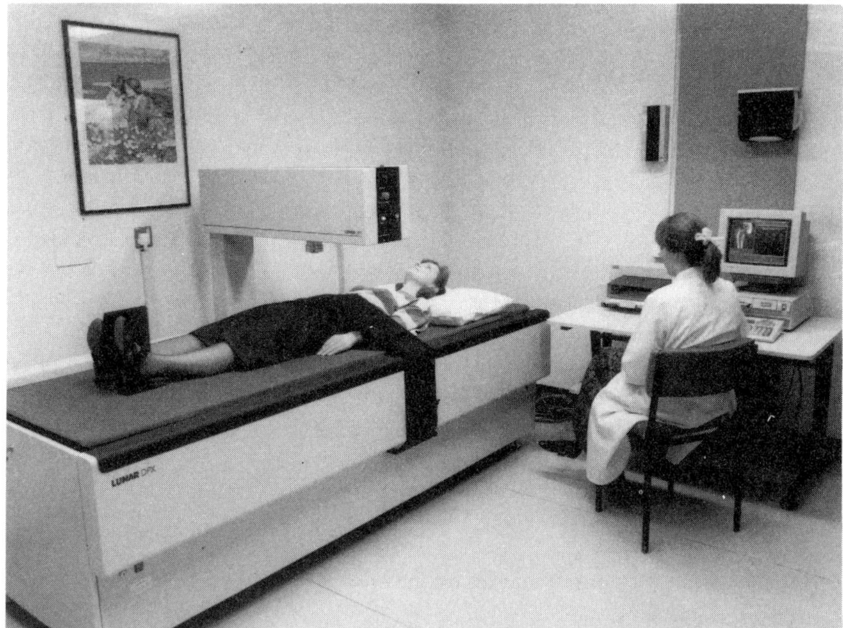

Fig. 20.1 Lunar DPX bone densitometer used for dual energy X-ray absorptiometry (DEXA).

should be noted that this is not a true density but an aerial density obtained from an image that is a two-dimensional projection. For orthopaedic images the pixel or sample size is 0.6 mm × 0.6 mm, providing a high resolution scan image. Scan length is a maximum of 240 mm.

The subject is placed supine on the scan table with the leg to be scanned placed in neutral rotation in a foot brace. A foam positioning device is placed behind the knee to allow a slight knee flexion. The leg should be parallel to the long axis of the table so that the transverse beam path is perpendicular to the femoral shaft. Careful positioning of the subject is important to ensure good reproducibility.

Data analysis

Using the standard automated analysis procedure, the program defines a reproducible origin point at the extreme medial border of the lesser trochanter. Four regions of interest are positioned relative to this reference point overlying areas where bone remodelling has previously been observed (Fig. 20.2). Two regions, C and D, are placed on the medial side of the implant, above and below the reference point. These overlie the calcar and lesser trochanter. Two regions, A and B, are placed on the lateral side of the implant

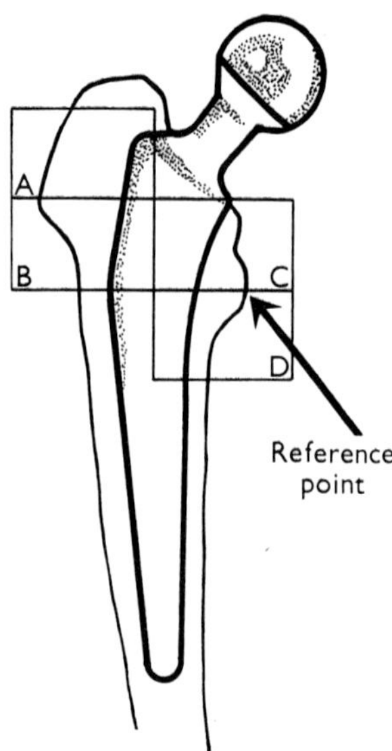

Fig. 20.2 Regions of interest on the proximal femur to be measured.

overlying the greater trochanter. These four regions can be mapped onto subsequent scans to identify changes in regional BMD. By following BMD results in these areas over time, trends or patterns in bone loss in specific regions around an implant can be evaluated.

In addition to the standard analysis procedure, two other methods of analysing and presenting the basic scan data have been evaluated. A perimeter region can be defined around the entire implant surface (Fig. 20.3). The width and position of this region can be varied allowing analysis of bone along the length of the implant at any distance from the implant surface outwards to the bone edges. In addition, BMD can be calculated for smaller areas within this perimeter region.

Cross-sectional profiles can also be drawn across the femur at any position along its length (Fig. 20.3). The profile is essentially a plot of density at each sample point of the cross-section. The profile region may be only one scan line deep or may cover large portions of the femoral shaft.

Validation

Accuracy was assessed by measurements of bone simulation phantoms. Scans were performed on phantoms of four known densities and repeated with an orthopaedic implant present. Precision of the method was assessed both *in vitro* and *in vivo* for the standard acquisition and analysis procedures. *In-vitro* precision was assessed on excized femora. Cadaveric femora with implants

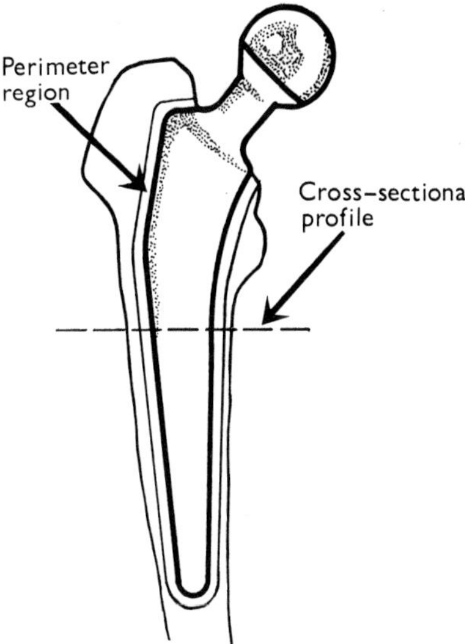

Fig. 20.3 Perimeter of implant and cross-section to be measured.

cemented in place were each scanned five times, with repositioning between scans.

The same bones were used to investigate sensitivity to positioning errors using a jig which allowed independent movements in rotation, vertical tilt, and horizontal angle. *In-vivo* precision was assessed on six subjects with asymptomatic hip replacements at least 2 years post-operatively.

Results

Validation of the technique

The radiation dose for all three scans is 0.15 mSv, so repeat measurements pose no hazard. Accuracy of the technique was assessed using bone simulation phantoms. The results shown in Table 20.1 indicate that the measured values of bone density do not differ significantly from the actual values at any of the four density values. Also, the presence or absence of the implant has no effect on the accuracy of the measured value.

Values of BMD measured on cadaveric femora are quite insensitive to changes in horizontal or vertical bone angles. However, changes in rotation of the femur had a significant effect on measured BMD values, particularly in regions C and D (Fig. 20.2) where the position of the lesser trochanter is of great importance. With internal rotation of 100 BMD, changes of up to 6 per cent were recorded for the calcar region compared to neutral rotation results.

Table 20.1 Accuracy of DEXA

	Measured density (mean and SD)	
Actual density	No implant	Implant
0.625	0.618 (0.006)	0.623 (0.004)
1.010	1.014 (0.005)	1.008 (0.002)
1.240	1.238 (0.004)	1.241 (0.003)
1.586	1.589 (0.002)	1.587 (0.004)

Table 20.2 Precision (CV%)

	Excised femora	Patients
CV	1.5–2.4%	1.9–4.6%
No. of specimens	4	6
No. of scans per specimen	5	5

Precision of the method can be described by the coefficient of variation. The percentage coefficients of variation (CV) for both *in-vitro* and *in-vivo* precision are shown in Table 20.2.

Assessment of analysis

Using the standard auto-analysis procedure, BMD results were calculated for the four regions shown in Fig. 20.2. Mapping these regions onto subsequent scans provides assessment of changes in regional BMD over time. Early clinical results show changes in regional BMD of between 6 and 20 per cent within 6 months from surgery, with the largest changes occurring below the calcar region. Larger losses occur in regions C and D than in regions A and B. In comparison, changes in BMD in the spine and contralateral hips were only 0.5–2.7 per cent in the same period. Figure 20.4 shows the pattern of bonc loss for two selected patients followed to 1 year post-operatively.

Alternative analysis approaches were adopted to examine the varying rates of bone resorption around the shaft of the implant on both medial and lateral sides within a perimeter region, and along cross-sectional profiles perpendicular to the shaft of the implant. A perimeter region could be defined along the entire length of the prosthesis (Fig. 20.3). Both the width and position of this region may be altered by the operator. Results obtained in a perimeter region 3-mm wide placed immediately adjacent to the implant surface are shown in Figs 20.5 and 20.6 for the medial and lateral sides. From the graph of the

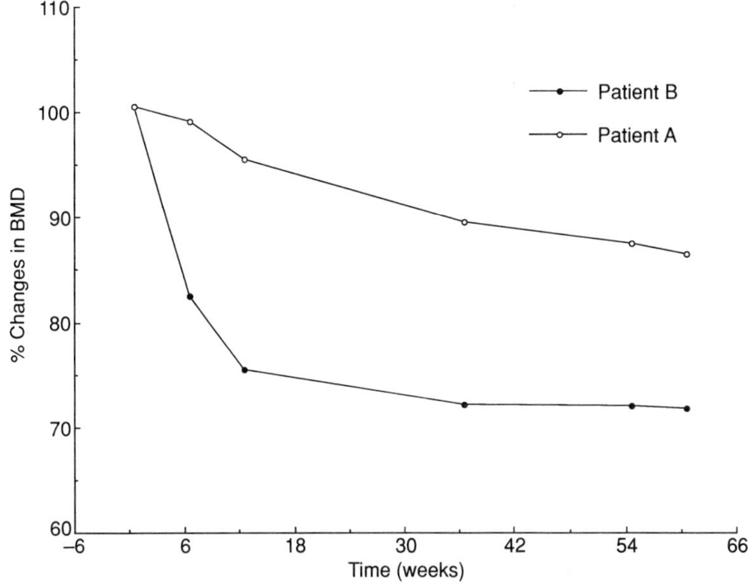

Fig. 20.4 Bone density in the calcar region followed over 1 year post-surgery.

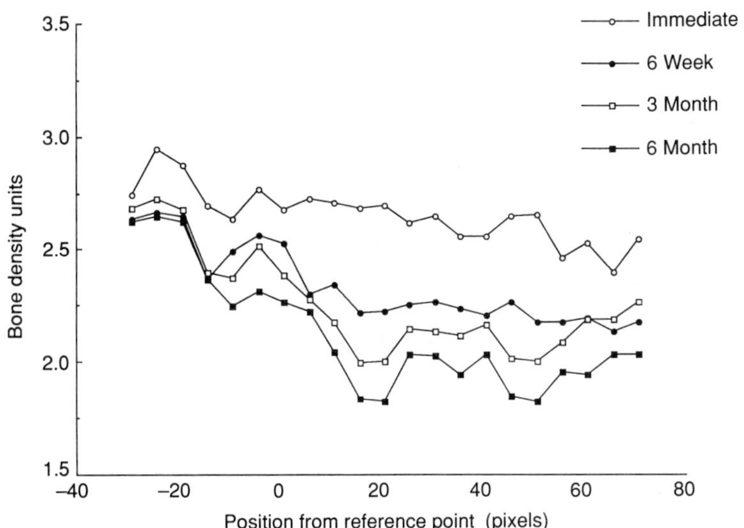

Fig. 20.5 Bone density in the lateral perimeter from 0 to 6 months.

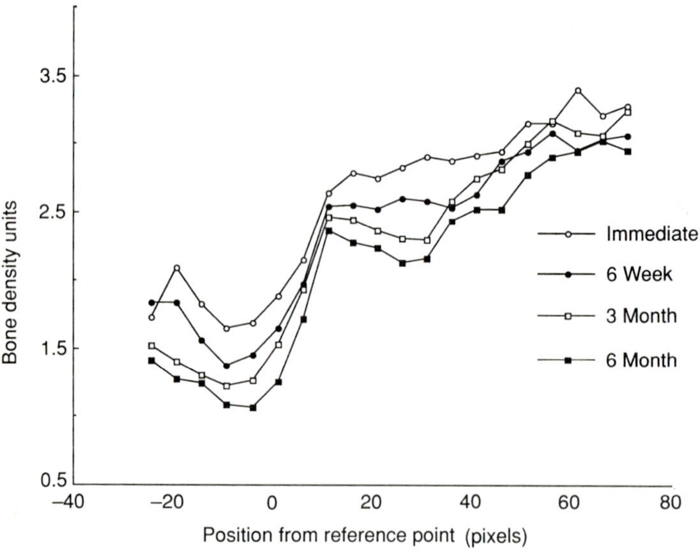

Fig. 20.6 Bone density in the medial perimeter from 0 to 6 months.

medial side (Fig. 20.5) it is clear that larger changes in bone density occur in the calcar region (between -40 and 0 on the graph) than further down the shaft of the implant.

Cross-sectional BMD profiles may provide similar spatial information about rates of bone loss at different points across the femur. Figure 20.7 shows cross-sectional profiles for an immediate post-operative scan and 6 month

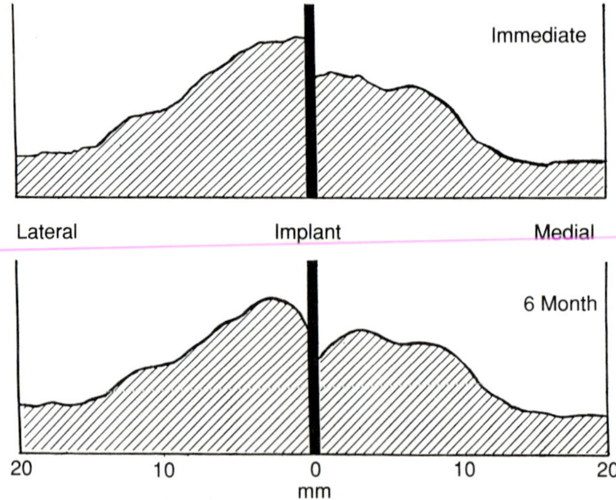

Fig. 20.7 Bone density profiles in two patients at 0 and 6 months post-surgery.

scan, with the cross-sectional profile at the same position with respect to the reference point. It is clear that there has been a decrease in BMD close to the implant surface at this position between these scans. Again the change is more pronounced on the medial side.

Discussion

This study demonstrates that DEXA systems provide an accurate, precise, and convenient means of monitoring bone loss in the proximal femur following total hip replacement. Precision error is small (2–4 per cent) compared with the changes in BMD observed (10–20 per cent), so the method is sensitive enough to detect changes occurring even in the early post-operative period. Good precision, however, is very dependent on reproducible positioning of the patient, with particular care required to ensure that the femur is placed in neutral rotation. One of the problems in assessment of cemented prostheses is demarcation between radiopaque cement and bone. The interfaces are generally interdigitated and the attenuation coefficients similar. Thus there will always be a region of uncertainty in the position of the inner bone surface. Further studies of this problem are in progress, but when interpreting data from cemented arthroplasties this factor must be considered.

Rates of bone loss between patients are highly variable and as yet do not provide sufficient information to identify patterns or trends in groups of patients. Results over the first 12 months post-operatively suggest that bone loss in the proximal femur is significant and progressive. Other studies using DEXA suggest a continuing slow bone loss after 1 year, even though serial radiography may be unable to resolve these small changes. The calcar region, C, shows the largest effect and perimeter regions indicate larger bone losses on the medial side. The standard analysis procedures (Figs 20.2 and 20.4) provide information on regional changes in BMD with time. Clinical results from this study concur with the observations of Kiratli *et al.* (1991), Richmond *et al.* (1990), and Stulberg *et al.* (1990).

The alternative analysis procedures (Figs 20.3, 20.5, 20.6, and 20.7) provide additional information about the local rates of bone loss around the implant. The perimeter display provides an assessment of BMD changes along the shaft of the femur, while the cross-sectional display may be useful in assessing the development of radiolucent zones. At present, numerical data from these two displays can only be obtained by subsequent processing using additional software packages. Until these procedures form part of the standard analysis software their use will be limited by the operator time required for their implementation.

Longer term prospective studies on larger groups of patients are necessary to further evaluate the natural history of bone remodelling. The technique may be of potential clinical use in assessing bone loss related to micromovement and loosening in hip joint femoral components.

Acknowledgements

The bone densitometer used in this work was purchased by the Glasgow Royal Infirmary Bone Densitometer Appeal Fund. This fund also covers the running costs of the facility. Thanks are due to the Bone Mineral Metabolism Group of the Department of Medicine for access to the facility for research work. The authors would like to thank Lunar Radiation Corporation for providing the orthopaedic software for use in this research.

References to Section 3

Albrektsson, B. J., Ryd, L., Carlsson, L. V., Freeman, M. A. R., Herberts, P., Regnér, L. *et al.* (1990). The effect of a stem on the tibial component of knee arthroplasty. A Roentgen stereophotogrammetric study of uncemented tibial components in the Freeman–Samuelson knee arthroplasty. *Journal of Bone and Joint Surgery*, **72B**, 252–8.

Amstutz, H. C., Ouzounian, T., Grauer, D., Flink, C., Kirkpatrick, J., and Bassett, L. (1986). The grid radiograph. A simple technique for consistent high resolution visualization of the hip. *Journal of Bone and Joint Surgery*, **68A**, 1052–6.

Aronson, A. S., Holst, L., and Selvik, G. (1974). An instrument for insertion of radiopaque bone markers. *Radiology*, **113**, 733–4.

Baldursson, H., Egund, N., Hansson, L. I., and Selvik, G. (1979). Instability and wear of total hip prostheses determined with roentgen stereophotogrammetry. *Archives of Orthopaedic and Traumatic Surgery*, **95**, 257–63.

Blane, C. E., Herzenberg, J. E., and Dipietro, M. A. (1991). Radiographic imaging for Ilizarov limb lengthening in children. *Pediatric Radiology*, **21**, 117–20.

Bohr, H. and Schaadt, O. (1983). Bone mineral content of femoral bone and the lumbar spine measured in women with fracture of the femoral neck by dual photon absorptiometry. *Clinical Orthopaedics and Related Research*, **179**, 240–5.

Brown, R. H., Burstein, A. H., Nash, C. L., and Schock, C. C. (1976). Spinal analysis using a three-dimensional radiographic technique. *Journal of Biomechanics*, **9**, 355–65.

Carlsson, A. S., Gentz, C., and Linder, L. (1983). Localized bone resorption in the femur in mechanical failure of cemented total hip arthroplasties. *Acta Orthopaedica Scandinavica*, **54**, 396–402.

Carlsson, L. V., Albrektsson, B. E. J., Freeman, M. A. R., Herberts, P., Malchau, H., and Ryd, L. (1992). A new radiographic method for detection of tibial component migration in total knee arthroplasty. *Journal of Arthroplasty* (in press).

Chafetz, N., Baumrind, N., Murray, W. R., Genant, H. K., and Korn, E. L. (1985). Subsidence of the femoral prosthesis. *Clinical Orthopaedics and Related Research*, **201**, 60–7.

Cornelissen, P., Cornelissen, M., van der Perre, G., Christensen, A. B, Ammitzboll, F., and Dyrbye, C. (1986). Assessment of tibial stiffness by vibration testing *in situ*—I. *Journal of Biomechanics*, **19**, 53–60.

Cornelissen, M., Cornelissen, P., van der Perre, G., Christensen, A. B, Ammitzboll, F., and Dyrbye, C. (1987). Assessment of tibial stiffness by vibration testing *in situ*—III. *Journal of Biomechanics*, **20**, 333–42.

Collet, C., Grossdidier, G., and Borrelly, J. (1985). Anatomical basis for study of acetabular migration of the prosthetic cup used in arthroplasty of the hip: a new technique for measurement of migration. *Anatomia Clinica*, **1**, 171–4.

Cunningham, L. and Kenwright, J. (1989). Mechanical measurements of fracture repair. *Proceedings of the Institution of Mechanical Engineers*, International Conference, pp. 133–8.

Dempsey, A. J., Finlay, J. B., Bourne, R. B., Rorabeck, C. H., Scott, M. A., and

Millman, J. C. (1989). Stability and anchorage considerations for cementless tibial components. *Journal of Arthroplasty*, **4**, 223–30.

Donarski, R. J. (1992). Bone fracture measurement using mechanical vibration. Ph.D. Thesis. University of Kent.

Edwards, J. (1981). Mechanical techniques for clinical assessment of fracture repair. *EEC Meeting on Measurements of Bone Healing*, 32–8.

Edwards, J. (1991). Bone: measurement of properties. In *Concise encyclopaedia of biological and biomedical measurement systems* (ed. P. A. Payne), pp. 74–81. Pergamon Press, Oxford.

Engh, C. A., Bobyn, J. D., and Glassman, A. H. (1987). Porous coated hip replacement: the factors governing bone ingrowth, stress shielding, and clinical results. *Journal of Bone and Joint Surgery*, **69B**, 45–55.

Friedman, R. J. (1981). Current status of total knee arthroplasty. *Postgrad Medicine*, **70**, 207–14.

Green, D. L., Bahniuk, E., Liebelt, R. A., Fender, E., and Markov, P. (1983). Biplane radiographic measurements of reversible displacement (including clinical loosening) and migration of total joint replacements. *Journal of Bone and Joint Surgery*, **65A**, 1134–43.

Gruen, T. A , McNeice, G. M., and Amstutz, H. C. (1979). Modes of failure of cemented stern type femoral components: a radiographic analysis of loosening. *Clinical Orthopaedics and Related Research*, **141**, 17–27.

Hammer, R., Edholm, P., and Lindholm, B. (1984). Stability of union after tibial shaft fracture, analysis of a non-invasive technique. *Journal of Bone and Joint Surgery*, **66B**, 529–34.

Hammer, R. R., Hammerby, S., and Linhdolm, B. (1985). Accuracy of radiologic assessment of tibial shaft fracture union in humans. *Clinical Orthopaedics and Related Research*, **199**, 233–8.

Hardinge, K., Porter, M. L., Jones, P. R., Hukins, D. W., and Taylor, C. J. (1991). Measurement of hip prostheses using image analysis. *Journal of Bone and Joint Surgery*, **13B**, 724–8.

Hvid, I. (1988). Trabecular bone strength at the knee. *Clinical Orthopaedics and Related Research*, **227**, 220–1.

Ilchman, T., Franzén, H., Mjöberg, B., and Wingstrand, H. (1992). Measurement accuracy in acetabular cup migration. A comparison of four radiologic methods versus Roentgen Stereophotogrammetric Analysis. *Journal of Arthroplasty*, **7**, 121–7.

Jones, P. R., Hukins, D. W. L., Porter, M. L., Davies K. E., Hardinge, K., and Taylor, C. J. (1992). Bending and fracture of the femoral component in cemented total hip replacement. *Journal of Biomedical Engineering*, **14**, 9–15.

Kay, P. R., Ross, E. R. S., and Powell, E. S. (1989). Development and Clinical Application of an External Fixator Monitoring System. *Journal of Biomedical Engineering*, **11**, 240–4.

Kiratli, B. J., Heiner, J. P., Wilson, M. A., McBeath, A. A. (1991). Bone mineral density of the proximal femur after uncemented total hip arthroplasty. (Abstract) *37th Annual Meeting, Orthopaedic Research Society*, Anaheim, CA.

Kirkpatrick, J. S., Clarke, I. C., Harlan, C. A., and Jinnah, R. H. (1982). Radiographic techniques for consistent visualization of total hip arthroplasties. *Clinical Orthopaedics and Related Research*, **174**, 158–63.

Laskin, R. S. (1988). Tricon-M uncemented total knee arthroplasty. A review of 96 knees followed for longer than 2 years. *Journal of Arthroplasty*, **3**, 27–38.

Lippert, F. G., Harrington, R. M., Veress, S. A., Fraser, C., Green, D., and Bahniuk, E. (1982*a*). A comparison of convergent and bi-plane X-ray photogrammetry systems used to detect total joint loosening. *Journal of Biomechanics*, **15**, 677–82.

Lippert, F. G., Veress, S.A., Harrington, R. M., and El-Garf, T. (1982*b*). Single X-ray photogrammetry system used to detect total joint loosening. *Proceedings of the Society of Photo-Optical Instrumentation Engineers*, Biostereometrics '82, **361**, 186–92.

Lunar Product Information Sheet (1990/91). *Dual energy X-ray absorptiometry and total hip arthroplasty*. Lunar Corporation.

Mazess, R B., Collick, B., Trempe, J., Barden, H., and Hanson, J. (1989). Performance evaluation of a dual energy X-ray bone densitometer. *Calcified Tissue International*, **44**, 228–32.

Mjöberg, B. (1986). Loosening of the cemented hip prosthesis. *Acta Orthopaedica Scandinavica*, **57** (*Suppl*. 221).

Mjöberg B., Selvik G., Hansson L. I., Rosenqvist R., and Önnerfalt R. (1986). Mechanical loosening of total hip prostheses. A radiographic and Roentgen Stereophotogrammetric study. *Journal of Bone and Joint Surgery*, **68B**, 770–4.

Nicholas, P. J. (1979). X-Ray diagnosis of healing fractures in rabbits. *Clinical Orthopaedics and Related Research*, **142**, 234–6.

Nokes, L., Mintowt-Czyz, W. J., Fairclough, J. A., Mackie, I., Howard, C., and Williams, J. (1984). Natural frequency of fracture fragments in the assessment of tibial fracture healing. *Journal of Biomedical Engineering*, **6**, 227–9.

Nunn, D., Freeman, M. A., Hill, P. F., and Evans, S. J. (1989). The measurement of migration of the acetabular component of hip prostheses. *Journal of Bone and Joint Surgery*, **71**, 629–31.

Oh, I. and Harris, W. H. (1982). Proximal strain distribution in the loaded femur. *Journal of Bone and Joint Surgery*, **64A**, 983–90

Panjabi, M. M., Lindsey, R. W., Walter, S. D., and White, A. A. (1989). The clinician's ability to evaluate the strength of healing fractures from plain radiographs. *Journal of Orthopaedic Trauma*, **3**, 29–32.

Pearcy, M. J. (1985). Stereo radiography of lumbar spine motion. *Acta Orthopaedica Scandinavica*, **56** (*Suppl*. 212).

Probst, K. J. (1980). *Stereo-Röntgen-Analyse (SRA)* von Lockerungsvorgängen alloplastischer Hüftgelenkimplantate. Enke Copythek, Ferdinand Enke, Stuttgart.

Richmond, B. J., Bauer, T. W., Stulberg, B. N., and Fox, J. S. (1990). Bone mineral density in patients undergoing uncemented total hip arthroplasty. *Calcified Tissue International*, **46**, 145.

Rosenstein, A. D., McCoy, G. F., Bulstrode, C. J., McLardy-Smith, P. D., Cunningham, J. L. and Turner-Smith, A. R. (1989). The differentiation of loose and secure femoral implants in total hip replacement using a vibration technique: an anatomical and pilot clinical study. *Proceedings of the Institution of Mechanical Engineers*, **203**, 77–89.

Russe, W. (1988). Röntgenphotogrammetrie der künstlichen Hüftgelenkspfanne. *Aktuelle Probleme in Chirurgie und Orthopädie*, **32**, 1–80.

Ryd, L. (1986). Micromotion in knee arthroplasty: a Roentgen stereo photogrammetric analysis of tibial component fixation. *Acta Orthopaedica Scandinavia*, **57** (*Suppl*. 220).

Ryd, L., Lindstrand, A., Rosenquist, R., and Selvik, G. (1986). Tibial component fixation in knee arthroplasty. *Clinical Orthopaedics and Related Research*, **213**, 141–9.

Selvik, G. (1989). Roentgen stereophotogrammetry—A method for the study of the kinematics of the skeletal system. *Acta Orthopaedica Scandinavica*, **60** (*Suppl*. 232).

Selvik, G., Alberius, P., and Aronson, A.S. (1983). A Roentgen stereophotogrammetric system. Construction, calibration and technical accuracy. *Acta Radiologica Diagnosis*, **24**, 343–52.

Shimagaki, H., Bechtold, J. E., Sherman, R. E., and Gustilo, R. B. (1990). Stability of initial fixation of the tibial component in cementless total knee arthroplasty. *Journal of Orthopaedic Research*, **8**, 64–71.

Sorenson, J. A., Hanson, J A., and Mazess, R. B. (1988). Precision and accuracy of dual energy X-ray absorptiometry (Abstract). *Proceedings of the 10th annual meeting of the American Society of Bone Mineral Research*, June 1988. New Orleans, Los Angeles.

Stulberg, B. N., Eberle, R. W., Deal, C. L., and Richmond, B. J. (1990). *X-ray absorptiometry for measuring peri-prosthetic bone mineral density*. Lunar Corporation Research/Technical Note (DOCSCl- E-191).

Sutherland, C. J., Wild, A. H., Borden, L. S., and Marks, K. E. (1982). A ten-year follow-up of one hundred consecutive Müller curved-stem total hip-replacement arthroplasties. *Journal of Bone and Joint Surgery*, **64A**, 970–82.

Takamoto, T. (1976). X-ray photogrammetric analysis of skeletal spatial motions. Ph.D. Dissertation. University of Washington, USA.

Thomas, B. J., Salvati, E. A., and Small, R. D. (1986). The CAD hip arthroplasty: five to ten year follow-up. *Journal of Bone and Joint Surgery*, **68A**, 640–6.

Tothill, P. (1989). Methods of bone mineral measurement. *Physics in Medicine and Biology*, **34–5**, 543–72.

Turner-Smith, A. R. (1990). X-ray photogrammetry of artificial joints. *The Photogrammetric Record*, **13**, 347–66.

Turner-Smith, A. R., White, S. P., and Bulstrode, C. (1990). X-ray photogrammetry of artificial hip joints. In *Close-range photogrammetry meets machine vision* (ed. A. Gruen and E. P. Baltsavias). *Proceedings of the SPIE*, **1395**, 587–94.

van der Perre, G., and Borgwardt-Christensen, A. (1985). Monitoring of fracture healing by vibration analysis and other mechanical methods. *Proceedings of specialist consensus meeting*, Leuven, 14–16 March 1985. Acco Leuven/Amersfoort.

van der Perre, G., Lowel, G., and Borgwardt-Christensen, A. (1989). *In vivo* assessment of bone quality by vibration and wave propagation techniques. *Proceedings of the 1989–90 COMAC BME* 11.2.6. (Monitoring of Fracture Healing).

Veress, S. A. (1989). Different considerations in analytical X-ray photogrammetry. *Photogrammetria*, **43**, 133–42.

Veress, S. A., Lippert, F. G., and Takamoto, T. (1977). An analytical approach to X-ray photogrammetry. *Photogrammetric Engineering and Remote Sensing*, **43**, 1503–10.

Wetherell, R. G., Amis A. A., and Heatley, F. W. (1989). Measurement of acetabular erosion. The effect of pelvis rotation on common landmarks. *Journal of Bone and Joint Surgery*, **71B**, 447–51.

Young, J. W. R., Kostrubiak, I. S., Resnik, C. S., and Paley, D. (1989). Sonographic evaluation of bone production at the distraction site in Ilizarov limb lengthening procedures. *American Journal of Roentgenology*, **154**, 125–8.

Section 4—Material Properties

21 Micromechanical aspects of normal and deformed cortical bone

A. ASCENZI AND A. BOYDE

Mechanical significance

The attention of investigators has long been attracted by the biomechanical significance of the osteonic structure of bone shafts. As is well known, 'osteons' (used here to stand for 'secondary osteon') are lamellar systems each concentrically arranged around a central canal whose axis runs parallel to the diaphyseal axis of a long bone. Under the polarizing microscope, osteonic lamellae may be divided into two types: some appear dark in cross-section and bright (or light) in longitudinal section, whereas others appear bright in cross-section and dark in longitudinal section. Lamellae of the first type have collagen fibres (and crystallites) showing the same orientation as the osteon axis; they are therefore referred to as 'longitudinal'. Lamellae of the second type have collagen fibres (and crystallites) all of which were, for a long time, supposed to be oriented only circularly, that is, orthogonally to the osteon axis; they were therefore called 'transverse'. This view previously discussed by Burkhardt (1929) is under review (Frasca *et al.* 1977; Ascenzi and Benvenuti 1986; Giraud-Guille 1988; Marotti and Muglia 1988) because oblique fibres have also been found beside the transverse ones. Although the appellation 'oblique' is more correct, we will continue to use the name 'transverse' for the second type of lamella.

The first authors to succeed in isolating individual osteon samples and in determining their tensile and compressive properties were Ascenzi and co-workers (Ascenzi *et al.* 1966; Ascenzi and Bonucci 1967, 1968). To be able to collect a homogeneous series of comparable isolated osteons, a detailed knowledge of their structural features and their degree of calcification is indispensable. On this basis, human fully-calcified osteons with three separate types of collagen fibre orientation were chosen. In the first or 'longitudinal type', fibres had a marked longitudinal spiral course in successive lamellae. In the second or 'alternate type', the osteonic structure was the result of an alternation of longitudinal and transverse lamellae. In the third or 'transverse type' the osteon was almost entirely made up of transverse lamellae.

The results obtained showed that ultimate tensile strength and modulus of elasticity under tension were highest in fully calcified longitudinal osteons. Conversely, ultimate compressive strength and modulus of elasticity under compression were highest in fully calcified transverse osteons. These findings indicate that the differences recorded between the biomechanical behaviour of the three osteon types depend essentially on the structural orientation of the lamellae. This inference has been confirmed by measurements using other micromechanical techniques to perform the shearing test (Ascenzi and Bonucci 1972) and 'pin-test' (Ascenzi and Bonucci 1976), and to record hysteresis loops (Ascenzi *et al.* 1985).

The results obtained by Ascenzi and co-workers were strengthened by a series of investigations carried out by Vincentelli and Evans (1971), Evans and Vincentelli (1974), and Evans (1974), using a quite different technique. Cross-sections of macroscopic longitudinal bone specimens were prepared from the level of the break area and loaded under tension and compression under the polarizing microscope. These authors were able to show that ultimate tensile strength has a significant negative correlation with the percentage of bright osteons in the break area, and an equally significant positive correlation with the percentage of dark osteons. Conversely, ultimate compressive stress has a significant positive correlation with the percentage of bright and intermediate osteons in that area.

When the results obtained by Ascenzi and co-workers and the findings of Evans and Vincentelli are considered together, it may be concluded that longitudinal osteonic lamellae are structures that are suitable for loading under tension and that osteonic transverse lamellae are suitable for loading under compression.

Distribution of osteonic lamellae in bone shafts

In the cortical bone of limbs, the continuous remodelling process creates a patchwork distribution pattern in which fragments of osteons (or interstitial bone) formed some time previously alternate with osteons which have just been laid down. In adult life the diaphyses of the long bones may be considered as consisting entirely of osteons and interstitial bone. When reference is made both to the patchwork distribution pattern of cortical bone and the mechanical properties of the two types of osteonic lamellae, it appears clear that an analysis of the distribution of the lamellae in bone shafts would provide the information needed for correlating stress distribution and bone microstructure.

Femur

The first research on the femur was carried out by Portigliatti Barbos *et al.* (1983). This was done using a Zeiss Videoplan, a semi-automatic apparatus designed for the acquisition and computation of geometric data and having as essential features a digitizer tablet, a computer, and a monitor. An examination of the colour-coded maps prepared from three sections obtained from the middle portion of the diaphysis revealed that the distribution of the two types of osteonic lamellae was not random. Transverse lamellae were mainly found at the boundary between the medial and posterior aspect of the bone. Conversely, longitudinal lamellae were mainly distributed in the lateral and anterior aspects (Fig. 21.1).

To be able to offer an explanation of this distribution pattern, it seemed useful to refer to stress distribution in the femoral diaphysis. After the pioneering investigations of Küntscher (1934, 1935*a*,*b*) and Evans and Lissner (1948) using the stresscoat technique, Blaimont (1968) was the first author to predict a rotational distribution of stresses from the proximal to the distal end of the femur by applying circumferential extensometry. Studies on *in-vivo* loading of the lower limb indicate significant antero-posterior and medio-lateral bending of the femur during gait (Paul 1966, 1971). A more recent model (Ruff 1981) shows that when compression is exerted in the femoral head

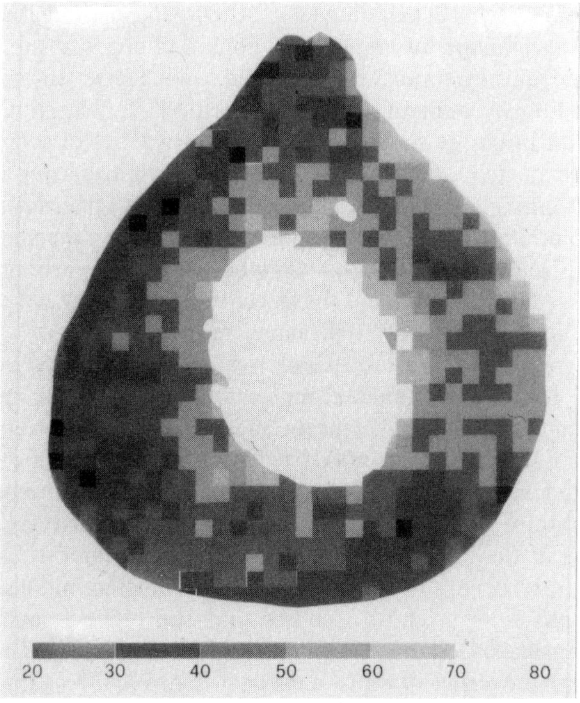

Fig. 21.1 Diagram of the proportion of transverse (bright) lamellae in a cross-section prepared from the middle of a human femoral shaft with the reference colour scale. The left and lower sides of the diagram correspond to the lateral and anterior aspects of the section, respectively. Semiautomatic method.

and condyles, the three-dimensional curvature of the femoral shaft will be such as to induce an angular shift in its bending planes, with progressive rotation from a medio-lateral slope to an antero-posterior one. In applying the engineering beam theory in a precise functional interpretation of variations in long bone cross-sectional geometry, Ruff and Hayes (1983) state that changes in the direction of greatest bending rigidity in the femur could be interpreted in terms of loading and stress distributions. It follows that the anticlockwise rotation of the main axes over almost the whole length of the femoral shaft, found by these authors, reflects a gradual shift from predominantly medio-lateral to predominantly antero-posterior bending in moving from the proximal toward the distal end. Thus the literature reports data that consistently indicate a distribution of bending forces in the femur that could account for the distribution of lamellae in the middle portion of bone, as observed by Portigliatti Barbos *et al.* (1983).

These stimulating results suggested to the same authors the advisability of extending their investigation to the whole of the femoral diaphysis and the other long bones, in seeking confirmation of their preliminary conclusions. This would have been a demanding task, because the method used by them to analyse the distribution of lamellae in bone sections was time-consuming (each section requiring about 2 months), and a much faster automatic method was looked for. By applying this later method (Boyde *et al.* 1984), the distribution of bright or transverse lamellae in a series of successive cross-sections (each located 1 cm from the next) was determined using a microscope with stage scanning, circularly polarized light, and a Quantimet 720 image-analysing computer. To show the results graphically, a scheme was adopted wherein the size of a square plotting symbol varied proportionally with the results averaged in that area: thus the larger the side (or surface) of the square, the greater the proportion of bright lamellae present.

Three diagrams of the most representative sections of a femur are illustrated in Fig. 21.2. They were obtained from sections taken from the proximal (a), middle (b), and distal third (c) of the diaphysis, respectively. Examination and comparison of these diagrams make it clear that the distribution of osteonic and interstitial lamellae is not random and that it changes progressively from the proximal third to the distal third. In Fig. 21.2(a), transverse lamellae are concentrated at the level of the medial wall of the femoral diaphysis. The diagrammatic recording of the section taken from the middle third of the femoral diaphysis (Fig. 21.2(b)) shows that the highest concentration of transverse lamellae is found in the posterior portion of the medial wall, at the boundary with the posterior wall—a result closely resembling that reported in the diagram in Fig. 21.1. In the lower third of the femoral diaphysis (Fig. 21.2(c)) a preferential distribution of transverse lamellae over the medial half of the posterior wall is seen, although they are also found over an adjacent small area of the medial wall.

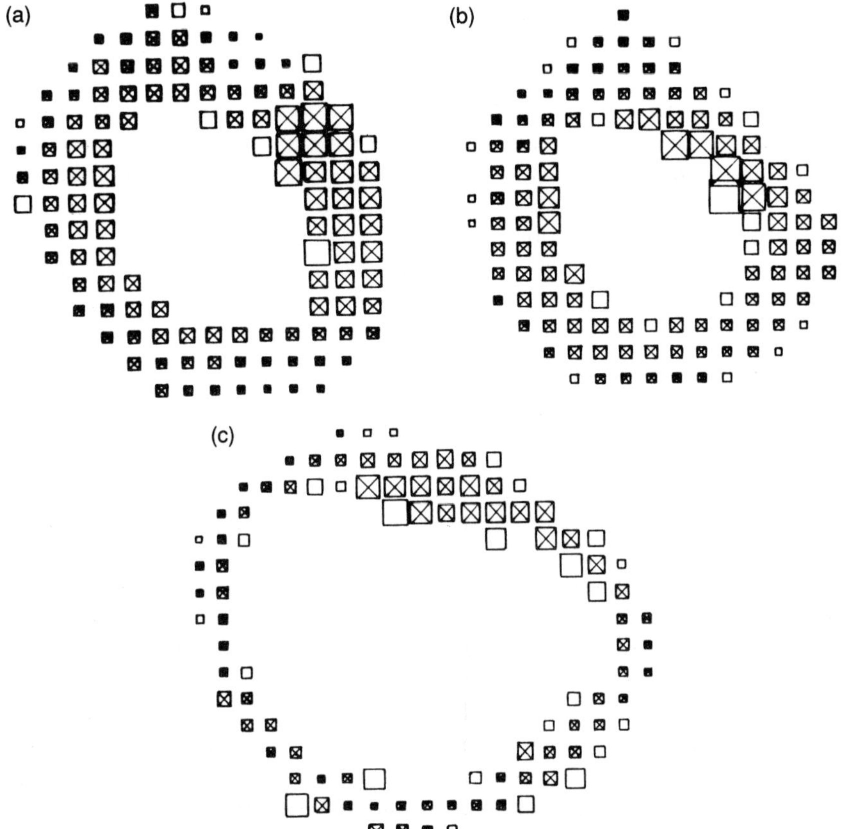

Fig. 21.2 Diagrams (a), (b), (c) display the proportion of transverse (bright) lamellae from the upper, middle, and lower third of the femoral diaphysis. The squares without diagonals indicate that only a fraction of them are covered by the section. The left and lower sides of the diagrams correspond to the lateral and anterior aspects of the sections, respectively. Automatic method.

From all these data it may be inferred (Portigliatti Barbos *et al.* 1984) that in going from the proximal to the distal end of the shaft, the distribution of regions with a preferred transverse orientation undergoes a rotation from the medial toward the posterior wall, whereas, as would be expected, zones with preferred longitudinal orientation undergo a rotation in the opposite direction from the lateral toward the anterior wall. Since this rotational distribution of the two types of osteonic lamellae in the femoral shaft is closely related to the bending stress distribution, there is evidence that, besides the mechanical significance of bone geometry, there is also a characteristic distribution of tissue texture at a microscopic level. The rotational arrangement of lamellae can be considered as symmetrically distributed in the two bones of the same individual, even if some differences may arise from an asymmetrical use of the lower limbs (Portigliatti Barbos *et al.* 1987).

The interpretation given here on the distribution of lamellae in the femoral diaphysis was previously accepted by all the investigators (Portigliatti Barbos *et al.* 1984). However, in more recent times, an alternative explanation has been favoured, based upon the findings of Boyde and Riggs (1990) and Riggs (1990). The latter studied macroscopic mechanical testing samples cut from contrasting segments of the equine radius. This bone shows a collagen orientation map which is clearer than any found in human bones so far studied. Here it was shown that bone with a preferred longitudinal collagen orientation was stiffer in both compression as well as, as expected, in tension. Conversely, bone with a more transverse collagen orientation was more deformable under the same load in both compresion and tension. According to this information, a new viewpoint arises in which one can predict deformability and bending patterns in bone in a different way: 'more longitudinal, less deformation: more transverse, more give'. The discrepancies between compressive tests on isolated osteones (Ascenzi and Bonucci 1968) and macroscopic bone samples with well-defined collagen orientation preferences (Riggs 1990) in respect of the behaviour of regions with a preferred longitudinal collagen orientation, may reflect the lack of support for the phase in isolation.

In accepting this new interpretation a problem arises: are osteons of the same type comparable in man and other mammals? On this topic the investigations are very scanty. X-ray diffraction shows that in bovine longitudinal osteons the crystallites reveal a higher degree of orientation in respect to the human osteons (Ascenzi *et al.* 1978). May this be the case also for the horse? An investigation on this matter would be desirable.

In all the femoral sections, the incidence of transverse lamellae falls from the inner toward the outer circumferential edge. None of the currently available data are capable of yielding a direct explanation for this finding, although it seems likely that the same arrangement may be related to other stresses operating in bone, such as dynamic stresses depending on muscular activity. This tentative interpretation may also be applicable to the other long bones examined, particularly tibia, fibula, and humerus, each of which reveal the finding noted for the femoral shaft.

Tibia and fibula

Going on to examine the distribution of transverse lamellae in the leg bones, tibia and fibula (Caiando *et al.* 1989), some minor technical modifications have been applied in recording the distribution of transverse lamellae corresponding to variations in the thickness of cortical bone.

For the tibia, the results are shown in Fig. 21.3, where only the most representative diagrams are reported, one for each quarter, and the ratio of the surface area of transverse lamellae is symbolized by simple squares. The proportion of areas rich in transverse lamellae is highest in the posterior tibial wall, and to a lesser extent, in portions of the anterior sector. This finding is especially obvious in the two proximal diaphyseal quarters (Figs 21.3(a,b)), while it changes progressively as the distal bone end is approached (Figs 21.3(c,d)). In this last portion, transverse lamellae are still found preferentially in the posterior wall, but the region in which they are concentrated tends to be displaced progressively toward the antero-medial wall with a clear pattern of

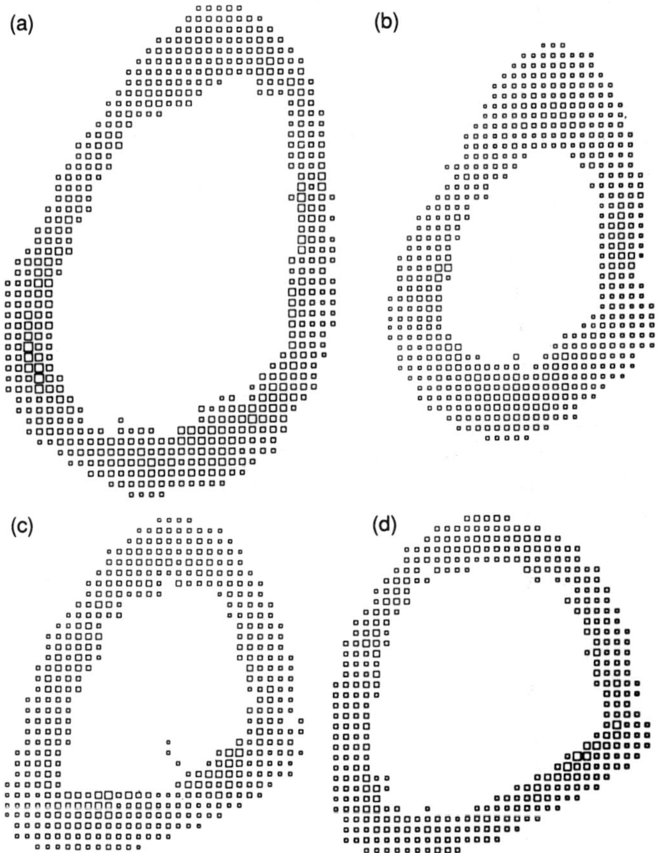

Fig. 21.3 Diagrams of the proportion of transverse lamellae in sections of the tibial diaphysis cut at the level of the proximal quarter (a), the intermediate quarters (b and c), and the distal quarter (d). The left and the lower sides of the diagrams correspond to the medial and posterior aspects of the sections, respectively. Automatic method.

decreasing frequency and more homogeneous distribution within the cross-section.

When all these data furnished by the tibia are compared with those supplied by the macroscopic shape and the biomechanical investigations, the following conclusion can be drawn. The fact that the highest concentration of transverse lamellae is found in the posterior sectors of the proximal half of the tibia may be accounted for, according to the initial hypothesis, by the backward oriented concavity of the bone. In addition, the elongation of the cross sectional shape in the antero-posterior direction suggests that the bone may be best able to resist potential buckling in that direction. This interpretation is supported by the measurements of Minns *et al.* (1975) who found that the mean bending stiffness in the antero-posterior plane was 9.989 N m/mm compared with 3.588 N m/mm in the medio-lateral plane in isolated human tibiae.

According to the data of Riggs (1990) and Boyde and Riggs (1990), however, the findings supplied by the distribution of transverse lamellae could be interpreted as indicating that there is greater resilience built into the bone structure along the antero-posterior axis. At the distal end of the tibia, an almost homogeneous distribution of transverse lamellae in the cross-section seems to be in accordance with the almost circular geometry of this bone portion. At the same time the concentration of transverse lamellae in the posterior wall, with their relatively higher frequency in the antero-medial wall, suggests that bending stiffness may be oriented in a somewhat different direction from that observed at the proximal end. This interpretation is strengthened by the investigations of Ruff and Hayes (1983), who applied the beam theory, and reported that, for the cross-sections located 80, 65, 50, 35, and 20 per cent from the distal end of the bone, the values of the angle measuring the direction of the greatest bending rigidity were 66.1, 69.3, 64.7, 51.6, and 44.3 per cent, respectively. A diagrammatic display of the distribution of transverse lamellae in the fibula appears in Fig. 21.4. Here too, only the most representative sections are shown—one for each diaphyseal quarter.

In the posterior wall of the proximal, second, and third quarters (Figs 21.4(a,b,c)) the proportions of transverse lamellae are actually much higher than in that of the tibia. Conversely, in the distal quarter of the fibula (Fig. 21.4(d)) the transverse lamellae are preferentially distributed in the medial and anterior walls, although patches are also seen in the lateral and posterior walls.

In discussing these results it should be recalled that until 1970 there was a consensus of opinion that during walking and running the full weight bore down on the dome of the talus via the tibial platform, so that the fibula played no role in weight-bearing or loading along the axis, acting at most as a lateral buttress. Lampert (1971) examined the anatomical features of the ankle in detail and simulated the weight bearing process to calculate strain distribu-

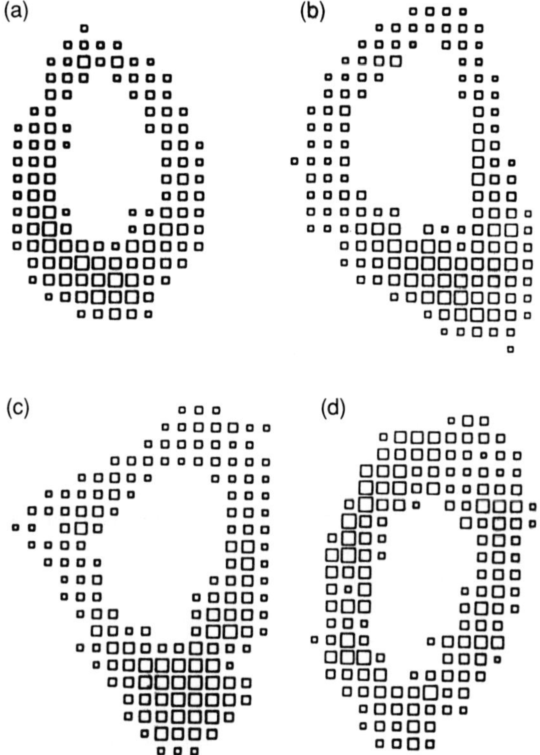

Fig. 21.4 Diagrams of the proportion of transverse lamellae in sections of the fibular diaphysis cut at the level of the proximal quarter (a), the intermediate quarters (b and c), and the distal quarter (d). The left and lower sides of the diagrams correspond to the medial and posterior aspects of the sections, respectively. Automatic method.

tion: he hypothesized that the force generated by the fibular articulation with the talus and perhaps also by the lower tibio-fibular ligaments, would push against the proximal tibio-fibular joint. The outcome would be that the fibula functions as an eccentrically loaded beam subjected not only to axial stresses but also to large bending moments. These results appear to be in agreement with those obtained in the present study, which seem to indicate that the maximum bending rigidity of the fibula, like that of the tibia, is oriented in an approximately antero-posterior direction.

At the distal end of the fibular shaft, the enhanced proportion of transverse collagen in the medial and anterior wall may be explained by considering that the articular surface with the talus is not vertical but obliquely angled, so that the pressure exerted on the fibula is transmitted through its medial side.

Humerus

Proceeding now to examine the distribution of osteonic lamellae in the humerus, it should first be stated that for the investigation on this bone two different methods were used. The first was almost identical with that applied to the tibia and fibula, while the second or single field, digitizing method was developed by Boyde and Riggs (1990). Using a broad uniform and stable light source, the bone section is viewed between two large sheets of circularly polarizing filter. In this way all the bone areas with more transverse lamellae appear brighter while the areas with more longitudinal lamellae appear darker. The circularly polarized light image is transferred via a CCD TV camera to an image analysing computer and the digitized image presented as a map using a topographic pseudocolour scheme.

Since the results obtained by applying the two methods were very similar, only those provided by the first are illustrated here (Fig. 21.5). In the proximal third, the humerus shows that transverse lamellae are most frequent within the medial wall and in much of the posterior wall (Fig. 21.5(a)). This concentration is probably connected with the change in the direction of the bone head, which is oriented medially, backwards, and upwards. In the middle third (Fig. 21.5(b)), the bone shows an almost elliptical outline due both to the torsional groove for the radial nerve and the prominence for the insertion of the deltoid muscle. Here, transverse lamellae are concentrated at the two poles of the ellipse crossing the major axis (i.e. in the sectors most remote from the neutral axis of bending) and roughly corresponding to the anterior and posterior walls. This is an indication that the bending of the bone tends to occur mainly in one plane—an interpretation apparently confirmed by the similar feature (noted above) observed in the tibial shaft (Minns *et al.* 1975; Carando *et al.* 1989). In the distal third of the humeral shaft (Fig. 21.5(c)), a lower incidence of transverse lamellae in both the posterior and anterior walls repeats the situation found in the middle third. It should, however, be borne in mind that the bone takes on a triangular outline at the distal level, which means that the anterior wall has a lower volume than the posterior wall.

Ulna and radius

Going on now to analyse the distribution of transverse lamellae in the ulna and radius, it will first be noted that only the second method used for the humerus was applied here. As regards the ulna, the shaft shows a microscopic distribution of transverse lamellae that is closely correlated with its macroscopically complicated shape. At the proximal level, where there is an anterior concave curvature associated with a bowing of the interosseous or lateral border, a higher incidence of transverse lamellae occurs in the anterior wall, with a clearly progressive tendency for this zone to expand towards the

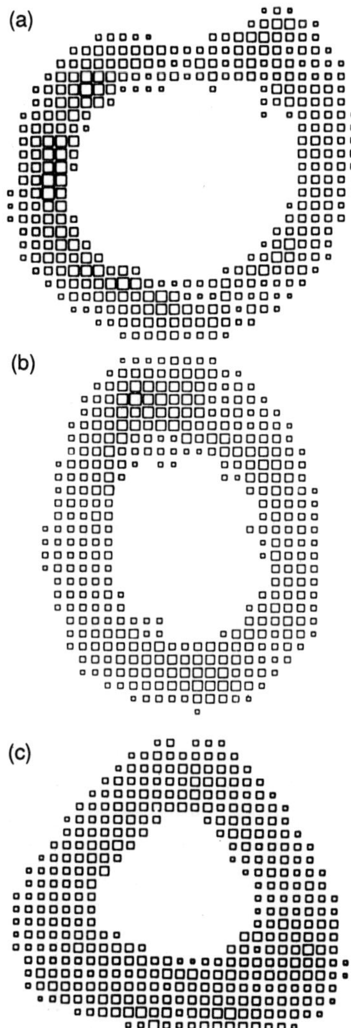

Fig. 21.5 Diagrams (a), (b), (c) display the proportion of transverse lamellae from the upper, middle, and lower third of the humeral diaphysis. The left and lower sides of the diagrams correspond to the medial and posterior aspects of the sections, respectively. Automatic method.

lateral border, indicating a structure able to resist (or to comply to) compressive stresses (Figs 21.6(d,e)). Only in the most distal portion of the ulna are the zones rich in transverse lamellae homogeneously distributed around the total bone wall—a finding in agreement with the macroscopic cylindrical shape of the bone at this level (Fig. 21.6(f)).

The complexity of the macroscopic shape of the radial shaft provides a further example of the characteristic distribution of transverse lamellae concentrations along a bone. The incidence of transverse lamellae is fairly uniform in only a small proximal portion of the shaft, just below the tuberosity for the insertion of the distal tendon of the biceps muscle, where the bone is cylindrical (Fig. 21.6(a)). In the remaining bone, where the shaft is curved with its convexity corresponding to the lateral surface and its concavity to the interosseous or medial border, a high incidence of transverse lamellae is found, indicating that a compressive load may be experienced at this site (Figs 21.6(b,c)).

Distribution of osteonic lamellae in pathologic bone shafts

As an example of the pathologic distribution of osteonic lamellae, the case chosen was that of a woman who died at the age of 63, and showed greatly bowed femurs, as a result of severe juvenile rickets. In presenting this case it is worth recalling that the first author to be interested in bowed bones was apparently Benninghoff (1927), who attempted to explain the macroscopic changes occurring in a bowed femur by adopting a descriptive morphological method. In this author's view, the femur loses its circular shape in acquiring an elliptical section whose main axis lies in the bending plane, so that the deformed bone is supported on its concave side.

A sounder and more rigorous explanation for the deformed shape of the section corresponding to the portion of maximum bowing was later put forward by Pauwels (1954, 1968, 1979) in developing the physico-mathematical bases governing the changes in long bones brought about by bending. These bases are derived from the general theory on bending and consider the case of a bent femoral diaphysis in which the bending plane and the plane of application of the force do not necessarily coincide, so that the bending is said to be 'distorted'.

Accepting these premises, cross-sections prepared from the portion of maximum bowing of the femurs pertaining to the case mentioned above, were analysed using the method applied to the tibia and fibula. The results are illustrated in Fig. 21.7. From these diagrams it appears clear that the transverse lamellae are preferentially distributed in the elongated lateral and medial walls of the bone. Here they occupy not only the sectors which have the greatest surface area, but also the remaining portions of the same walls. In contrast with the transverse lamellae, the longitudinal ones are mainly distributed in the short anterior wall and in the posterior portion of the lateral wall.

In offering an explanation for this finding, the results of a recent investigation on single osteons by Ascenzi *et al.* (1990) can be taken into

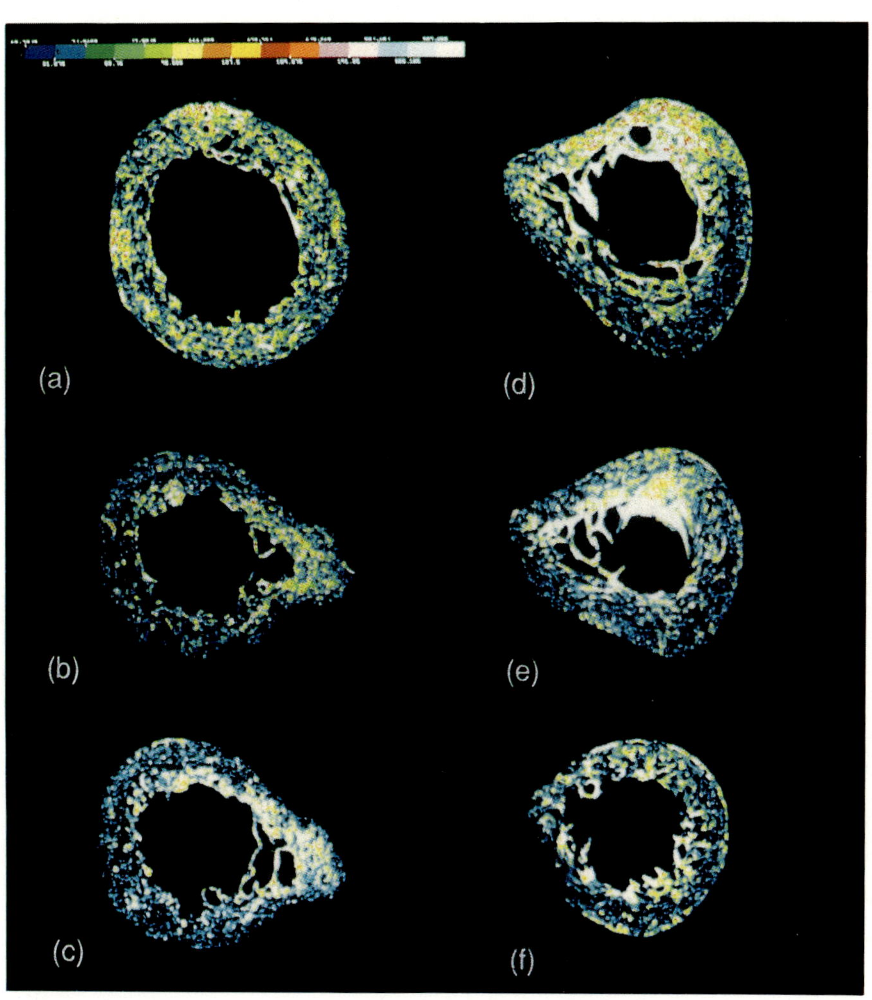
(a)
(d)
(b)
(e)
(c)
(f)

Fig. 21.7 Diagrams of the proportion of transverse lamellae in the cross-sections of two bowed femurs at the level of the maximum bending concavity. R, right and L, left. Automatic method.

account. By working on fully calcified cylindrical units, having longitudinal and alternate lamellae respectively, and testing them in bending, these authors succeeded in providing evidence that isolated osteons with alternate lamellae are those best able to resist bending. Thus the high concentration of transverse lamellae in the elongated walls of the bowed femurs is in accordance with the need to compensate the deformity produced by the bending of the bones and, at the same time, reveals that an obvious relationship exists between the macroscopic features of the deformed femurs and their modified microscopic structure.

Conclusions

Analysis of secondary compact bone performed to determine the function of lamellae in a single osteon and their distribution in bone sections, allows the following main conclusions to be drawn:

1. In osteons, lamellae are of two types—longitudinal and transverse. Longitudinal lamellae, which appear dark in cross-section when examined under the polarizing microscope, withstand loading by tension, whereas

transverse lamellae, which appear bright in cross-section under the polarizing microscope, withstand loading by compression in isolated osteons.

2. In cross-sections obtained from the shafts of long bones, the distribution of the lamellae found in osteons and interstitial bone is not random.
3. An accurate comparison of the lamellar distribution and geometry of the bone shaft indicates the existence of a close relationship between these two conditions.
4. In general terms, it may be stated that there is a high incidence of transverse lamellae in bone sectors loaded by compression, whereas longitudinal lamellae are especially distributed in sectors loaded by tension.
5. The progressive increase in the concentration of longitudinal lamellae from the inner toward the outer circumferential edge of a section may be tentatively explained by supposing that this finding is related to other stresses operating in bone, such as dynamic stresses depending on muscular activity.
6. The rearrangement in the distribution of osteonic lamellae in a bone shaft is a very delicate and refined process, as proved by the slight structural lack of symmetry found in the bone tissue of subjects who had used their limbs asymmetrically in life.
7. The high concentration of transverse lamellae in the elongated sides of the elliptical section in a pathologically bowed shaft increases the bending stiffness of the deformed bone.

Acknowledgements

The series of personal investigations reported in this paper has been supported by grants from the National Research Council of Italy, the Medical Research Council, and the Science and Engineering Research Council, UK. The authors acknowledge the participation of Drs P. Bianco, S. Carando, M. Portigliatti Barbos, and C. M. Riggs in the joint work reviewed in this paper; and particularly thank Dr Lidia Salvadori, the late Elaine Maconnachie, and Lucio Virgilii for helpful assistance.

Debate in the meeting

It may not be true for all animal species that longitudinal osteons are less stiff than transverse osteons. The orientation of collagen is different in different animal species and so studies have to be conducted on single species.

22 Age-dependent changes of the rat femur

K. Indrekvam and N. Langeland

Introduction

External forces produce changes in the mechanical properties of the skeleton during growth. Knowledge about the structural and material qualities of bone during normal activity at different ages is therefore needed to understand better the mechanical demands in clinical situations. Methods for measuring the surface deformation of bones *in vivo* in large animals with implanted strain gauges have been described in several reports (Evans 1953; Lanyon and Smith 1969; Cochran 1972; Rybricki *et al.* 1977; Carter *et al.* 1980), and a few studies in small animals have also been presented (Keller and Spengler 1982, 1989; Biewener *et al.* 1986, Husby *et al.* 1988). Changes in mechanical properties of rat femora during growth have been reported after *in-vitro* mechanical testing (Vogel 1979; Ekeland *et al.* 1982; Keller *et al.* 1986), and recently also by *in-vivo* measurements (Keller and Spengler 1989).

The present investigation was planned to further explore the correlation and relevance of *in-vivo* and *in-vitro* mechanical variables, and to study the changes in material composition and mechanical qualities that occur during growth.

Materials and methods

A total of 72 male Wistar rats were used (Mol:WIST, Møllegaard Breeding Center, Ejby, Denmark). There were nine age groups with eight rats each: 6, 12, 20, 26, 32, 38, 44, 52, and 56 weeks old. The animals were given standard maintenance rat diet (RM1 expanded, Special Diets Services, Witham, England) and water *ad libitum.*

Strain measurement in vivo

A single-element strain gauge (0.6/120 LY 11, Hottinger Baldwin Messtechnik, Darmstadt, Germany) was implanted on the anterior surface of the left femoral diaphysis of each rat. Construction of the strain gauge units, implantation, and recording of strain have been described in detail earlier

(Husby *et al.* 1988). The animals were operated on under anesthesia with a mixture of fentanyl/fluanisone and midazolam. Post-operative analgesia was managed by buprenorphine-hydrochloride injected subcutaneously.

Strain in the axial direction was recorded from the second post-operative day and for 2 days while the rats were running on a treadmill at a speed of 10 m/min. Peak values of strain (maximum deformation) of 30 walking cycles were measured, and the mean value was calculated. Stride frequency was measured as well. The rats were killed on the last day of strain recording. *In-vivo* stiffness was estimated from the body weight/strain ratio, because actual load on each femur during running was unknown.

Geometry

Excised femora were mechanically cleansed of soft tissue, and the femur length was measured from the caput to the medial condyle with a sliding caliper. The cross-sectional area at the location of the strain gauge was estimated from measurements of the outer and inner antero-posterior and medio-lateral diameters (Engesæter *et al.* 1978). The area moment of inertia (I) was calculated using the equation applicable for a hollow ellipse (Roark and Young 1976):

$$I = \frac{\pi}{4}(R_o r_o^3 - R_i r_i^3)$$

where R_o and R_i are the outer and inner medio-lateral radius, and r_o and r_i the outer and inner antero-posterior radius.

Bending tests

The excized femora were kept moist, and three-point bending tests were performed within 3 h with an Instron® 1193 machine (Husby *et al.* 1988). Each femur was deflected at a speed of 10 mm/min until fracture occurred. Stiffness was calculated from the linear portion of the load-deflection curves. Ultimate load was recorded. Maximum bending stress (σ) was calculated from the load data using the equation for an elastic, isotropic, homogeneous beam with a hollow elliptical cross-section (Roark and Young 1976).

$$\sigma = \frac{Mxr_o}{I}$$

where M is the bending moment defined for this test as 1/4 (load × length) where load is ultimate load and length is the distance between the supporting bars of the jig (14 mm); r_o is the distance from the neutral axis (half the external antero-posterior diameter of the femoral diaphysis); and I is the area moment of inertia. Maximum bending stress was taken as the strength of bone material.

Collagen, calcium, and phosphorus analyses

Excised femora were mechanically cleansed of all soft tissues and hydrolysed in 6 M hydrochloric acid at 110°C for 18 h. The amount of hydroxyproline was used as a measure of the collagen content of the bone (Firschein 1969). A sample from each hydrolysed femur was therefore analysed for hydroxyproline by an amino acid analyser (Biotronik Aminoacid Analyzer LC 7000, Wissenschaftliche Geräte GMBH, Frankfurt, Germany) with an integrator (Spectra Physics SP 4200 computing integrator, San Jose, USA).

To measure collagen synthesis, ^{14}C-proline (15 μCi/100 g body weight) was injected intraperitoneally 24 h before killing the rats. The amount of radioactive hydroxyproline in the bone was measured in a β-counter (Wallac 1217 Rackbeta Liquid Scintillation Counter, Turku, Finland) after separation from proline by columns of Dowex 50W × 8. The specific activity, expressing incorporation of collagen during 24 h, was calculated from the ^{14}C-hydroxyproline/total hydroxyproline ratio *ad modum* Firschein (1969).

Calcium and phosphorus analyses were carried out by means of continuous-flow spectrophotometry (Technicon SMAC system, New York, USA; Gitelman 1967; Daly and Ertinghausen 1972; Amador and Urban 1972). Mineralization was expressed as the calcium/hydroxyproline ratio.

Statistical analysis

Median values with 0.25 and 0.75 fractiles were used to express the average and the variance of the measurements. The Kruskal–Wallis test was used to test differences in medians between groups (Minitab, Ryan *et al.* 1985), and was considered significant when $p<0.05$. The Kendall rank correlation procedure was applied to test for correlation between two variables (BMDP P3S, Dixon 1985). Multiple linear regression analysis was used to identify the variables with the highest explanation ratio of strain values (BMDP P1R, Dixon 1985).

Results

Body weight, femur length, cross-sectional area, and area moment of inertia are shown in Table 22.1.

On the third post-operative day, 8 out of 72 strain gauges were excluded because of electrical malfunction. Peak strain was of the same magnitude for all the age groups (Fig. 22.1). The median stride frequency was 1.7 (1.5–1.8) stride/s, and there were no significant differences between the age groups. There was, however, a tendency towards a slower stride frequency and lower strain values for older animals.

Table 22.1

Age (weeks)	Bodyweight (g)	Femur length (mm)	Cross-sectional area (mm^2)	Area moment of inertia (mm^4)
6	199 (182–204)	30.1 (29.7–30.5)	1.2 (1.1–1.3)	4.3 (3.9–4.7)
12	324 (309–332)	34.9 (34.5–35.2)	1.7 (1.5–1.8)	5.5 (5.0–6.5)
20	376 (371–409)	37.5 (36.8–37.9)	2.0 (1.9–2.0)	6.9 (6.1–7.4)
26	454 (418–481)	39.1 (38.1–39.3)	2.3 (2.0–2.3)	9.0 (7.3–9.8)
32	489 (452–510)	38.8 (38.4–39.0)	2.3 (2.1–2.5)	9.9 (8.7–10.5)
38	490 (482–521)	39.8 (38.6–41.0)	2.5 (2.3–2.7)	11.0 (9.1–11.9)
44	476 (452–523)	40.1 (39.6–40.5)	2.6 (2.5–2.6)	11.7 (10.7–12.0)
52	495 (463–527)	39.9 (39.2–40.4)	2.5 (2.3–2.6)	11.6 (10.5–12.4)
56	544 (531–618)	40.9 (39.3–41.9)	3.0 (2.7–3.1)	15.0 (13.2–15.6)

Medians (0.25–0.75 fractiles).

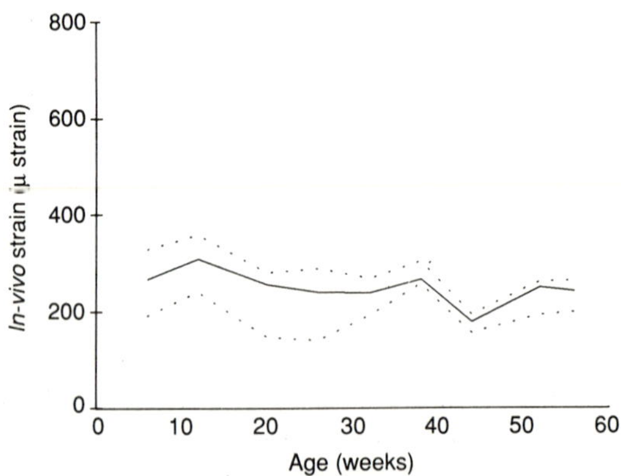

Fig. 22.1 *In-vivo* strain values. Medians (0.25–0.75 fractiles).

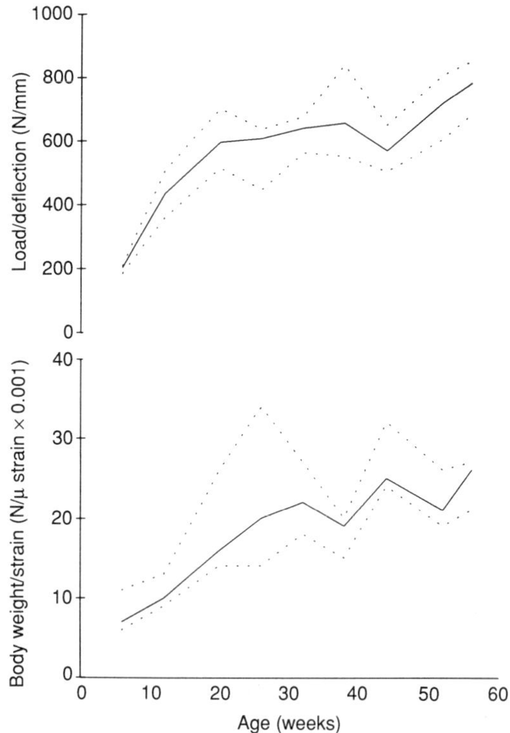

Fig. 22.2 Stiffness calculated as (body weight)/(*in-vivo* strain) (N/μstrain) and *in-vitro* load/deflection (N/mm). Medians (0.25–0.75 fractiles).

In-vivo stiffness expressed as (body weight)/(*in-vivo* strain) ratio was correlated with the more conventional load/deflection ratio (Kendall's $\tau=0.40$, $p<0.005$). Stiffness increased with age ($p<0.001$; Fig. 22.2).

Maximum bending stress exhibited an increase during maturation, whereas the difference among adult age groups was insignificant. Ultimate load, however, increased steadily from 6 to 56 weeks of age ($p<0.001$; Fig. 22.3).

Total contents of hydroxyproline, calcium, and phosphorus in the femora naturally increased with age ($p<0.001$). Also the calcium/phosphorus ratio increased during growth from 1.47 at 6 weeks to 1.66 at 1 year of age ($p<0.001$).

Mineralization calculated as the calcium/hydroxyproline ratio had higher values in older bones ($p<0.001$; Fig. 22.4). When the biochemical measurements were compared to mechanical measurements, there was correlation between *in-vivo* stiffness calculated from the strain gauge recordings (body weight/*in-vivo* strain ratio) and mineralization (calcium/hydroxyproline ratio) (Kendall's $\tau=0.41$, $p<0.005$).

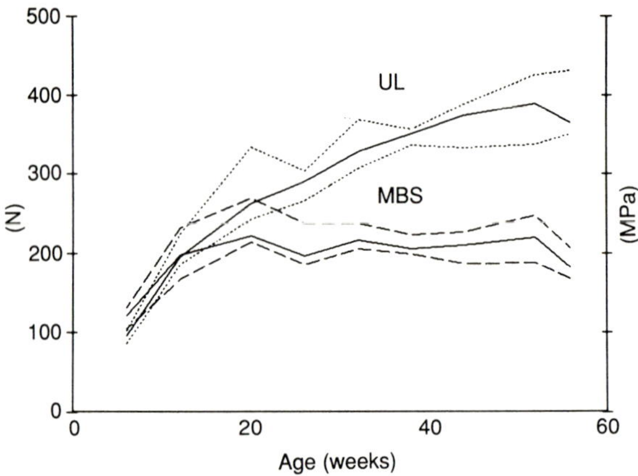

Fig. 22.3 Ultimate load (N) and maximum bending stress (MPa). Medians (0.25–0.75 fractiles).

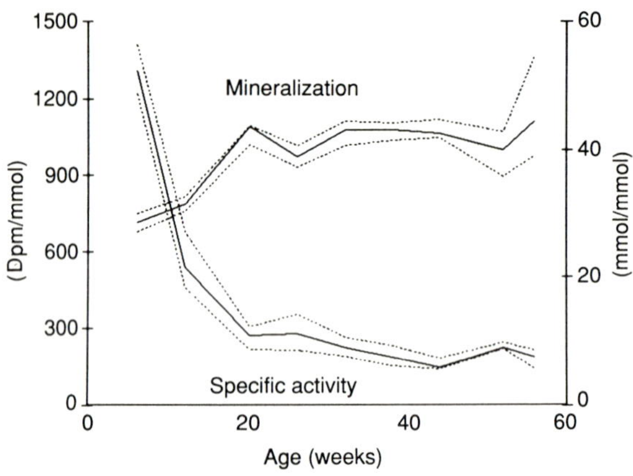

Fig. 22.4 Mineralization and collagen synthesis. Medians (0.25–0.75 fractiles).

The specific activity of ^{14}C-hydroxyproline, representing the rate of collagen synthesis, decreased with age ($p<0.001$; Fig. 22.4).

Linear regression analysis indicated that mineralization, cross-sectional area, and femur length separately had an effect on strain ($p=0.04$, $p=0.01$, and $p=0.04$, respectively). However, none of these variables seemed to be the major determinant for strain in a multiple regression analysis.

Discussion

The present study demonstrates that the main trend for bone dimensions, mechanical properties, and chemical composition is an increase with age. The exceptions are *in-vivo* strain values which were remarkably stable with age, and maximum bending stress which increased mostly during maturation, but thereafter was fairly constant in adult life. The *in-vivo* body weight/strain and *in-vitro* load/deflection calculations of stiffness correlated well, as did body weight/strain and mineralization.

Strain during running exercises in the present study remained remarkably stable at different animal ages. This observation is in agreement with Wolff's law, and implies that the various mechanical characteristics of bone are well balanced during growth and are optimized on the basis of the functional demands. Reports from a variety of species also suggest that strain is nearly independent of age (Lanyon *et al.* 1979; Biewener *et al.* 1986; Keller and Spengler 1989).

The authors' measured maximum strain result for the anterior mid-diaphyseal rat femur came close to that of Keller and Spengler (1989). Biewener *et al.* (1986), recording strain from the tibiotarsus of exercized chicks, reported about a 2–4-fold higher strain value for the proximal and distal midshaft. Higher strain values in their paper probably reflect species differences, gauge location, gauge type, or running speed. The validity of the present study's *in-vivo* recordings was confirmed by their reproducibility and the correlation between *in-vivo* body weight/strain and *in-vitro* load/deflection.

A single-element gauge recording of strain in the axial direction was used in the present study. The gauge was designed to accept a bending radius down to 0.3 mm, which is well below the smallest radius of the bones. Strain distribution remains unknown; such information requires the implantation of rosette gauges around the diaphyseal circumference.

Increased bone stiffness with age was to be expected when testing the bone as a composite structure (Torzilli *et al.* 1981, 1982; Ekeland *et al.* 1982; Jonsson *et al.* 1984; Keller *et al.* 1986). Increased stiffness was also demonstrated when calculated from strain-gauge recordings. Structural stiffness depends on both geometry and material properties. Increased area moment of inertia clearly shows that bone dimensions are important in this respect.

An interesting finding in the present study was that ultimate load increased from 12 to 56 weeks of age, whereas maximum bending stress (bone material strength) did not reveal any significant change during adult life. A constant level for bone material strength has been reported from about 4 months of age in rats (Vogel 1979; Ekeland *et al.* 1981; Keller *et al.* 1986). Thus, the increase in load bearing capacity (e.g. the strength of bone as a structure) with age in adults rats must be a result of larger cross-sectional dimensions. Consequently, this indicates that young bones rely upon elastic deformation to avoid fractures, whereas old bones are more dependent upon size and shape of the bone. This interpretation agrees with tests of human femora, where both a lower stiffness, lower bending strength, and greater deflection before fracture have been reported for children when compared with adults (Vinz 1970; Currey and Butler 1975).

In the present study, the total content of minerals increased with age as reported by Ekeland and Underdal (1983), but a disproportionate rise in calcium content relative to phosphorus occurred ($p<0.01$). It has been reported that bones of young individuals contain a greater proportion of amorphous calcium phosphate than bones of old rats in which most calcium appears as crystalline hydroxyapatite (Termine 1972; Eanes *et al.* 1973; Lees and Davidson 1977). The calcium/phosphorus (Ca/P) ratio for amorphous calcium phosphate has been reported to be 1.47, and for hydroxyapatite 1.67 (Grynpas *et al.* 1984). This correlates with the finding that the Ca/P ratio in the present study was rather low in the youngest animals and then increased with age.

The intimate association of minerals with collagen is essential to the formation of hard connective tissues like bone. The proportion of minerals in bone tissue, expressed as the calcium/hydroxyproline ratio was found to be greater in older animals than in younger ones ($p<0.005$). This is in agreement with Urry's (1974) suggestion that the binding of calcium to protein in bone tissue is a continuous process that progresses with age.

Both increased mineralization and a higher proportion of crystalline hydroxyapatite in the older bones theoretically make the bone tissue of these animals stiffer (Currey 1975, 1988*a*; Lees and Davidson 1977). There was correlation between mineralization and stiffness calculated from the strain recordings. The older animals also had the highest maximum bending stress compared with the younger ones. Currey (1975, 1988*b*) reported that greater mineralization in specimens of the bovine femur produced greater ultimate tensile strength and a higher modulus of elasticity. Burstein *et al.* (1975) found that with progressive decalcification ultimate stress decreased. All these studies therefore emphasize the importance of mineralization for the mechanical properties of bone.

Torzilli *et al.* (1981) reported from a canine bone study that geometry and material contributed equally to structural changes, whereas the present study, as do several others, supports the view that structural properties of rat bones are dependent to a greater extent upon geometry than on material (Ekeland *et al.* 1981; Keller *et al.* 1986).

In conclusion, the functional demands of physical activity caused strain of the same magnitude in the age groups tested. *In-vivo* recordings of stiffness corresponded to *in-vitro* measurements. Although the *in-vivo* recordings primarily refer to the measuring point, the correlation with *in-vitro* measurements indicates a general validity of the results. Increased dimensions, increased mineralization, and lower rate of collagen synthesis contributed to stiffer bone with age. The results are compatible with the theory that the mechanical behaviour of bone is regulated by changes in both geometry and material properties. Nearly constant strain compared to greater alterations of other variables, indicates that strain might be a measure of different bone properties being adjusted to each other to meet the changing mechanical demands during growth.

Debate in the meeting

Bone marrow density and measures of calcium do not relate clearly to strain, so anthropomorphic information needs to be added to describe the shape of the bone, and histomorphometric information to describe the microscopic organization of the bone. All these factors together are required to derive the strength of bone.

23 Can the strain transduction mechanism in bone influence the development of osteoporosis?

J. VANDER SLOTEN, R. VAN AUDEKERCKE, AND G. VAN DER PERRE

Introduction

Osteoporosis is a frequently observed bone metabolic disease, mostly seen in the elderly and especially in post-menopausal women. Actually, one can distinguish between three types of osteoporosis (Burr and Martin 1989), all of them characterized by a reduction in total bone volume and, hence, a reduction of the mechanical strength of the bones.

The first type of osteoporosis is the post-menopausal osteoporosis, characterized by a rapid bone loss in women within 10–15 years after menopause. This disorder mostly affects trabecular bone in which the combination of an enhanced bone resorption rate and an unchanged bone formation rate produce a net bone loss within the cancellous bone. Senile osteoporosis is the second type and is the normal consequence of ageing, affecting both men and women. A net reduction in bone formation causes bone losses in cortical as well as in trabecular bone. Finally, disuse osteoporosis is the third type of osteoporosis, caused by a reduction of mechanical stimuli to the skeleton. It is observed in astronauts and in patients staying in bed for a long time. This type of osteoporosis can be reversed if the person involved returns to his or her normal level of activity. An excellent and detailed description of the effects and treatment of these types of osteoporosis has been given by Burr and Martin (1989).

Functional loading of bones is required to maintain a certain bone volume. Overloading will result in bone deposition and underloading in bone resorption. These effects have been clearly demonstrated in a number of animal experiments by Lanyon and his group (Lanyon 1987). Deficiency of functional loading is the cause of disuse osteoporosis and probably also plays a role in senile osteoporosis. The development of post-menopausal and senile osteoporosis is undoubtedly influenced by a multitude of biological, hormonal, and perhaps mechanical factors.

Osteoporosis can, in a severe degree, lead to collapsing vertebrae and femoral neck fractures. Within the research that is presented here, the authors have concentrated on the trabecular structure in the proximal femur and investigated whether failures in the self-regulating mechanism of maintaining and changing bone mass by functional strains, can influence the way in which osteoporosis develops in the proximal femur.

Method

Geometrical model

The trabecular structure in a medio-lateral cross-section of a proximal femur was modelled by means of plane truss finite elements (i.e. one-dimensional elements, interconnected by hinges). These truss elements, whose parameters are the Young's modulus, the Poisson's ratio, and the cross-sectional area, can only be loaded by axial forces. Provided these truss elements are aligned with the main material directions at each location within the trabecular structure, the mechanical behaviour of the trabecular structure can be imitated accurately (vander Sloten and van der Perre 1989). The density of the mesh was uniform and the directions of the plane truss elements were according to the main material directions in the trabecular structure. The mesh is shown in Fig. 23.1. Preliminary calculations showed that the stability of this plane truss network was assured.

The cross-section of each of the truss elements was proportional to the density of the trabecular structure it represented, i.e. the denser the bone was

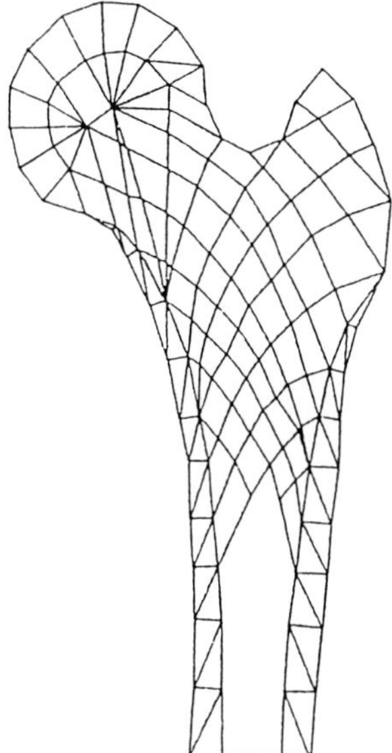

Fig. 23.1 Plane truss mesh of the trabecular structure within the proximal femur.

at a certain location, the larger was the cross section of the corresponding truss element. The density was estimated by measuring, in a direction perpendicular to a truss element, the amount of bone material per unit of length. Material properties of the elements were $E = 15$ GPa (Young's modulus) and $\mu = 0.3$ (Poisson ratio).

The structure was loaded according to one-legged stance, with load vectors at the femoral head and at the trochanter major. According to literature (Brown and Di Gioia 1984) the load on the femoral head was distributed on six nodes. The nodes at the distal end were assumed to be fixed as a boundary condition.

Mathematical model for the development of osteoporosis

A failure in the self-regulating mechanism of bone mass vs functional loading was assumed, and the hypothesis that trabeculae that are loaded less than average would gradually vanish and that trabeculae that are loaded more than average would gradually become thicker was simulated. Trabeculae loaded in compression and trabeculae loaded in tension were differentiated between and the average compression and average tensile stress over all elements was calculated. The cross-sectional area of each of the plane truss elements was adapted proportionally to the difference between the actual stress within the element and the average stress. In this way, an iterative algorithm was constructed by applying the mathematical model to the modified finite element mesh.

Table 23.1 shows the mathematical expressions that were used to simulate the development of osteoporosis. Of course this is just a working hypothesis and one of many possible mechanisms that could be evaluated. The recursive

Table 23.1 Overview of the formulas used to adapt the cross-section of the truss elements

Tension	σ^{+}_{max}	
		$A' = A\left(1 + 0.5\,\dfrac{\sigma - \sigma^{+}_{avg}}{\sigma^{+}_{avg} - \sigma^{+}_{max}}\right)$
	σ^{+}_{avg}	
		$A' = A\left(1 - 0.5\,\dfrac{\sigma^{+}_{avg} - \sigma}{\sigma^{+}_{avg}}\right)$
	0	
		$A' = A\left(1 - 0.5\,\dfrac{\sigma - \sigma^{-}_{avg}}{-\sigma^{-}_{avg}}\right)$
Compression	σ^{-}_{avg}	
		$A' = A\left(1 + 0.5\,\dfrac{\sigma^{-}_{avg} - \sigma}{\sigma^{-}_{avg} - \sigma^{-}_{max}}\right)$
	σ^{-}_{max}	

A = initial area of truss element; A' = area after adaptation; σ = stress.

algorithm was programmed in PASCAL. The finite element software used was SYSTUS by FRAMATOME, France, and was executed on a VAX 3600 computer.

Results

Eight iteration steps were made after which the solution stabilized. Stabilization does not mean that the area did not change any more; it means that the tendencies in each of the truss elements had become stable. The evolution of each of the truss elements was followed as a function of the iteration step. Figure 23.2 shows the tendencies that were observed in the truss elements within the head, neck, and inter-trochanteric region.

Discussion

The results of this simulation were compared with the so-called 'Singh-index' which assigns a number to the degree of osteoporosis, based upon the

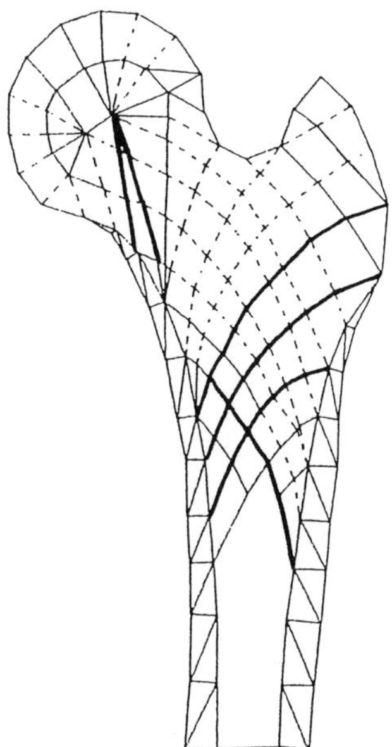

Fig. 23.2 Schematic representation of the evolution within the truss elements: bold line means that the area increases steadily; normal line means that the change in area is minimal (± 30 per cent of the initial value); dashed line means a steady decrease of the area.

appearance of the trabecular density on a standard frontal radiograph of the proximal femur (Fig. 23.3, Dambacher and Rüegsegger 1985).

Stages 4 and 3 of the Singh index are characterized by the disappearance of the tensile trabeculae, which was also observed in this simulation (Fig. 23.2). It is clear that this has severe consequences for the strength of the neck of the femur: the major trabecular system in this region has disappeared and this may lead to femoral neck fractures at low loading on the femoral head, which is indeed frequently observed in people suffering from osteoporosis. Stage 5 of the Singh index was not modelled by the simulations, on the contrary, the simulation predicted a thickening of these trabeculae. The trabeculae that were most stable in Stages 2 and 1 of the Singh index were also amongst the most stable trabeculae predicted by the simulations. Although this work was limited, and no absolute agreement between the authors' simulations and all stages of the Singh-index could be established, a correlation between simulation and clinical experience was observed as to the trabeculae which disappear first when osteoporosis develops. This may be influenced by an uncontrolled weakening of those trabeculae which are loaded less. This failure in the self-regulating and self-stabilizing mechanism of maintaining and changing bone mass is likely to be one of the causes for osteoporosis in the proximal femur.

It is therefore not unreasonable to assume that mechanical factors play a role in the development of all types of osteoporosis. The relative ease of these simulations make it possible to investigate a number of working hypotheses for the development of osteoporosis and compare the simulation results with clinical data.

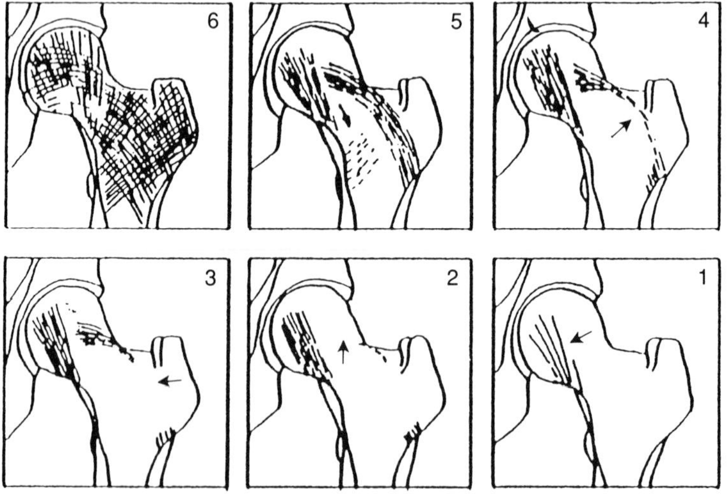

Fig. 23.3 The stages of osteoporosis according to the 'Singh-index'.

Debate in the meeting

There was concern that the scale of the trusses in this geometric model was quite different from the scale of the trabecular bone and therefore that its answers might not relate accurately to trabecular arrangements. The author (J.V.S.) pointed out that the cross-section of each beam element was adjusted so that it stood for the amount of bone that is present in a certain direction at a certain point. The trusses cannot change size or direction. Might it not be better to build each truss as a super-element containing many smaller elements? The authors are working in this direction but the current model already consists of more than 20 000 elements and the computational load is severe.

A number of workers have shown that there is a thickening of the neocortical bone of the head. As it thickens it becomes more mineralized and brittle. Is remodelling of cortical bone due to the resorption of trabeculae, or are the trabeculae disappearing because the cortical bone is remodelling? Which comes first? This dichotomy could be incorporated into the model.

24 The influence of axial distraction upon muscles

A. H. R. W. SIMPSON, P. J. KYBERD, P. E. WILLIAMS, J. KENWRIGHT, AND G. GOLDSPINK

Introduction

Little is known about the response of muscle to the distraction that occurs during leg lengthening or during the correction of deformity, though the behaviour of the muscles often limits the ultimate gain in leg length. Temporary dysfunction is usual and permanent disability can result, particularly if lengthening is performed in patients with poliomyelitis.

Patient studies

Soft tissue problems are notoriously under-reported by orthopaedic surgeons yet despite this, in his review of leg lengthening techniques, Paley (1988) concluded that muscle was the major remaining problem in leg lengthening. In the reported series, the incidence of contractures varies considerably: no muscle problems or joint contractures were recorded in the 454 segments that Ilizarov lengthened (1989). Mezhenina *et al.* (1984) reported 4 cases of knee stiffness out of 78 patients, Monticelli and Spinelli (1981) reported 9 joint contractures out of 43 patients undergoing lengthening, and Paley (1988) reported 6 cases of joint contracture in 18 patients undergoing lengthening. As indicated by this study, the cases of knee stiffness and joint contracture almost certainly represent an inadequate lengthening response of the muscle in these patients. Kawamura *et al.* (1968) has shown that biochemical abnormalities can occur under experimental conditions in normal muscles when there is only a 10 per cent increase in limb length, suggesting that damage occurs with distraction. Ilizarov (1989) has shown that under experimental conditions parts of the distracted muscle develop the same appearance as seen in embryonic tissue, which suggests that there is a proliferative response of muscle tissue with adaptation to an increase in length.

There is therefore some evidence that muscle adapts to permanent change in length though its capacity to do so has not been defined, nor has the relationship of the adaptation to different rates of distraction.

Muscle structure and function

A muscle belly consists of many bundles of muscle fibres. The muscle fibres are grouped together into bundles which are surrounded by perimysium. The whole muscle is surrounded by epimysium. There are also connective tissue elements in the spaces between the muscle fibres within a bundle and this is known as endomysium. It has been described by Rowe (1974) to consist of alternating sheets of parallel collagen fibres. The collagen fibres lie at an angle of between 45 and 60° relative to the long axis of the muscle depending on whether the muscle belly is lengthened or contracted. Each muscle fibre consists of a number of myofibrils which in turn consist of a number of myofilaments. The myofilaments are made up of actin and myosin molecules. Movement between the actin and myosin molecules results in the force of muscle contraction. The actin and myosin molecules are arranged into a collective unit called a sarcomere which is the contractile unit of the muscle. Williams and Goldspink (1978) demonstrated that when the resting length of a muscle is altered, the muscle adds or subtracts sarcomeres so that the pre-existing sarcomeres are returned to their optimum working length. Williams and Goldspink also showed that when the sarcomeres are added, they are added at the ends of the muscle in the region of the musculotendonous junction.

Vandenburgh and Karlisch (1989), using his cell culture system of myoblasts on a collagen gel matrix, had shown that the growth rate is dependent on the mechanical environment. The relationship of this to the *in-vivo* situation has not been investigated; in particular, the interrelationship between the connective tissue and contractile elements was impossible to study in his model.

Animal studies

In this present study, the hypothesis is investigated that muscle will act like other tissues such as bone and skin, and form new tissue on distraction if this is applied at the appropriate rate and rhythm. The ability of muscle to adapt to changes in length is studied as well as its sensitivity to sustain damage under different distraction regimes.

Method

In 20 adult New Zealand white rabbits, the muscles in the lower leg were subjected to distraction. An external fixation device (Orthofix M100) was applied to the medial surface of the tibia and a mid diaphyseal osteotomy was created. Lengthening was commenced at rates which varied between 0.7 and

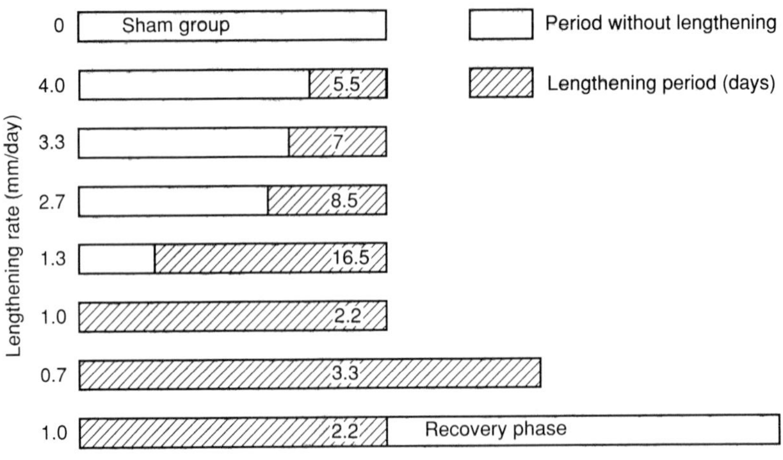

Fig. 24.1 Lengthening rates in the experimental groups.

4 mm per day, daily increases being divided into two increments. The groups lengthened are shown in Fig. 24.1.

Before sacrifice, animals were given an anaesthetic (which did not include any neuromuscular blocking agents) and physiological measurements of muscle function were made. The tendon of tibialis anterior was divided distally and the muscle belly was dissected as far as its proximal attachment. The tendon was then attached with a suture to the lever arm of a recording device that measured the active and passive force generated by the muscle. A single 1 ms square wave pulse produced a single twitch contraction of the tibialis anterior which had been set to various lengths between the extremes of its range. The active force produced during the twitch contraction for each length was measured and length/tension curves were then plotted (Fig. 24.2). The passive force exerted by the muscle as it was stretched was measured for each preset length, and then length/passive force curves were plotted for the lengthened and unlengthened control tibialis anterior and soleus muscles.

After these recordings the animals were sacrificed and samples collected for histology. To assess the sarcomere number and length the tibialis anterior and the soleus were excized from each leg having been fixed by sutures to a rigid wooden spatula at the exact length corresponding to the 90° position of the ankle joint.

Results

No abnormalities were found in the control limbs or sham group in range of joint movement, muscle stiffness, or in the responses to single twitch and

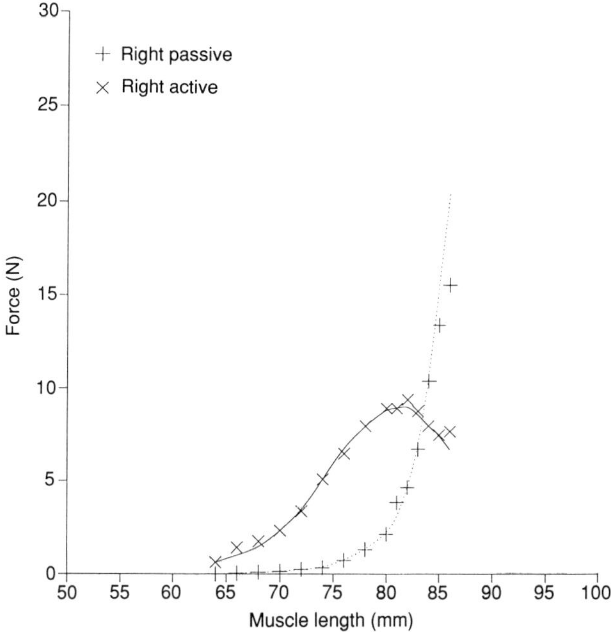

Fig. 24.2 Forces developed by tibialis anterior—control group.

tetanic contraction. A single twitch response, Fig. 24.3, shows the functional adaptation of the muscle which has been distracted at 1.0 mm/day; the level of the maximum force generated is very similar in the lengthened and control legs and the peak for the lengthened leg has moved to the right by an amount very close to the lengthening increment (i.e. 2 cm). Similar adaptations were seen in all specimens in which distraction rates were at 1 mm per day or less. At higher rates of distraction, Fig. 24.4, there was evidence of dysfunction with smaller peaks recorded in the lengthened muscle and little shift of the peak to the right; these changes reflect inadequate adaptation to the new length. In the muscles from animals which were not sacrificed until 1 month after the end of distraction there was recovery of function on the single twitch test even at higher rates of distraction up to 1 mm/day. The passive tension/length curves showed no adaptation at the faster rates but moderate adaptation at 1 mm/day. At the faster rates of distraction greater damage was seen on histological examination.

Figure 24.5 shows the number of sarcomeres per muscle fibre for various rates of distraction. At faster rates a greater percentage of the increase in overall muscle length was produced by stretching each individual sarcomere such that it exceeded the optimum length for the remaining sarcomere.

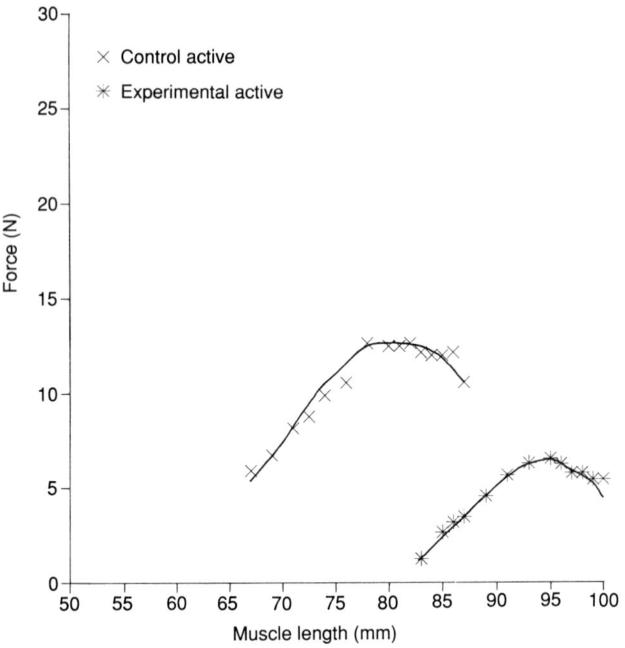

Fig. 24.3 Forces developed by tibialis anterior—20 per cent @ 1 mm/day.

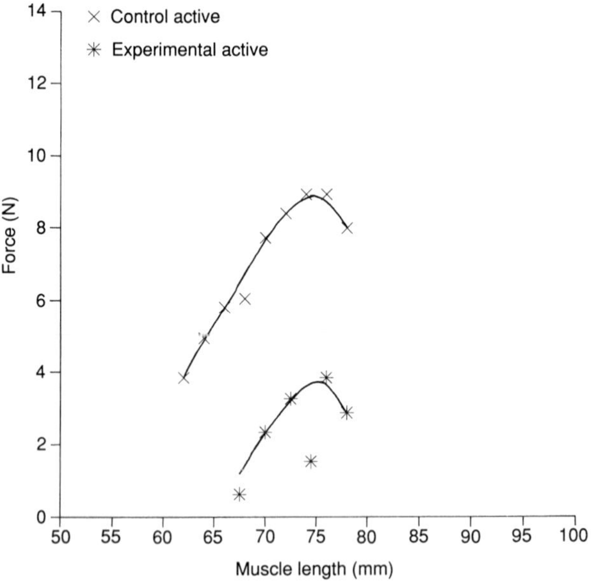

Fig. 24.4 Forces developed by tibialis anterior—20 per cent @ 2.7 mm/day.

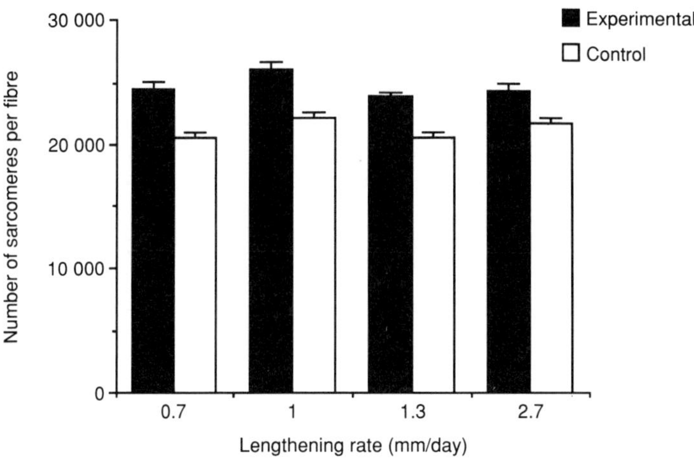

Fig. 24.5 Increase in serial sarcomeres with distraction.

Discussion

The studies on normal muscle show that, even in the adult, the number of sarcomeres from one end of a muscle fibre to the other is adjusted in response to the normal working length of the muscle so that at any time, the sarcomeres are working at their optimum length with the optimum overlap of the actin and myosin molecules. During leg lengthening, the authors in this study have shown that muscle fibres increase the number of sarcomeres running from one end to another. At the more rapid rate of distraction however, the muscle is unable to add on sarcomeres at a fast enough rate so that the individual sarcomeres become over-stretched.

The histological studies show that abnormalities develop at all but the slowest rates (0.7 mm/day) of distraction. Whorled fibres and centralization of nuclei were seen. Although the exact significance of these changes is not known, they are common findings in many types of muscle damage. The length/passive tension curves also demonstrated that there was a relative increase in muscle stiffness. The increase in stiffness will result in the antagonist using more energy in order to produce the same ankle movement and therefore the system will be of reduced efficiency. The increased stiffness may also result in a reduced speed of muscle response and this may be important in the more testing physiological conditions of natural existence such as running and jumping. The cause of fibrosis is not clear; similar changes are seen in many types of muscle damage (Engel *et al.* 1966; Sjostrom *et al.* 1978).

In the present studies, normal muscles have been distracted but in clinical practice abnormal muscle may be present within the lengthened limb. In congenitally short limbs, the muscles may have an inherently higher stiffness which may also be present in muscles affected by poliomyelitis. Similarly in patients undergoing bone transport who are elderly or who have had a serious injury to the compartments of the leg, there may be a change in the mechanical properties of the muscles making them acutely sensitive to any lengthening procedures.

It would seem that the rate of lengthening of a muscle should be considered very carefully when embarking upon leg lengthening, as the critical rate of muscle distraction may be equally important to that of the bone.

Debate in the meeting

Earlier work by Goldspink and Williams suggests that when muscle is lengthened sarcomeres are added at each end, so it is the absolute lengthening rate which is important rather than the proportional rate. In contrast, nerve is added along its whole length, so in this case percentage strain is the important factor.

In view of the delayed accommodation by soft tissues, the meeting wondered whether it was appropriate to allow patients to develop equinus during the elongation phase which could be removed during the consolidation phase. The author (A.H.R.W.S.), however, considered that the anterior muscles may become over-stretched, unable to overcome the passive forces generated by the posterior muscles, and so be unable to correct themselves. In addition, the common peroneal nerve may be more prone to damage due to the weaker anterior structures. In summary, an equinus should not be allowed to develop but the posterior structures should be encouraged to lengthen at the same time as the limb is being lengthened. If necessary a slow rate of distraction should be used. (See 'Debate' in Chapter 25, p. 226 for further discussion.)

25 The response of peripheral nervous tissue to limb lengthening

A. H. R. W. Simpson and J. Kenwright

Introduction

Limb lengthening procedures and operations performed to correct deformity are frequently associated with soft tissue problems. Permanent nerve palsy is a well recorded complication. This most commonly occurs when there has been an acute correction of deformity or when lengthening with the faster, Wagner-type distraction regimes. Although complete and permanent nerve palsies are rarely recorded, it is likely that temporary or incomplete nerve lesions are a lot more common. Indeed, Galardi *et al.* (1990) found electrophysiological evidence of neuromuscular damage in 10 out of the 10 segments that they were lengthening. The response of peripheral nerves to acute lengthening and gradual lengthening will be discussed.

Basic structure

Peripheral nerves consist of myelinated and unmyelinated fibres which are collected into bundles referred to as fascicles. Each of these bundles of axons is surrounded by a tube of tissue called perineurium which consists of a regular structure of flattened laminae of fibroblasts alternating with collagenous sheets. The fascicles are enclosed by a further fibrous tube, the epineurium, which has relatively little organization compared to the perineurium. The spaces between the nerve fibres are filled with a loose connective tissue termed the endoneurium which contains lymphatics and blood vessels of the peripheral nerve.

The vasculature of a nerve can be divided into intrinsic and extrinsic systems (Lundborg and Rydevik 1973). The extrinsic system consists of regional vessels of varying diameter which approach the nerve at different levels along its length. These vessels are spiralled and tortuous. The extrinsic vessels pierce the nerve and anastamose with the intraneural network. The intraneural systems consist of plexuses in the epi-, peri-, and endoneurium. In the endoneurial spaces, the vessels are primarily capillaries and so constitute the nutrative vascular supply to the nerve fibres.

Mechanical properties

Haftek (1970) performed a study on fresh post-mortem samples of rabbit tibial nerves which were tested in an Instron machine. He described peripheral nerves going through three phases: (a) elastic; (b) plastic; and (c) viscoelastic. Haftek considered the epineurium to be the main structure responsible for the elasticity of the nerve, however the data of Sunderland and Bradley (1961) suggests that the perineurium is responsible for the ultimate strength of the nerve. This is supported by the study by Kwan *et al.* (1988). This study also puts these mechanical results into the context of the living animal and shows that nerves are under tension even in the resting position of the limb.

The arrangement of a tight limiting membrane such as the perineurium enclosing a tissue compartment supplied by small blood vessels would allow a compartmental syndrome to occur, actually within the perineurium.

Response to acute strain

Background

Sunderland and Bradley (1961) in their studies on fresh post-mortem samples of human radial, ulnar, and median nerves found that the greatest elongation at the elastic limit of a nerve trunk was 20 per cent, although for some nerves it was as low as 8 per cent. Complete mechanical failure occurred at 30 per cent strain. However, the base length was taken as the length at which the nerve started to resist tension. As already discussed, Kwan *et al.* (1988) demonstrated that the nerve is under tension *in vivo*, therefore the strains compared to the *in-vivo* length of the nerve, at the elastic limit and rupture point, will be considerably lower (approximately 3 and 13 per cent beyond *in-vivo* strain respectively).

Lundborg and Rydevig (1973) studied the effect of acute stretch on the vasculature. They exposed the rabbit tibial nerve, transsected it proximally to the ankle joint and attached sutures to this end. At 8 per cent strain there was a reduction in flow particularly in the venules.

In 1872 Mitchell stretched the sciatic nerves of rabbits and found that at 25 per cent strain the nerves stopped conducting immediately. It was not until over a century later that Brown *et al.* (1989) stretched a peripheral nerve and then observed it for the following 3 h. Their model necessitated dividing the tibial nerve at the knee and clamping of the distal stump. The nerve was stimulated proximally and distally and recordings obtained from the abductor hallucis. They found that at 15 per cent strain with the ankle dorsiflexed the nerves would conduct for 30 min and then stop. At 10 per cent strain they found no change in conduction over the period that they monitored (3 h).

This study

In this longer-term study at lower acute strains the author have studied nerve function. Strain was measured on the tibial nerve of the adult New Zealand White rabbit by means of small metallic rings attached by a 9-0 epineurial suture. Plain radiographs corrected for magnification were taken during and after the lengthening regime and measurements were taken on an X-ray digitizing table allowing an accuracy of <0.1 mm. Cutaneous electrodes were used to stimulate the tibial nerve posterior to the medial malleolus and also just distal to the knee joint. A 1 ms square wave pulse was used throughout. The amplitude was increased until the maximum response was obtained. The muscle action potential from the abductor hallucis was measured in the sole of the foot by means of cutaneous electrodes. The data was recorded in digital form and an average of 16 stimulations were obtained.

The latency is the time between the stimulus artefact and the first appearance of the muscle action potential. The difference between the latencies from stimulation at the knee and at the medial malleolus reflected the time taken for the nerve impulse to travel the known distance between the two stimulation points. From these measurements the conduction velocity was derived. With maximum stimulation at the knee joint and at the medial malleolus the muscle action potential should have been similar unless there had been loss of conduction in some of the fibres between the two stimulation points. Any reduction in this ratio from the post-operative values therefore indicated nerve dysfunction caused by the lengthening process.

The animals that had undergone a single stage distraction showed an increase in strain throughout their length between the sciatic notch and the medial malleolous. The strain appeared to be evenly distributed throughout the length of the nerve although there may have been a very slightly higher value at the level of the osteotomy. The animals that had had a strain of less than 8 per cent (mean 6.00 per cent) showed no perceptible change in conduction velocity or amplitude ratio. Those animals that had undergone an acute distraction resulting in strains between 9.7 and 12.1 per cent (mean 10.2 per cent) continued to function for 48–72 h and then ceased functioning.

Histological examination correlated with the electrophysiological findings: nerves that had undergone strains between 9.7 and 12 per cent showed oedema and an increase in intraneural fibrosis.

Highet and Holmes (1943) in their study of eight patients and Highet and Sanders (1943) in their study of the external popliteal nerve of seven dogs found that when the nerves were sutured under tension to bridge a large nerve defect the degeneration occurred in the proximal stump for up to 9 cm as well as throughout the distal stump. There was extensive intraneural fibrosis at strains of the order of 12 per cent. Similar findings were obtained by Simpson and Kenwright in this study of the rabbit tibial nerve: animals subjected to an acute strain of 10–12 per cent had delayed loss of function and on histological examination had extensive intraneural fibrosis.

Response to gradual lengthening

Background

In an elegant series of experiments using a specially designed nerve puller, Bray (1970) used a microelectrode attached to the cell surface or to an axon or dendrite to demonstrate that the nerve processes would elongate in response to tension. The experiments were performed on dissociated chick sensory ganglia neurons. He obtained a maximum length of 960 μm by pulling on an axon at 40 μm/h for 24 h. At rates of 100 μm/h the axons elongated for several hours with no obvious thinning. At rates of 170 μm/h the axons elongated for about 1.5 h but with accompanying thinning. At rates of over 1000 μm/h the axons were immediately pulled into a thin strand which then broke. In another series of experiments Bray investigated the axon membrane during normal growth by grinding up glass and carmine. Particles 1–5 μm in diameter attached themselves to the plasma membrane of isolated rat sympathetic neurons which were growing in culture. The particles were observed to remain the same distance from the cell body during growth of the axon suggesting that the new membrane was being added in the region of the axon tip.

When the tensile force is exerted throughout the length of the axon, the site and mechanism by which the axon elongates is not known nor is the maximum rate known that an axon can accommodate lengthening *in vivo*.

Highet and Sanders (1943) conducted experiments on bridging nerve defects. They mobilized the nerve and approximated the ends by deflection of the accompanying joints, then applied plaster casts to allow them to gradually straighten out the joints. The lengthening rates for the nerve were of the order of 3.5 mm/day and at this rate they were unable to detect any new nerve formation and in fact noted retrograde degeneration of the proximal stump.

Milner (1989) used mini-tissue expansion devices implanted beneath the sciatic nerve in rats. He found that nerves tolerated slow distraction better than rapid distraction and that strains of 50 per cent produced a significant reduction in conduction velocity. When the tissue expanders were deflated the nerves contracted down to a length that was still greater than their original length. However, whether this was due to a proliferative response or whether it was due to a viscoelastic creep response was not determined. The large reduction in conduction velocity at strain of 50 per cent would be against this being a proliferative response.

This study

In this study the same animal model as for single stage distraction to investigate gradual distraction is used. In contrast to the one-stage distraction group, the animals that had undergone gradual distraction demonstrated a maximum strain at the level of the centre of the bone gap. The animals in the gradual distraction group were all lengthened to amounts in excess of the 10 per cent strain that had resulted in loss of function in the previous group. At the lower strain rates the nerve continued to function with no significant change in conduction velocity or amplitude ratio. There appeared to be a critical strain rate of the order of 1.5–2 per cent/day above which nerve function ceased. Lengthening was continued in one animal to see whether there was a maximum total strain above which function ceased. The limb of this animal was lengthened a total of 8 cm (80 per cent). The bulkier posterior structures lengthen less than the anterior structures which causes the ankle to drop into equinus, but despite this the nerve strain of the central segment was 60 per cent. This nerve continued to function. A maximum limit for the total strain at these lower lengthening rates was therefore not found in this study.

Electron microscopy of this animal showed good preservation of the myelin sheaths and no change in the diameter of the axons.

Discussion

Lundborg and Rydevik (1973) have shown that at 15 per cent strain there is a complete cessation of flow in the blood vessels in the nerve. At 8 per cent strain they found that the blood flow started to reduce, particularly on the venous side. Brown *et al.* (1989) have shown that after 30 min at 15 per cent strain the nerves cease to conduct. This time-period is similar to the delay in loss of function after ischaemia from a tourniquet, suggesting that the loss of function that Brown *et al.* encountered could be secondary to the ischaemia. During the course of their experiment they found no loss of function at lower strain values, however, their model precluded them following lower strain values for a prolonged period. In this study strains of the order of 10 per cent where found to result in a delayed loss of function after 48–72 h. This could be explained by delayed ischaemia resulting from an intraneural compartmental syndrome. This delay in loss of function may explain why some patients who had a correction of deformity initially appeared to have normal nerve function but then developed a delayed loss of function.

Peripheral nerves are often described as having a wavy pattern of the axons on longitudinal section. These are said to straighten out first during lengthening. However Simpson and Kenwright (1992) found that nerves fixed at their *in-vivo* length and then sectioned longitudinally were straight; the wavy appearance is an artefact of preparation with the reduction of tension on the nerve when it is transsected. The nerves do not simply straighten out to accommodate the increase in length. Results from the gradual distraction group indicate that the nerves are not simply stretching or bow stringing to accommodate the increase in limb length.

The electrophysiological tests used by Galardi *et al.* (1990) indicate loss of nerve, muscle or neuromuscular junction function in all of the limbs that they monitored. However, except in the two patients with changing conduction velocity of the common peroneal nerve, it is not possible to tell whether the loss of function was due to the nerve, the muscle, or the junction. This study suggests that the tibial nerve of the rabbit has the ability to create new nerve tissue in response to lengthening. The loss of function seen in the common peroneal nerve may be as a result of the associated increase in pressure on the nerve as it wraps around the fibular neck. Neural transport mechanisms are known to be very sensitive to pressure, and dysfunction in these transport mechanisms may prevent the necessary building blocks being transported down the nerve to allow creation of the new axoplasm.

Debate in the meeting

At the cellular level, the difference between nerve and muscle in susceptibility to damage on limb lengthening could be explained by the size of the cells. In the case of nerve we are considering axons which are continuous from the spinal cord right down to the foot and so in percentage terms they may undergo a smaller strain than the cells of muscle or other connective tissue. No relationship has yet been shown with the physiological state at insult but it is speculated that, as the damage seen is similar to the eccentric damage sometimes seen in fell runners, it may be that damage during limb lengthening could be reduced by the right physiotherapy regime for pre-training. At the moment this is still speculation.

26 Functional and structural adaptation of canine knee articular cartilage to physiological joint loading

J. S. Jurvelin, I. Kiviranta, A.-M. Saamanen, J. Arokoski, M. Tammi, T. Räsänen, A. Vaarnamo, and H. J. Helminen

Introduction

Our musculoskeletal system is made up of bony levers with points of support at the joints. In synovial joints, articular cartilage shields bone ends against surplus mechanical stress during muscular work and motion. The cartilage tissue together with the synovial fluid give the necessary lubrication and low friction to the articulating surfaces.

In spite of the important functional role of cartilage in joints, it is inadequately known how mechanical loading modulates composition, structure, and properties of the cartilage. During decreased or increased joint loading, several factors related to both quality and quantity of the stress pattern may contribute significantly to the cartilage response. It is now clear that unloading, immobilization, and continuous compression of high degree between the articulating surfaces are deleterious to the cartilage matrix (Langenskiold *et al.* 1979; Palmoski *et al.* 1979, 1980; Salter and Field 1960; Sood 1971). These cartilage alterations appear to mimic typical arthrotic changes or, at least, indicate atrophy of the matrix.

Articular cartilage is injured after experimental, repetitive impulsive loading (Radin *et al.* 1978). Running is probably a more physiological way to enhance articular cartilage load. Experimental, vigorous exercise may injure articular cartilage (Krause 1969; Vasan 1983), but there are several reports supporting the anabolic effects of running (Holmdahl and Ingelmark 1948; Lanier 1946; Sääf 1950). Differences in the experimental design, species, and age of the trained animals make the comparison between the previous studies difficult. For example, one of the factors that may injure cartilage is a rapid, forced start to a running training programme. It remains unclear to what extent articular cartilage shows functional adaptation analogous to Wolff's law in bone (Wolff 1892) or the Woo–Akeson–Amiel–USCD-hypothesis in tendons and ligaments (Woo and Akeson 1987). In most cases, functional characteristics of articular cartilage after altered joint loading have been discussed on the basis of known interdependency between proteoglycan content and mechanical properties of the tissue.

The authors have conducted a series of experiments where alterations in the canine knee joint loading were produced, on the one hand, by immobilizing the joint for 11 weeks and, on the other hand, by training the dogs on a treadmill according to 4 or 20 km daily running programmes (Jurvelin *et al.* 1986*a*,*b*, 1988). In this paper, local changes in the mechanical properties of knee joint cartilage are summarized. New results from the 40 km daily running programme are also included. In addition, indentation and shear tests were incorporated in histochemical, biochemical, and transmission electron microscopic analyses of the cartilage matrix to reveal the basic structure–function relationships of the tissue. Special attention was paid to controlling the validity of the theoretical and experimental methods of the study.

Materials and methods

Articular cartilage from the knee joints of female beagle dogs of pure breed ($n=63$) was used in the experiments. Further, fresh articular cartilage of bovine knees ($n=10$) was tested in the methodological analyses.

The dogs were divided into different experimental groups (Fig. 26.1). Immobilization and unloading of the knee joint was obtained by casting the right hind limb in 90° flexion for 11 weeks (Jurvelin *et al.* 1986*a*). The increased joint loading was produced by three running exercise programmes which were started at the age of 15 weeks. By gradual increase of the daily running distance and speed, the dogs were accustomed to running on a treadmill adjusted to 15° uphill inclination. Ultimately, the first training programme group (TP1) ran 4 km/day, 5 days a week for 15 weeks (Jurvelin *et al.* 1986*b*). The TP2 (Jurvelin

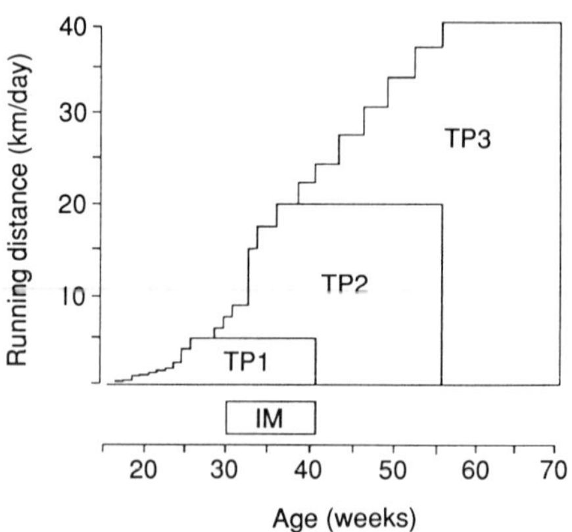

Fig. 26.1 Schedule of the experiments with dogs. Running on a treadmill 4 km/day (TP1), 20 km/day (TP2), or 40 km/day (TP3). Immobilization of the hind limb in 90° knee (stifle) joint flexion (IM).

et al. 1990) and TP3 ($n = 10$) groups continued running for additional 15 or 30 week periods, respectively, and the final daily running distances were 20 or 40 km, respectively (Fig. 26.1). All groups had their own age- and sex-matched control dogs ($n = 6$–12) which lived normally in their cages throughout the experiment.

Mechanical testing of articular cartilage

Creep tests

A microcomputer-controlled creep apparatus was constructed for indentation of canine knee articular cartilage at several predefined locations of femur and tibia (Fig. 26.2(a)). The apparatus is, in principle, a mechanical oscillograph traditionally used for rubber testing (Fig. 26.2(b)). Deformation of articular surface was measured by an inductive displacement transducer of differential transformer type. After amplification, the displacement signal was fed through a 12-bit A/D converter to a microcomputer which recorded the data. Application of the test load by a magnetic release started the actual indentation test and data acquisition by the computer. Using test stresses of 0.122–2.5 MPa and indenters of radii 0.32–0.81 mm, creep data was collected for 15–7200 s (Jurvelin *et al.* 1987).

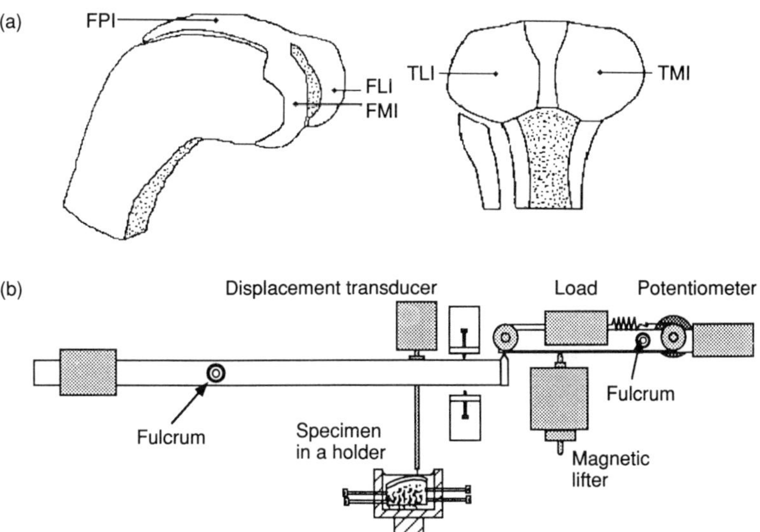

Fig. 26.2 (a) Indentation points of the canine knee articular cartilage. FPI = point on the patellar surface of femur, FMI = point on the medial femoral condyle, FLI = point on the lateral femoral condyle, TMI = point on the medial tibial plateau, TLI = point on the lateral tibial plateau. (b) Block diagram showing the mechanical structure of the apparatus used in the indentation creep tests See text for design and operational details.

Shear tests

The loading machine designed for indentation and shear stress relaxation tests consisted of a steel frame, a step-motor driven linear movement unit and a 20 kg load cell (Fig. 26.3(a)). The loading function was controlled by a PC/AT computer through a 12-bit acquisition board. The position of the indenter was monitored by an optoelectronic displacement transducer. Cylindrical osteochondral plugs from the patellar surface of bovine femur were first indented. After that the same plug (radius = 6 mm) was sandwiched between two smooth metallic discs and the shear test was conducted on the specimen (Fig. 26.3(b)). The loading protocol was identical for indentation and shear.

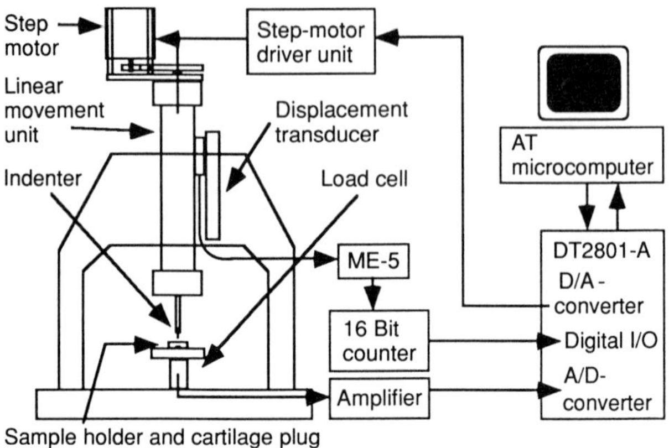

(a)

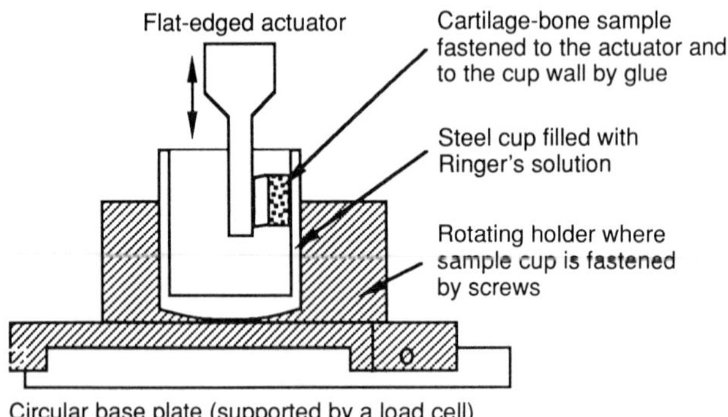

(b)

Fig. 26.3 (a) Block diagram showing the structure of the apparatus used in the indentation and shear stress relaxation tests. (b) Configuration of the shear measurements. See text for design and operational details.

After a 2 per cent shear prestrain, 10 per cent shear strain was imposed on the specimen using a fast (150 ms) ramp. The strain was maintained at a constant value throughout the test (3600 s) and the computer sampled the force and displacement signals. Thickness of uncalcified cartilage, necessary for determination of the strain and calculation of the shear modulus, was measured using a stereomicroscope (Jurvelin *et al.* 1987) or a penetrating needle technique (Hoch *et al.* 1983).

Analysis of the indentation measurements

According to the biphasic KLM model, cartilage behaves like an equivalent incompressible elastic material (i.e. $v = 0.50$) instantly after the step-load application ($t \geqslant 0$). At equilibrium, the fluid flow ceases and all the load is carried by the elastic solid matrix. Consequently, the equilibrium deformation is governed by the intrinsic elastic properties (shear modulus, μ_s and Poisson ratio, v_s) of the solid matrix (Mak *et al.* 1987; Mow *et al.* 1989). To test the suitability of the single phasic elastic model of articular cartilage for indentation (Hayes *et al.* 1972) the numerical solution of this elastic contact problem was extended to small Poisson ratios (Jurvelin *et al.* 1990). Taking the boundary conditions into account, the mathematical problem satisfying the field equations of the linear theory of elasticity for homogeneous, isotropic material reduces into inversion of the Fredholm integral equation of the second kind with symmetric kernel. The numerical solution of the problem produces a theoretical scaling function, K, which gives a correction for the effect of variable cartilage thickness on the deformation (Fig. 26.4).

A general viscoelastic model of articular cartilage (Parsons and Black 1977) was utilized for the analysis of the rate of time-dependent creep phase. In this model the creep is recorded as a function of time and the retardation time spectrum, $L(\tau)$, indicating that the rate of creep is determined using the first derivative of deformation, $\omega(t)$, with respect to ln t (Fig. 26.5).

Histochemical analysis

The samples for histochemical analyses were taken from the opposite (contralateral) knee at the sites corresponding to those of indentation points. Tissue sections were stained with Safranin O, a cationic dye that binds in a stoichiometric way to glycosaminoglycan polyanions (Kiviranta *et al.* 1985). The local absorbance in stained sections was analysed with a microspectrophotometer (Fig. 26.6). The value of Safranin O concentration was calculated in different (superficial, intermediate, and deep) cartilage zones of the uncalcified cartilage, and in the calcified cartilage layer.

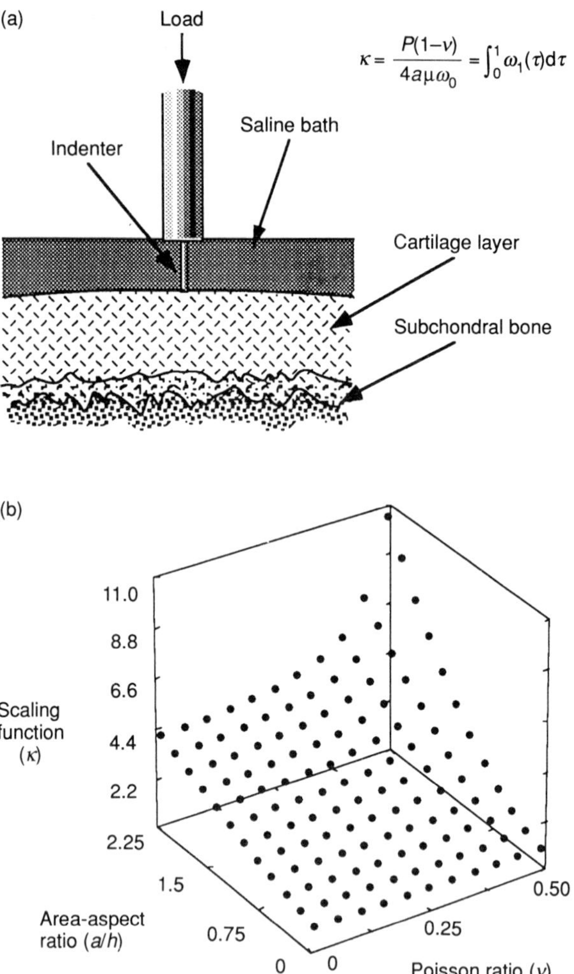

Fig. 26.4 (a) Indentation testing of articular cartilage. An impervious, plane-ended indenter is applied perpendicularly to the cartilage surface. The specimen is immersed in physiological saline solution. (b) Three-dimensional plot of the scaling function (κ) as a function of area-aspect ratio (a/h, a = indenter radius, h = cartilage thickness) and Poisson ratio (ν_s). P is the indenter load, ω_0 is deformation.

Biochemical analysis

The cartilage samples for biochemistry, collected from the adjoining sites to those for the histochemical analyses, were frozen and powdered in liquid nitrogen under pressure and freeze-dried. The powder was extracted with 4 mol/l guanidium chloride (GuCl) in 50 mmol/l sodium acetate buffer (pH

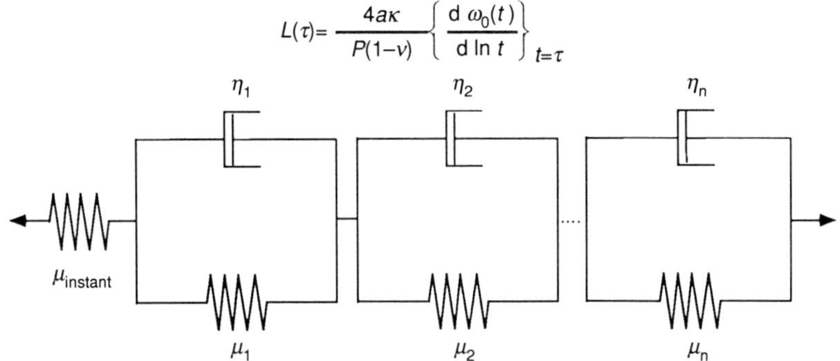

Fig. 26.5 The generalized viscoelastic model for articular cartilage and the formula for calculation of retardation time spectrum $L(\tau)$. For explanation of the symbols, see Fig. 26.4.

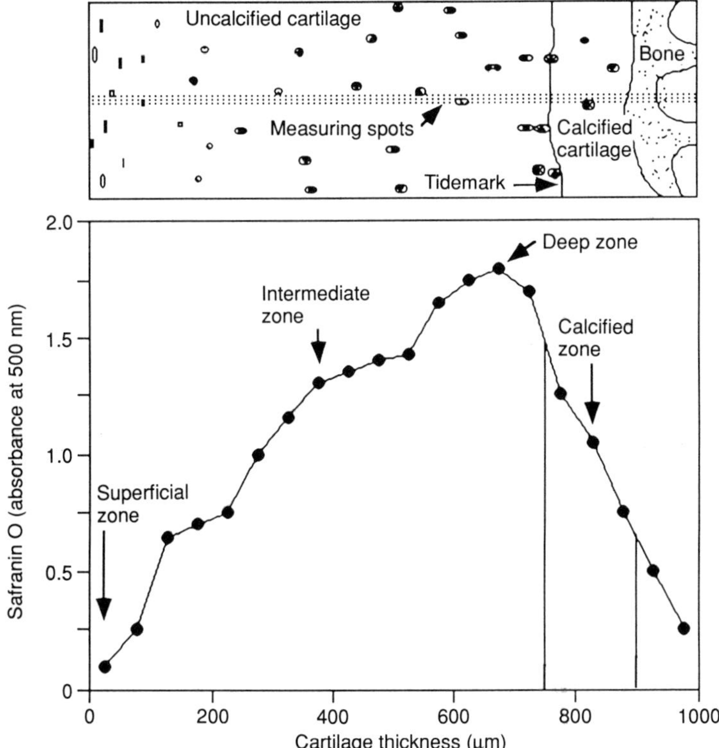

Fig. 26.6 The microspectrophotometric analysis system. The content of Safranin O stain in different zones of articular cartilage was determined at consecutive measuring spots in three lines shown in the schematic picture of articular cartilage (above). The area covered by small and large dots represents respectively the total content of Safranin O in the uncalcified and in the calcified cartilage (below).

5.8) in the presence of protease inhibitors. Extract and residue were analysed separately. Uronic acid and hydroxyproline contents were determined. After purification of prostaglandins (PGs) from the extract, the ability of the PGs to aggregate with exogeneous hyaluronan was tested. In summary, wet weight, dry weight, uronic acid (PG) content, hydroxyproline (collagen) content, PG extractability, and aggregation with exogenous hyaluronan were determined in the biochemical analyses (Jurvelin *et al.* 1987).

Transmission electron microscopic (TEM) analysis

Articular cartilage samples for TEM analysis consisted of the intermediate part of the lateral condyle of femur (FLI) and the intermediate area of the lateral condyle of the tibia (TLI). Samples were sawn according to the predominant direction of artificial split lines, perpendicularly to articular surface using a dentist drill. Blocks were fixed for 24 h in 2 per cent glutaraldehyde in 0.1 M sodium cacodylate buffer and decalcified in buffered 4.13 per cent EDTA solution for 1 day. Samples were post-fixed in buffered 1 per cent osmium tetroxide, dehydrated, and embedded in Epon. Thin vertical sections parallel to split lines were cut using ultramicrotome, stained with muranyl acetate and lead citrate, and examined in a JEM-1200 EX transmission electron microscope at 80 kV.

The electron micrographs were taken systematically from articular surface up to about 54 μm depth at a magnification of ×25 000. Collagen fibrils were assayed by superimposing a lattice over the photomicrographs (a final magnification of ×75 000). Volume density (V_v) of collagen fibres was determined as a relation $V_v = P_c/P_t$, where P_c is the number of test points hitting collagen fibres and P_t is the total number of test points in the area.

Results

Indentation and shear characteristics of articular cartilage

Indentation creep measurements of the canine knee articular cartilage showed typically an instantaneous deformation followed by creep to the equilibrium state. By focusing in the first 500 ms after a sudden load application, normally at least three total cycles of damped free oscillations were recorded (frequencies 15–50 Hz, Fig. 26.7(a)).

The stress relaxation curves for a typical indentation and shear test of patellar groove cartilage were analogous to creep curves with an instant high stress followed by strong relaxation (Fig. 26.7). Even for the shear test, where no significant fluid flow is supposed to occur (Mak *et al.* 1987), the relaxation was typically fivefold. Instant indentation and shear responses showed

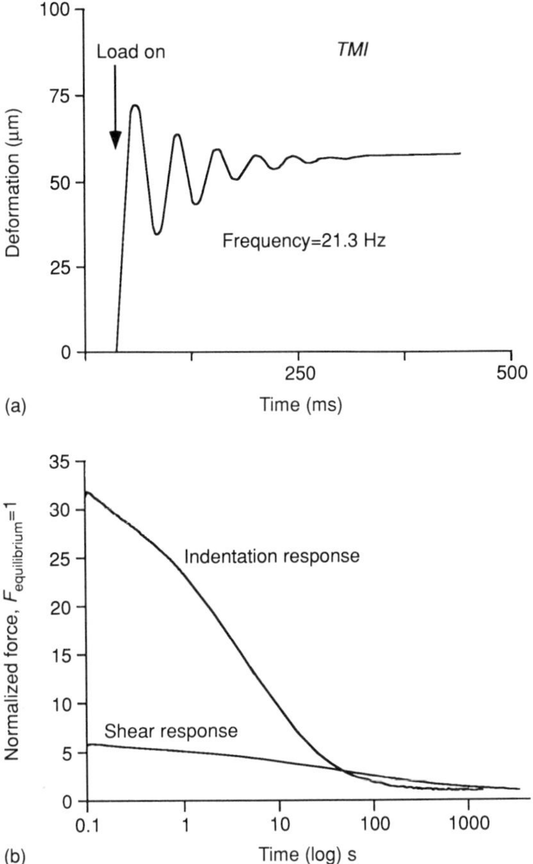

Fig. 26.7 (a) Typical short-term indentation creep curve of canine knee articular cartilage. Test point = TMI, test stress = 0.39 MPa, indenter radius = 0.40 mm. (b) Typical indentation and shear stress relaxation of bovine knee articular cartilage. The stress is normalized with the equilibium response.

nonlinearity at strains of 0–20 per cent. However, with small strains, the linear elastic model was valid for canine knee articular cartilage giving mechanical parameters that were independent of the indenter radius or stress (Table 26.1). When testing with several indenters, the values of instant modulus were independent of the indenter radius if the Poisson ratio (ν_s) was assumed to be 0.50 (Jurvelin *et al.* 1987). This suggested that, instantaneously after load application, both canine tibial and bovine femoral articular cartilage behaved like incompressible elastic materials. At equilibrium, both creep and stress relaxation measurements suggested a Poisson ratio of 0.40 for bovine femoral and canine tibial cartilages (Jurvelin *et al.* 1987).

Table 26.1 Instant shear modulus, equilibrium shear modulus, and retardation time spectrum, $L(\tau)$, as determined with three indenters of different radii. Test point *TMI*, test stress = 0.31 MPa

Indenter radius (mm)	Area-aspect-ratio (a/h)	Shear modulus (MPa) Instant[1]	Shear modulus (MPa) Equilibrium[2]	$L(\tau)$ (m^2/N ln t) $\times 10^7$)
0.32	0.31 ± 0.02[3]	0.62 ± 0.17	0.32 ± 0.08	4.15 ± 1.22
0.40	0.39 ± 0.02	0.58 ± 0.10	0.31 ± 0.05	4.73 ± 1.80
0.80	0.78 ± 0.04	0.62 ± 0.06	0.33 ± 0.05	4.11 ± 1.31

[1] $\nu = 0.50$; [2] $\nu = 40$. [3] Mean $\pm$ SD ($n = 6$)

Composition, structure, and properties of the canine knee articular cartilage

Canine knee articular cartilage showed distinct topographical variations in the biochemical composition, thickness, and biomechanical properties (Table 26.2). PG content of the tibial cartilage was higher than that of the femoral cartilage. In the surface zone of tibial cartilage the volume fraction of collagen was particularly low (Fig. 26.8). In general, the tibial cartilage was thicker but softer than the femoral cartilage. The instant and equilibrium modulus of the femoral cartilage (FMI, FLI) were 65 and 11 per cent higher, respectively, than those of the tibial cartilage (TMI, TLI). Typical frequencies of damped free

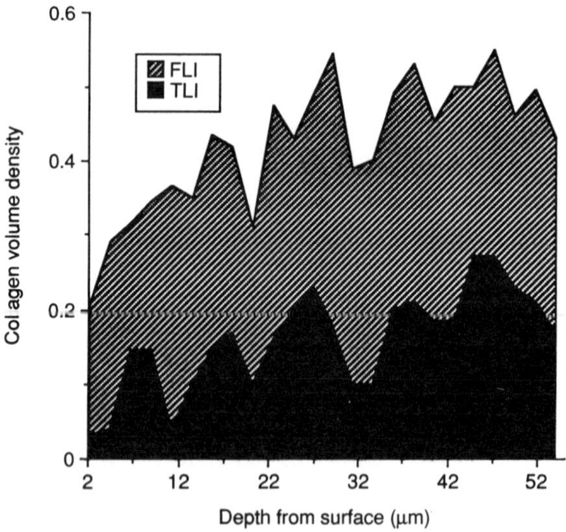

Fig. 26.8 Volume density of collagen fibres as a function of depth from the surface in the articular cartilage of canine femur (FLI) and tibia (TLI).

Table 26.2 Biomechanical and biochemical properties of articular cartilage in the canine femur and tibia. Test stress = 0.39 MPa, indenter radius = 0.40 mm

	Femur			Tibia	
	FPI	FMI	FLI	TMI	TLI
Biomechanical properties					
Thickness (mm)	0.49 ± 0.12[1]	0.84 ± 0.09	0.52 ± 0.07	1.18 ± 0.20	0.85 ± 0.13
Instant shear modulus (MPa)	1.40 ± 0.58	1.02 ± 0.26	0.94 ± 0.17	0.54 ± 0.20	0.65 ± 0.17
Equilibrium shear modulus (MPa)	0.50 ± 0.09	0.36 ± 0.03	0.32 ± 0.04	0.29 ± 0.06	0.32 ± 0.03
Rt-spectrum ($\times 10^{-7}$ m^2 N ln t)	2.72 ± 1.10	3.08 ± 0.66	3.73 ± 0.74	3.49 ± 0.96	2.60 ± 0.32
Biochemical properties					
Uronic acid/wet wght (nmol/mg)	31.1 ± 4.9	34.3 ± 7.47	28.8 ± 5.1	48.7 ± 7.2	46.0 ± 6.5
Uronic acid/ hydroxyproline (nmol μg)	200 ± 27	242 ± 66	185 ± 32	492 ± 66	300 ± 46
Proteoglycan extractability (%)	89.1 ± 16.1	89.4 ± 11.9	87.4 ± 14.0	89.3 ± 19.5	89.8 ± 15.0
Water content (%)	78.7 ± 5.0	73.3 ± 1.7	80.3 ± 4.0	77.1 ± 3.5	77.1 ± 2.1

[1] Mean ± SD (n = 10–12)

oscillations were 45.5 ± 8.2 Hz for femoral cartilage and 27.6 ± 5.0 Hz for tibial cartilage. The mean equilibrium stiffness of canine knee articular cartilage, as based on the five test points, was 0.36 MPa (Table 26.2).

Significant correlations between the biochemical and mechanical parameters in the analysed material have been presented in detail previously (Jurvelin *et al.* 1987). In brief, no significant correlation was found between the instantaneous modulus and PG content of the tissue. Instead, the instantaneous modulus correlated negatively with the percentage of proteoglycans extracted by GuCl. The equilibrium modulus correlated positively with the PG content. The creep deformation was affected by the the PG content as indicated by significant correlations between PG content and the retardation time spectrum, $L(\tau)$ (Table 26.3).

Effects of altered joint loading on articular cartilage

All tested canine cartilage samples of the control or experimental animals were normal with smooth, intact surface showing no evidence of fibrillation or

Table 26.3 Indentation points where significant ($p<0.05$) correlation coefficients were found between the biomechanical and biochemical parameters in the tested canine knee material

	Shear modulus		Retardation
	Instant	Equilibrium	time spectrum
Uronic acid/wet weigh		FPI, FLI	FPI, FLI, TMI
Uronic acid/hydroxyproline		FPI, FLI	FPI, FLI, TMI
Proteoglycan extractability	FPI, FLI		

$N=11$–12

osteophyte formation. Also, synovial fluid and synovial membrane looked normal in all joints.

Immobilization of the knee joint for 11 weeks significantly altered the mechanical properties of the canine knee articular cartilage. The cartilage thickness decreased by 9 per cent and deformation was systematically greater during testing in the immobilized dogs than in the controls (Jurvelin *et al.* 1986*a*). The rate of fluid flux under compression, as indicated by the values of the retardation time spectrum, increased in immobilized dogs. The Safranin O concentration of the matrix decreased in the immobilized knees (Kiviranta 1987). The reduction of the staining was most prominent in the superficial zone.

Neither the thickness nor the instant shear modulus of the cartilage was significantly altered by increased joint loading. After TP1 (Jurvelin *et al.* 1986*b*), the increase in stiffness was significant in the medial tibial condyle (TMI). In general, cartilage showed a trend to increased thickness and stiffness after TP1. When the degree of training was increased to 20 km/day, cartilage stiffness at TLI, as indicated by the equilibrium shear modulus, increased significantly (by 13 per cent, $p<0.05$). Otherwise, stiffness was at the control level or even slightly below it (Jurvelin *et al.* 1990). Running 40 km/day led to significant softening of the cartilage of femoral and tibial lateral condyles (Table 26.4). The changes in cartilage stiffness were closely related to the alterations in the PG content (Fig. 26.9). Moderate running exercise (TP1) elevated the Safranin O concentration of the uncalcified cartilage on the summits of the femoral condyles by 28 per cent. PGs were increased in the intermediate, deep, and calcified cartilage, while the superficial zone did not react (Kiviranta 1987). More strenuous training (TP2) decreased Safranin O concentration of the uncalcified and calcified cartilage. The reduction was most prominent in the superficial zone (Kiviranta 1987). Safranin O content was decreased after TP3 in the lateral and medial condyle of femur and in the

Table 26.4 Stiffness (shear modulus) of articular cartilage after immobilization (*IM*) and physical training programs (4 km/day, 20 km/day, 40 km/day). Ratios between mean values of experimental and control (*CO*) groups

	Shear modulus				
Indentation point	IM	CO	4 km	20 km	40 km
FEMUR					
FPI	1.04	1.00	1.15	0.98	1.06
FMI	0.79*	1.00	0.99	0.95	1.00
FLI	0.77*	1.00	1.11	0.91	0.88*
TIBIA					
TMI	0.87	1.00	1.40*	1.00	0.96
TLI	0.71**	1.00	0.98	1.13*	0.86*

*$p<0.05$, **$p<0.01$ (Mann–Whitney U-test or Wilcoxon matched-pairs signed rank test).

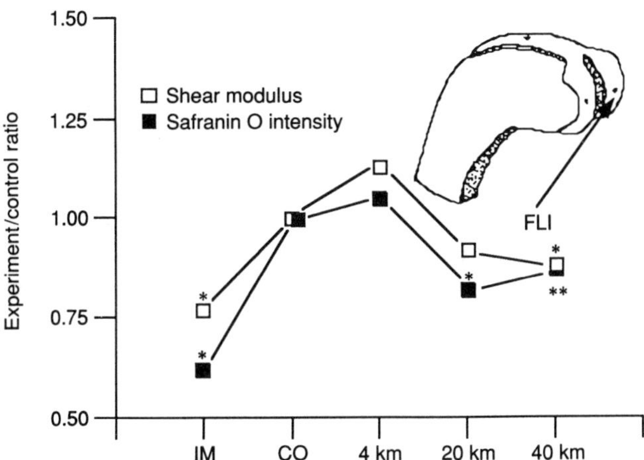

Fig. 26.9 Shear modulus and Safranin O concentration of uncalcified FLI cartilage during different joint loading conditions. Ratios are between mean values of experimental and control groups. *$p<0.05$, **$p<0.01$ (see Jurvelin *et al.* 1986*a*,*b*, 1988 for calculation of the shear modulus).

lateral condyle of tibia. PG content was decreased mainly in the superficial zone (Arokoski *et al.* 1992).

Discussion

Methodological and theoretical considerations

Indentation is the most suitable technique for the mechanical testing of small-sized canine articular cartilage samples. Mechanically, the step-load is difficult to obtain in a creep mesurement. Although the damped free oscillations are characteristic of materials showing elastic properties they are rarely revealed in the tests of articular cartilage with 'dead-weight' loading apparatus. Due to possible inertial and frictional problems of the loading apparatus, the oscillations may be over-damped and the true instant deformation is impossible to determine. In principle, they can be used to quantify the material properties of articular cartilage by the appropriate theoretical modelling and by measuring the frequency and damping of the oscillations

When the indentation creep technique was used with the impervious, plane-ended indenter, instantaneous and equilibrium cartilage responses after application of the indentation load could be quantified using a linear elastic model for the cartilage layer bonded to rigid bone. However, this was only possible if very small strains were utilized. A strong analogy between instantaneous indentation and shear behaviour was discovered. Instantaneously, indented cartilage behaved like an incompressible ($\nu = 0.50$) elastic material. At equilibrium, the elastic model suggested a Poisson ratio of 0.40 for the canine TMI cartilage. This value agrees with the ν-values of 0.40–0.50 determined and used by previous investigators to calculate the elastic moduli from indentation tests (Kempson 1971; Hori and Mockros 1976; Parsons and Black 1977; Armstrong *et al.* 1980; Lane and Ralis 1983; Hoch *et al.* 1983; Gershuni and Kuei 1984). However, from the biphasic indentation analysis, ν-values equal to zero have been determined for bovine knee articular cartilage (Athanasiou *et al.* 1991). The equilibrium modulus obtained from the shear measurements was typically five times higher than that obtained from indentation measurements. The disparity was attributed to different load-carrying mechanisms at indentation and shear equilibria. The shear balance is primarily generated through the elastic stiffness of the collageneous fibres (Myers *et al.* 1988). Due to buckling, collagen fibres have a smaller role during indentation equilibrium whereas the osmotic forces generated by PGs are of great importance. Our values for the shear moduli are consistent with the values reported previously for the cartilage of canine femur (Altman *et al.* 1984).

Structure function relationships of articular cartilage

PGs are important in resisting water flow within the loaded cartilage (Kempson *et al.* 1970; Armstrong and Mow 1982; Parsons and Black 1987). This was confirmed by significant correlations of the equilibrium modulus and retardation time spectrum with the PG content. While the rate of creep and the total equilibrium were significantly dependent on the PGs, the instantaneous arrangement of the matrix after load application was not directly related to the PG content. Instead, the enhanced entrapment of the PGs in the collagen network, as shown by the low extractability with guanidium chloride, was found in the cartilage with the high instant modulus values. Obviously, the interactions responsible for the cohesiveness of the solid matrix contribute significantly to instant stiffness of the tissue. The tibial plateau cartilage, while rich in PGs, was instantly softer than the cartilage of the opposing femoral condyle. As a novel finding, we measured a very low collagen content in the superficial tibial cartilage. In the case of indentation, the instantaneous shape change of the cartilage leads to stretching of the superficial collagen fibres (Askew and Mow 1978). The tensile stiffness of cartilage matrix correlates better with the collagen/PG ratio than with the collagen alone (Akizuki *et al.* 1987). It is concluded that there is a strong analogy between tensile behaviour and instant compressive behaviour as suggested by Mizrahi *et al.* (1986).

Adaptation of articular cartilage to physiological joint loading

As far as the authors know, there are no experimental or theoretical investigations describing the stress pattern in the canine knee joint during running. Based on X-ray analysis (Kiviranta 1987) it is anticipated that in the canine knee, as in the human knee (Brown and Shaw 1984), the summits of the femoral condyles carry a heavy load during running. In the tibia, the compressive load is transmitted partially through the menisci (Kettelkamp and Jacobs 1972, Shrive *et al.* 1978; Seedhom 1979; Seedhom and Hargreaves 1979; Ahmed and Burke 1983). The femorotibial force remains low in immobilized animals due to unloading of the limb, whereas considerable patellofemoral force arises from the pull of quadriceps and the patellar tendon (Buff *et al.* 1988).

Cartilage areas regularly subjected to high levels of stress show a higher PG content (Bjelle 1975; Sweet *et al.* 1977; Slowman and Brandt 1986; Kiviranta 1987; Manicourt and Pita 1988) and are stiffer (as estimated from two second indentations) than cartilage in the areas of low stress levels in normal joints (Kempson 1971; Swann and Seedhom 1986). The low stiffness of the cartilage in the central area of tibial plateaus may reflect a low weight-bearing role for those areas in the canine knee (Shrive *et al.* 1978; Seedhom and Hargreaves

1979; Swann and Seedhom 1986). Cartilage shows higher tensile stiffness at the periphery of highly loaded areas (Akizuki *et al.* 1987). The result is consistent with the idea of functional adaptation of cartilage, since large tensile hoop stresses most likely exist at those areas and a higher tensile stiffness is required. It is evident that, not only the degree of stress but also the type of stress (compression, tension, shear) significantly affects the modulation of cartilage structure and properties.

Immobilization of the canine knee in flexion for 11 weeks decreased the PG content and caused softening of the articular cartilage without any signs of tissue degeneration. Although the biomechanical properties of cartilage after immobilization resembled those of soft and permeable arthrotic articular cartilage, this structural and functional atrophy is proposed to be typical of cartilage adaptation due to unloading. In the immobilized knee, articular cartilage under femoropatellar contact preserved biomechanical characteristics close to the control level. This was assumed to be a consequence of small compression and movement between femur and tibia.

Moderate training of dogs (i.e. TP1: 4 km/day for 15 weeks) increased PG content and caused a slight stiffening of the cartilage. The results after strenuous training (TP2, TP3) indicated that if the intensity of running training was raised, cartilage no longer responded with a PG increase or improved mechanical properties. In the highly loaded summits of femoral condyles particularly, the beneficial effects of TP1 declined to the control level or even below it by TP2. In the central areas of the tibial plateaus (points TMI and TLI), the stiffness of exercized cartilage increased by 7 per cent as compared with controls and the positive effects of TP1 were preserved better during TP2. PG content also remained high. However, after TP3 the TLI cartilage was also significantly softer.

Due to the repetitive mechanical interaction of the articular surfaces, increased joint motion and stress requires the cartilage to resist friction and wear effectively. The lubrication mechanisms of surfaces and the material properties of the tissue must be adapted to these increased functional demands. In this light, adaptation of the tissue may not lead to improved mechanical properties but this is reflected by the ability of cartilage to retain its properties at the normal (control) level. The authors' results suggest that modulation of mechanical properties by joint loading follows from the concomitant changes in the PG content and quality. Although even very strenuous running training does not lead to the cartilage injury which is typical of impulsive joint loading (Radin *et al.* 1978), the results after 20 or 40 km daily training imply that young canine knee articular cartilage may not be totally adapted to the high mechanical demands. This conclusion is in harmony with previous reports on degenerative changes in cartilage after heavy running training in guinea-pigs (Stofft and Graf 1983), mice (Krause 1969), and dogs (Vasan 1983).

Debate in the meeting

The meeting wondered whether the activity of the animals outside their normal exercise period was influencing the cartilage development. Even among the animals on the treadmill for 7 h/day, movement outside the exercise period was similar to the matched control dogs. Blood tests showed that extra exercise was inducing no extra strain.

The calcified zone at the base of the animal's articular cartilage formed between their entry into the trial at 15 weeks and the end of the trial at 40 weeks. This experiment therefore showed the ability of young immature and growing cartilage to adapt to exercise and injury. The author (J.S.J) agreed that a different result might be found in more mature animals.

27 Transmission of rapidly applied load through articular cartilage: the mechanics of osteoarthrosis

J. J. O'Connor and I. Johnston

Introduction

It is difficult to deduce from inspection of an arthritic hip how the degenerative changes started and progressed. In 1973, Bullough *et al.* speculated that 'osteoarthritis is that sequence of events which follow from failure of the articular cartilage within an habitual contact area', i.e. a region of the joint which is frequently loaded. The terms osteoarthritis and osteoarthrosis are used interchangeably in this text and are intended to describe that form of degenerative joint disease for which there is not a clear inflammatory or traumatic cause. Radin (1990) defined osteoarthrosis as 'a deterioration of an articular joint occurring primarily from mechanical factors'. It manifests as full-thickness cartilage loss, at least 1 cm in diameter, in the habitually load-bearing areas of a joint. In 1991, Goodfellow *et al.* described antero-medial osteoarthritis of the knee, based on examination of tibial plateaux excised during unicompartmental arthroplasty. Unicompartmental disease is found in about 50 per cent of osteoarthritic knees requiring arthroplasty and occurs medially in 80–90 per cent of those cases. The lesion is found on the distal femur and on the anterior facet of the tibial plateau. The posterior facets of the femoral condyle and the tibial plateau remain undamaged. The lesion is only rarely associated with osteonecrosis of the medial condyle. All the ligaments of the knee are still intact. Since such patients are younger than those in which ligament damage was found (Goodfellow and O'Connor 1992, Table IV), it seems reasonable to deduce that antero-medial osteoarthritis of the knee is an example of early osteoarthrosis, starting with failure of the cartilage, and to seek a mechanical explanation for that failure (White, personal communication). Why does the lesion start medially, why on the anterior tibia, why on the distal femur? In this paper, the authors will suggest answers to the second and third questions.

Figure 27.1 shows the authors' computer model of the knee with its cruciate ligaments (O'Connor *et al.* 1990; Collins and O'Connor 1991; O'Connor 1992). The model reproduces the rolling action of the femur on the tibia first described by Weber and Weber in 1836. The femur rolls to the back of the

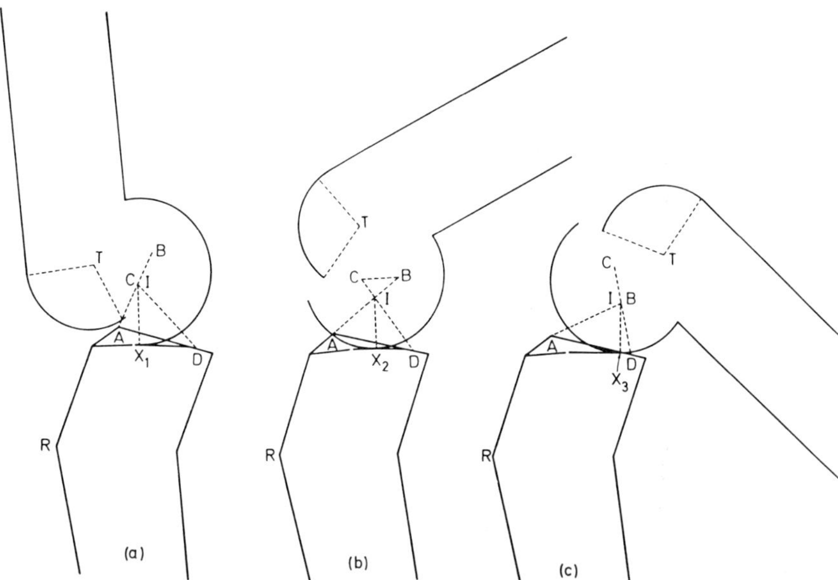

Fig 27.1 Computer-drawn model of the knee. The femur and the tibia are connected by the cruciate ligaments represented by the lines *AB* and *CD*. The two bones and the two ligaments form a four-bar linkage. The instant centre of the linkage lies at the intersection of the cruciate ligaments, *I*, and moves backwards during flexion as the ligaments rotate about their points of origin and insertion. As a result, the femur rolls as well as slides on the tibia, the contact point between them moving from the anterior tibial plateau, X_1, in extension to the posterior tibial plateau, X_3, in flexion.

tibial plateau while flexing and to the front while extending. In full extension, Fig. 27.1(a), the distal femur makes contact with the anterior tibia. The sites of the early arthrotic lesions are therefore surfaces which make contact in extension. They are habitual contact areas.

Failure of the cartilage cannot merely be the effect of large loads, since the knee transmits very large loads in flexion as when going up and down stairs (Andriacchi *et al.* 1982). What is peculiar then about the surfaces which make contact at full extension? To answer that question, we need to move to the gait laboratory.

In the gait laboratory, the forces transmitted between the foot and the ground during the stance phase of walking are measured with a force plate. The traces of the vertical component of the foot/ground reaction from two subjects are shown in Fig. 27.2. In addition to the effects of gravity, the curves include the dynamic effects needed to accelerate the body up and down. During the 10 ms or so immediately after heelstrike, some subjects exhibit a very high initial loading rate, up to 110 times body weight per second (i.e.

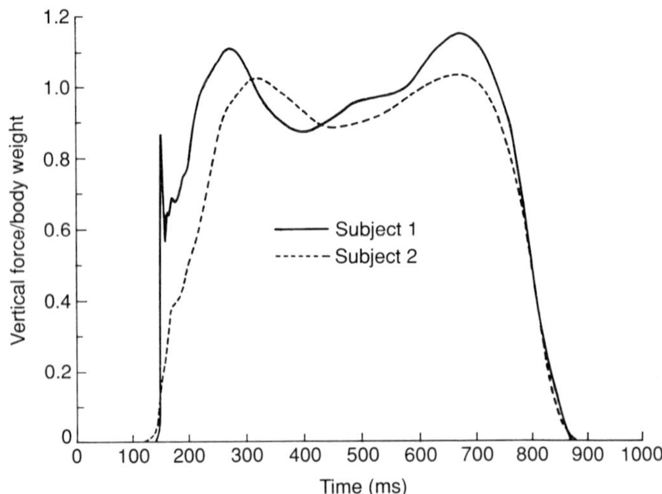

Fig 27.2 The vertical reaction between the foot and the ground for two subjects during the stance phase of walking. Subject 1 exhibits a high initial loading rate after heelstrike, with a distinct impulsive peak called the heelstrike transient. Subject 2 exhibits a lower initial loading rate, with no discernible peak.

about 10 t/s) with a characteristic impulsive peak called the heelstrike transient. Only about one-third of the population exhibit such high loading rates (Radin *et al.* 1986).

At heelstrike, the knee is nearly straight. The early arthrotic lesions therefore occur on those surfaces which transmit the heelstrike transient. This observation adds powerful clinical support to the Radin hypothesis (Radin and Paul 1971) that cartilage failure results from the transmission of repeated impulsive loads. Radin *et al.* (1991) examined the gait of a group of young adult subjects complaining of intermittent knee pain and found their loading rates at heelstrike to be significantly higher than those of a pain-free control group. High loading rates may be the characteristic of the group most at risk from osteoarthrosis.

Meachim (1972) and Meachim and Bentley (1980) inspected the cartilage in hundreds of joints in the Liverpool morgue. They found some specimens with a clear horizontal split near the tide mark, the cartilage/bone interface. Radin and Paul (1971) succeeded in reproducing cartilage failure in the laboratory *in vitro* with a regime of repeated impulsive loads. In living animals, subjected to such a loading regime, early signs of cartilage failure were found which were similar to Meachim's slides (Radin *et al.* 1984). Near the tidemark, there were distinct splits or cracks.

How can cartilage, a tissue containing 80 per cent fluid, 20 per cent solid, exhibit a mode of failure more typical of a brittle material, especially when it is carrying *compressive* loads on its surface? In brittle engineering materials, cracks such as those found by Meachim and as seen in Radin's rabbits are usually propagated by *tensile* stress. In cartilage in the unloaded state, the swelling pressure of the proteoglycan gel is balanced by tensile stress in the fibres of the collagen network (Maroudas 1976). Weightman and Kempson (1979) suggested that 'any mechanical failure of articular cartilage is likely to be a tensile failure of the collagen meshwork, either as a result of a single application of a load of great magnitude or as a result of a cyclic application of load'.

How can cartilage, when loaded in compression, generate tensile stresses? The purpose of the work reported in this paper was to answer that question by calculating the distribution of stress within a layer bonded to a substrate in response to pressure applied to its free surface.

How should the cartilage be modelled for the purposes of stress analysis? Mow and his colleagues (Mow *et al.* 1980; Lai *et al.* 1990; Mow *et al.* 1990) have developed mixture theories which take explicit account of the fluid and solid constituents of cartilage. However, fluid flow takes time and for rapidly applied loads 'the instantaneous behavior of the layer of biphasic tissue under a Heaviside load is the same as that of an equivalent incompressible ($v_s = 0.5$) elastic layer' (Mow *et al.* 1990). The authors will therefore model cartilage under rapidly applied loads, such as during the heelstrike transient, as a layer of linear elastic material bonded to a rigid substrate and subjected to surface pressure applied over a finite area, Fig. 27.3(a). Such an approach was used by Hayes *et al.* (1972) to analyse the response of cartilage to indentation. Anisotropic nonhomogeneous elastic layers were analysed by Askew and Mow (1978) and more recently by Eberhardt *et al.* (1991) to explain age-dependent surface fibrillation and the development of vertical splits at the surface of cartilage. Freeman (1972) had suggested that these phenomena were the result of fatigue failures arising from alternating tensile stresses in the surface layer.

Before describing the calculations and results in detail, a qualitative argument from elementary strength of materials theory (Case and Chilver 1971) can be used to infer how tensile stresses can arise at the tide-mark. If cartilage is doing its job, load applied to the free surface over a finite area is spread over a larger area of the subchondral bone plate, Fig. 27.3(a). Directly under the load, the surface pressure is greater than the interface pressure. Outside the loaded region, the interface pressure is bigger than the surface pressure. In each case there is a vertical pressure gradient which would normally promote fluid flow within the cartilage but, if the loads are applied sufficiently quickly, there is no time for flow. A balance of forces in the vertical direction for any vertical slice of the layer, Fig. 27.3(b), then requires that there should be shear stresses acting on vertical faces. If there are shear stresses on vertical faces, there have to be complementary shear stresses on horizontal

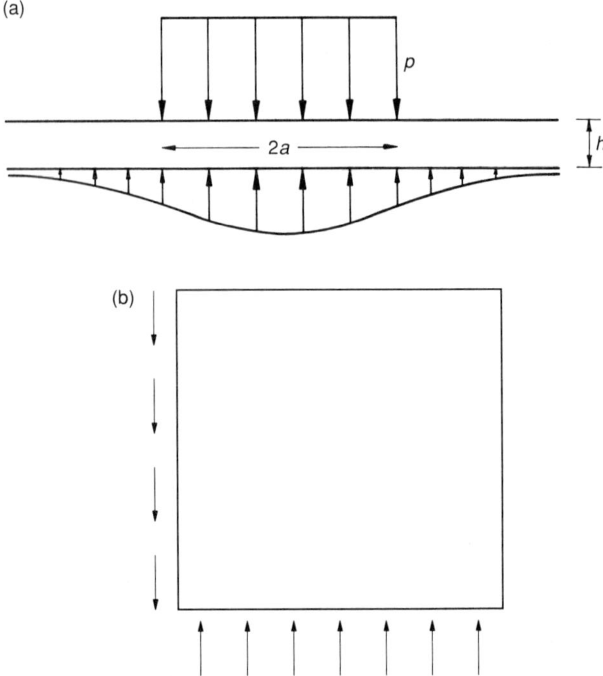

Fig 27.3 (a) A model of a cartilage layer with a uniform distribution of pressure applied to its upper surface. The cartilage spreads the load over a larger area of the subchondral bone at the interface with its lower surface. At any point under the load, the surface pressure is larger than the subchondral pressure. At any point outside the loaded region, the subchondral pressure is higher than the surface pressure. (b) Equilibrium of a block of cartilage. When there is a difference in pressure between the upper and lower surfaces, there have to be shear stresses on the vertical faces.

faces (Case and Chilver 1971). This argument differs slightly from that used by Armstrong (1986) who used an approximation valid for thin elastic layers to deduce the distribution of stress within rapidly loaded cartilage and concluded that 'the maximum shear stress is at the subchondral bone junction beneath the point of maximum pressure gradient' (measured *along* the articular surface).

Figure 27.4 shows a state of pure shear stress, in the upper diagram. The shear stresses on the upper left triangle pull it upwards and to the left and are balanced by a pure tensile stress on the 45° face. The shear stresses on the upper right triangle push it downward to the left and are balanced by a pure compressive stress on the 45° face. Thus, a state of pure shear stress is statically equivalent to a pure tensile stress on one 45° face and a pure compressive stress on the other. Wherever there are shear stresses, there will be tensile and

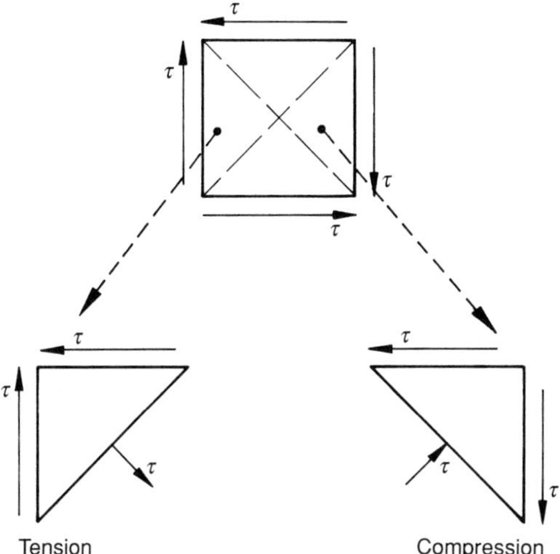

Fig 27.4 Pure shear stress on the sides of a square block. The shear stresses are statically equivalent to a pure tensile stress on one plane at 45 degrees to the sides of the block and a pure compressive stress on the other 45 degree plane.

compressive stresses on 45° planes. In general, the stresses on planes at 45° to the planes of maximum shear stress are called the *principal stresses*. They may both be tensile or compressive; one may be tensile and the other compressive, as in Fig. 27.4. The more tensile principal stress is called the *maximum* principal stress. The authors have reduced the problem to the question: are the shear stresses which arise in cartilage because of vertical pressure gradients large enough to overcome the compressive effects of the load and produce maximum principal stresses which are tensile? If there are tensile stresses, it may be possible to define the circumstances in which they are large enough to initiate and/or propagate Meachim cracks.

Method and materials

The authors considered the two-dimensional state of stress in an ideally elastic layer bonded to a rigid substrate and subjected to pressure loads with no friction applied over a strip of width $2a$ of its surface, the remainder of the surface being stress free, Fig. 27.3(a). The layer thickness is h. The aspect ratio, half the width of the contact strip to the thickness, is A. The bond at the interface is imposed by ensuring that horizontal and vertical displacements of the layer at the interface are zero.

The methods of the classical mathematical theory of elasticity were used to evaluate the stress distribution. The stresses from the appropriate form of the

Airy stress function were worked out (Sneddon 1951). The effects of different surface pressure distributions and of different aspect ratios were considered. Calculations for various values of Poisson ratio were carried out, approaching the value of 0.5 appropriate for cartilage under rapidly applied loads. Similar methods were used by Hayes *et al.* (1971), Askew and Mow (1978), and Eberhardt *et al.* (1991).

Results

Figure 27.5 shows the distributions of the three possible components of stress on the subchondral bone, vertical pressure, horizontal pressure, and shear stress, for a Hertz (elliptical) distribution of pressure, an aspect ratio of unity,

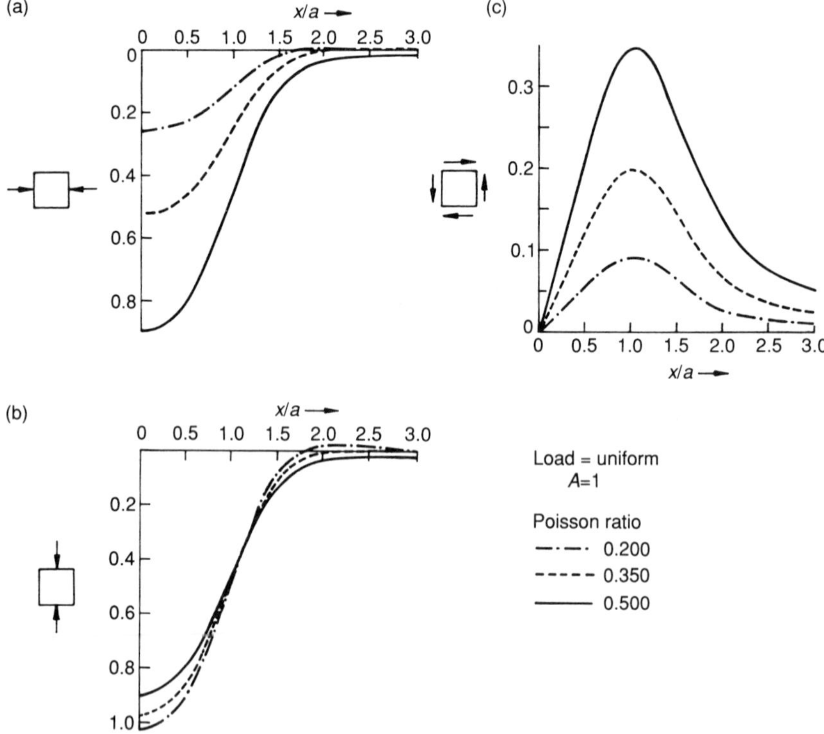

Fig 27.5 Distributions of stress along the cartilage/bone interface in response to a uniform distribution of pressure applied to the free surface over a strip $-1 < x/a < 1$ for an aspect ratio of unity and different values of Poisson's ratio. (a) Horizontal pressure, (b) vertical pressure, and (c) shear stress are plotted against distance along interface from centre of contact area. All stresses expressed as proportions of the applied uniform pressure.

and for different values of Poisson ratio. The shear stress acting on horizontal and vertical planes is largest near the edge of the contact area and largest of all for a Poisson ratio of 0.5, the value appropriate to rapid loading of cartilage, when the vertical and horizontal pressures are nearly equal. The shear stresses along the interface are then the maximum shear stresses at each point (Case and Chilver 1971) and the principal stresses act on planes at 45° to the interface. The distributions of shear stress shown in Fig. 27.5 are very similar to those described by Hayes *et al.* (1972).

Figure 27.6 shows contours of the maximum principal stress in a layer of aspect ratio 1.0 but with the vertical scale exaggerated to show the detail. On the contours 2–9, under the load, the maximum principal stress (and the minimum) is compressive. The maximum principal stress is tensile and the minimum principal stress is compressive outside contour 9, outside the embrace of the load. The tensile principal stress is largest under contour 10 at the interface, where Meachim found his cracks. The remaining results therefore describe effects at the interface.

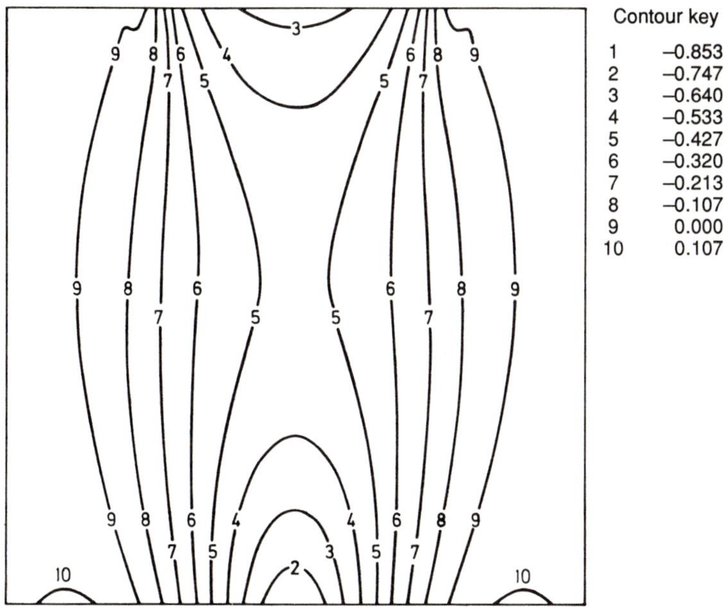

Contour key	
1	−0.853
2	−0.747
3	−0.640
4	−0.533
5	−0.427
6	−0.320
7	−0.213
8	−0.107
9	0.000
10	0.107

Fig 27.6 Contours of the algebraically maximum principal stress within a cartilage layer of aspect ratio unity in response to an Hertzian distribution of pressure applied to the surface. The vertical thickness of the layer has been multiplied by a factor of ten relative to the horizontal scale to show detail of the stress distribution. On contours 2–8, the maximum principal stress is compressive; outside contour 9, it is tensile. The largest tensile stresses are on the interface under contour 10. The stress values in the table are expressed as proportions of the average applied surface pressure.

In each of the following figures, the distribution of vertical pressure perpendicular to the interface and the distribution of the maximum principal stress are shown. All stresses are expressed as proportions of the average applied pressure. Compressive stresses are shown as negative, tensile stresses as positive. The region $-1<x/a<1$ lies under the load. All results are for a Poisson ratio of 0.45, close to the value appropriate to rapid loading of cartilage.

Figures 27.7 and 27.8 show the effects of load distribution, Fig. 27.7 for an aspect ratio of 4, Fig. 27.8 for an aspect ratio of 1, a smaller contact area or a thicker layer. The calculations were performed for distributions of pressure which were uniform, Hertzian and parabolic, for triangular and cosine squared distributions. Figure 27.7(a) shows that, for thin layers, large contact areas, the distribution of vertical pressure on the interface mirrors the surface distribution and that there is relatively little redistribution of load, except near the edge of the contact area, $x/a=\pm 1$. Similar effects were described by Hayes *et al.* (1972) and Askew and Mow (1978). Under the load, the general form of the distribution of the maximum principal stress, Fig. 27.7(b), also mirrors the applied load distribution. However, near the edge of the contact area and outside the embrace of the load, $x/a>1$, the value of the maximum principal stress goes positive, tensile, and its largest value is remarkably independent of the details of the way the load is applied. For an aspect ratio of unity and a thicker layer, Fig. 27.7, both the vertical pressure distribution on the interface and the distribution of maximum principal stress are very similar for all load distributions, the largest value of the tensile stress occurring outside the embrace of the load, $x/a>1$. The largest value is only slightly sensitive to load distribution both in magnitude and position. Further calculations show that the differences between different load distributions virtually disappear as layer thickness increases and contact area diminishes.

Figure 27.9 shows the effect of aspect ratio. As aspect ratio increases, the contact area on a layer of constant thickness grows. For small contact areas, thick layers, the pressure on the interface spreads out until, at an aspect ratio of 0.125, the pressure on the interface is almost uniform, Fig. 27.9(a). The corresponding maximum principal stress distribution is almost uniform, Fig. 27.9(b), going tensile only at some remove from the edge of the contact strip. For instance, calculations show that the maximum principal stress goes tensile only for $x/a>2.5$ for an aspect ratio of 0.25, and its largest value is only one twentieth the average applied pressure. For thin layers, large contact areas, the vertical pressure on the interface mirrors the surface distribution, as in Fig. 27.7(a). As layer thickness diminishes or contact areas grow, the site of the largest tensile stress moves in towards the edge of the contact area, Fig. 27.9(b), and the value grows and then diminishes. The peak value at aspect ratio 2.0 is bigger than the peaks at 4.0 and 1.0. The absolute maximum tension occurs for an aspect ratio of about 1.3 where it has the value of 0.13 times the average applied pressure.

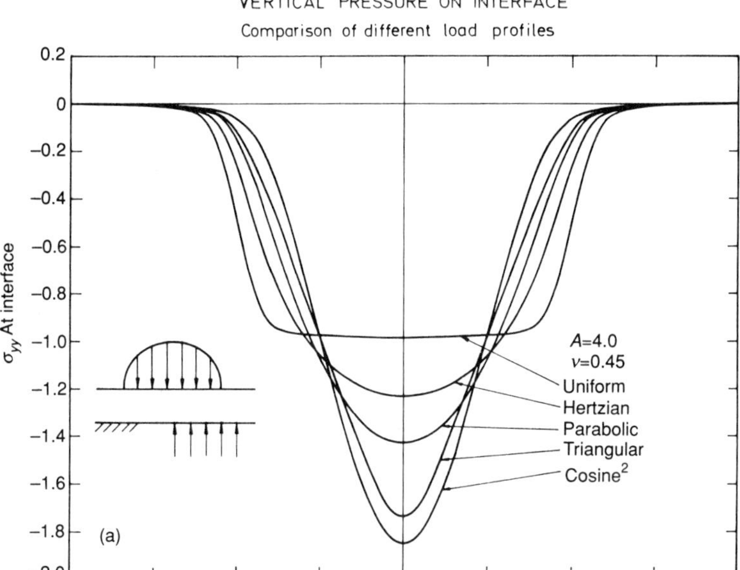

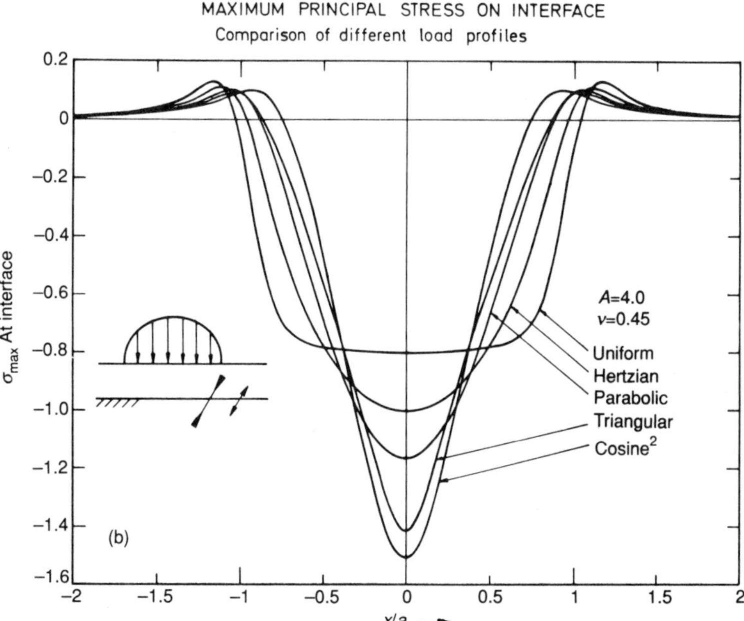

Fig 27.7 Distributions of (a) vertical pressure and (b) maximum principal stress on the interface plotted against distance along interface for an aspect ratio of four, Poisson ratio = 0.45 and different distributions of pressure applied to the free surface over a strip $-1 < x/a < 1$. All stresses are expressed as proportions of the average surface pressure. Compressive stresses are negative, tensile stresses positive.

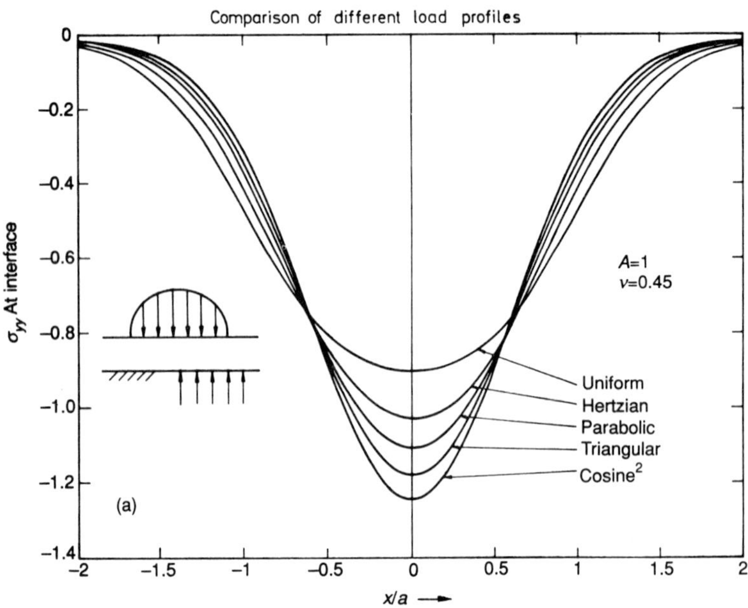

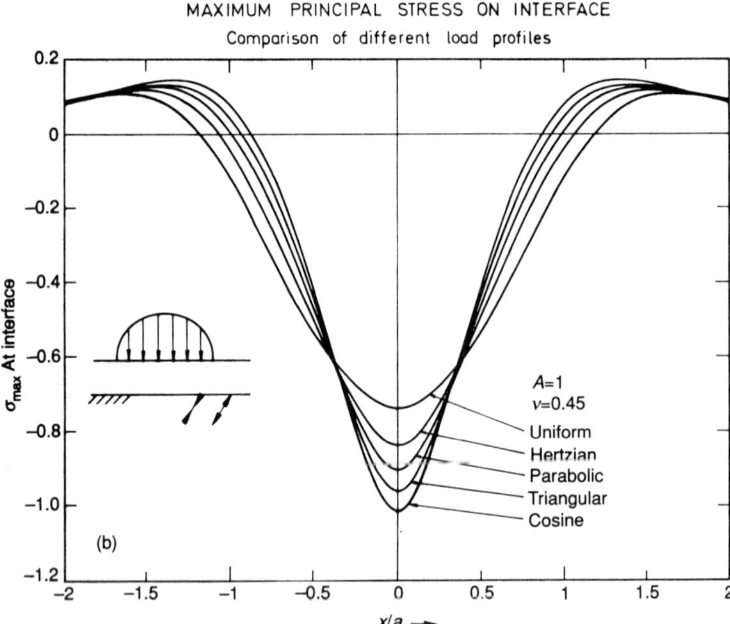

Fig 27.8 Distributions of (a) vertical pressure and (b) maximum principal stress on the interface plotted against distance along interface for an aspect ratio of unity, Poisson ratio = 0.45 and different distributions of pressure applied to the free surface over a strip $-1 < x/a < 1$. All stresses expressed as proportions of the average surface pressure. Compressive stresses are negative, tensile stresses positive.

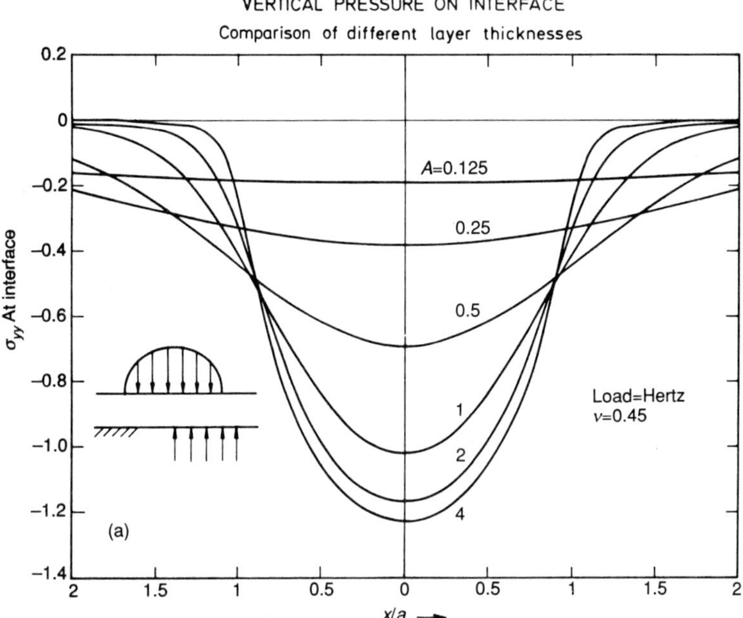

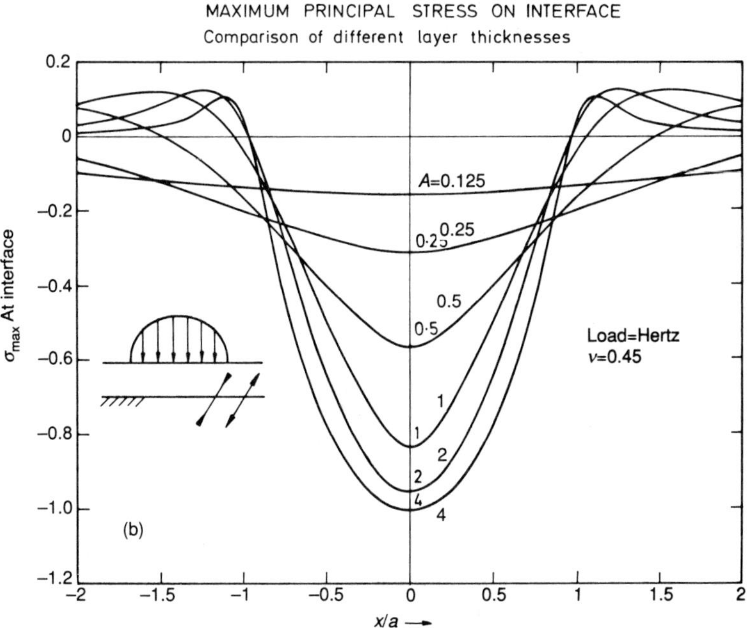

Fig 27.9 Distributions of (a) vertical pressure and (b) maximum principal stress on the interface plotted against distance along interface for a Hertzian distribution of pressure applied to the free surface over a strip $-1 < x/a < 1$, Poisson ratio $= 0.45$ and different values of aspect ratio. All stresses expressed as proportions of the average surface pressure. Compressive stresses are negative, tensile stresses positive.

Discussion

The stress distribution in cartilage is only one of several factors which, acting singly or together, could lead to its failure. The effect of the tensile stresses found in the present work may be exacerbated by weakening of the cartilage due to other factors.

Hypotheses regarding the onset of osteoarthrosis should explain the incidence of the disease in only a proportion of the population. The heelstrike transient is not a universal feature of gait, the loading rate varying over the population by a factor of nearly 4 (Radin and Paul 1991, Table 3).

While Hayes *et al.* (1972), Askew and Mow (1978), and Eberhardt *et al.* (1991) showed distributions of shear stress along the layer/substrate interface similar to those in Fig. 27.5 above, they did not calculate the principal stresses. Eberhardt *et al.*(1991) suggested that the shear stresses might be responsible for horizontal splits but the authors prefer the mechanism of Weightman and Kempson quoted above which requires tensile stress for failure of the collagen fibres in cartilage.

The authors' calculations show that tensile stresses are generated at the interface between an elastic layer and a rigid substrate when pressure loads are applied over part of the free surface of the layer. The distribution of tensile stress and its largest value, as a proportion of the average applied pressure, is not at all sensitive to the distribution of surface pressure and hence to the detailed geometry of the articular surfaces. These conclusions differ somewhat from Armstrong's (1986), whose approximate analysis suggested that the maximum shear stress on the interface would be largest under the point at which the surface pressure gradient was largest. The present more exact analysis shows that the maximum tensile stress on the interface is usually found outside the embrace of the surface loads, regardless of load distribution, and that its largest value occurs when the aspect ratio is about unity, circumstances not covered by Armstrong's thin layer approximation.

Since the tensile stress distribution is insenstive to the distribution of surface loads, the results may have application to articular cartilage in all joints (or to various models of joints) when loads are applied rapidly, as during the heelstrike transient. The tensile stresses could be responsible for the disruption of the collagen network near the cartilage/bone interface and for the initiation of Meachim cracks. Propagation of such cracks under repeated applications of the load could lead eventually to separation of a layer of cartilage from the bone, the onset of osteoarthrosis.

Stress levels needed to fracture articular cartilage in uniaxial tension have been found to be smaller in the deeper layers of cartilage and to diminish systematically with increasing age, with reported values in the range 0–30 MPa (Kempson 1979). Although the level of tensile stress found in this investigation is quite small, at most about 0.15 times the average applied pressure, impulsive loads could initiate and propagate cracks at the cartilage/bone interface in cartilage at the lower end of Kempson's strength range.

As suggested in Fig. 27.9(b), the maximum tensile stress at the interface as a proportion of the applied pressure increases and then diminishes as the total load and the contact area grows, the position of maximum tension moving outward ahead of the advancing contact boundary. As the contact boundary spreads, a crack could find itself initially in a tensile, crack-propagating, field and then in a compressive, crack-arresting, field. The methods of elastic fracture mechanics may allow more precise investigation of the circumstances in which such cracks could lead to cartilage loss in a region of habitual contact.

Conclusion

Pressure loads applied rapidly to cartilage generate tensile stresses at the bone/cartilage interface. The largest tensile stress occurs just outside the embrace of the loaded region, its value as a proportion of the applied pressure depends on the aspect ratio but is not very sensitive to the distribution of applied pressure. It is conceivable that, if a crack is initiated during one application of the load, it could be propagated a little by each subsequent application of the load. This mechanism would explain Meachim's cracks and why repeated application of the heelstrike transient could lead to anteromedial osteoarthritis of the knee.

Acknowledgements

The work reported here was supported by a grant from the Arthritis and Rheumatism Council. Revealing discussions with Mr Goodfellow and Dr Radin are gratefully acknowledged.

Debate in the meeting

The meeting wondered whether, since the turnover of collagen fibres in cartilage is so slow with the enormous number of loading cycles, fatigue failure during human lifetime was inevitable. The author (J.J.O.C.) disagreed, pointing out that the site of the failures would lead one to deduce that it is rapid loading that does the damage. If this is so, maybe one should change the regime of loading, possibly by giving up running or wearing soft shoes, and so avoid the peak load.

Why do some develop osteoarthritis and others not? The author suggested that there may be a bimodal distribution of the way people walk. About one-third of the population slam their feet to the ground while 'the rest of us are slobs; and the slobs win'.

In answer to concern about stress waves the author pointed out that with the speed of sound in cartilage of the order of 3000 m/s, loading rates during normal human motion are relatively very slow and a static analysis is quite adequate.

References to Section 4

Ahmed, A. M. and Burke, D. L. (1983). *In-vitro* measurement of static pressure distribution in synovial joints. Part I: tibial surface of the knee. *Journal of Biomechanical Engineering*, **105**, 226–36.

Akizuki, S., Mow, V. C., Muller, F., Pita, J. C., Howell, D. S., and Manicourt D. H. (1987). Tensile properties of human knee joint cartilage. II. Correlations between weight bearing and tissue pathology and the kinetics of swelling. *Journal of Orthopaedic Research*, **5**, 173–86.

Altman, R. D., Tenebaum, J., Latta, L., Riskin, W., Blanco, L. N., and Howell, D. S. (1984). Biomechanical and biochemical properties of dog cartilage in experimentally induced osteoarthritis. *Annals of the Rheumatic Diseases*, **43**, 83–90.

Amador, E. and Urban, J. (1972). Simplified serum phosphorus analyses by continuous-flow spectrophotometry. *Clinical Chemistry*, **18**, 601–4.

Andriacchi, T. P., Galante, J. O., and Fermier, R. W. (1982). The influence of total knee replacement design on walking and stair climbing. *Journal of Bone and Joint Surgery*, **64A**, 1328–35.

Armstrong, C. G. (1986). An analysis of the stresses in a thin layer of articular cartilage in a synovial joint. *Engineering in Medicine*, **15**, 55–61.

Armstrong, C. G. and Mow, V. C. (1982). Variations in the intrinsic mechanical properties of human articular cartilage with age, degeneration, and water content. *Journal of Bone and Joint Surgery*, **64A**, 88–94.

Armstrong, C. G., Bahrani, A. S., and Gardner, D. L. (1980). Changes in the deformational behaviour of human hip cartilage with age. *Journal of Biomechanical Engineering*, **102**, 214–20.

Arokoski, J., Kiviranta, I., Jurvelin, J., Tammi, M., and Helminen, H. J. (1992). The effects of long-term running (40 km/day) on glycosaminoglycan content and thickness of articular cartilage in young canine knee and shoulder joints. *14th Symposium of ESOA*, 24–27 May 1992. Noordwijkerhout, The Netherlands.

Ascenzi, A. and Benvenuti, A. (1986). Orientation of collagen fibers at the boundary between two successive osteonic lamellae and its mechanical interpretation. *Journal of Biomechanics*, **19**, 455–63.

Ascenzi, A. and Bonucci, E. (1967). The tensile properties of single osteons. *Anatomical Record*, **158**, 375–86.

Ascenzi, A. and Bonucci, E. (1968). The compressive properties of single osteons. *Anatomical Record*, **161**, 377–92.

Ascenzi, A. and Bonucci, E. (1972). The shearing properties of single osteons. *Anatomical Record*, **172**, 499–510.

Ascenzi, A. and Bonucci, E.(1976). Relationship between ultra-structure and 'pin test' in osteons. *Clinical Orthopaedics and Related Research*, **121**, 275–94.

Ascenzi, A., Bonucci, E., and Checcucci, A. (1966). The tensile properties of single osteons studied using a microwave extensimeter. In *Studies of the anatomy and bone and joints* (ed. F. G. Evans), pp. 121–41. Springer-Verlag, Berlin.

Ascenzi, A., Bonucci, E., Ripamonti, A., and Roveri, N. (1978). X-ray diffraction and electron microscope study of osteons. *Calcified Tissue International*, **25**, 133–43.

Ascenzi, A., Benvenuti, A., Mango, F., and Simili, R. (1985). Mechanical hysteresis loops from single osteons: Technical devices and preliminary results. *Journal of Biomechanics*, **18**, 391–8.

Ascenzi, A., Baschieri, P., and Benvenuti, A. (1990). The bending properties of single osteons. *Journal of Biomechanics*, **23**, 763–71.

Askew, M. J. and Mow, V. C. (1978). The biomechanical function of the collagen fibril ultrastructure of articular cartilage. *Journal of Biomechanical Engineering*, **100**, 105–15.

Athanasiou, K. A., Rosenwasser, M. P., Buckwalter, J. A., Malinin, T. I., and Mow, V. C. (1991). Interspecies comparisons of *in-situ* intrinsic mechanical properties of distal femoral. *Journal of Orthopaedic Research*, **9**, 330–40.

Benninghoff, von (1927). Ueber die Anpassung der Knochenkompakta an geanderte Beanspruchungen. *Anatomischer Anzeiger*, **63**, 289–99.

Biewener, A. A., Swartz, S. M., and Bertram, J. E. A. (1986). Bone modeling during growth. Dynamic strain equilibrium in the chick tibiotarsus. *Calcified Tissue International*, **39**, 390–5.

Bjelle, A. (1975). Content and composition of glycosaminoglycans in human knee joint cartilage. Variation with site and age in adults. *Connective Tissue Research*, **3**, 141–7.

Blaimont, P. (1968). Contribution à l'étude biomécanique du fémur humain. *Acta Orthopadica Belgica*, **34**, 665–844.

Boyde, A. and Riggs, C. M. (1990). The quantitative study of the orientation of collagen in compact bone slices. *Bone*, **11**, 35–9.

Boyde, A., Bianco, P., Portigliatti Barbos, M., and Ascenzi, A. (1984). Collagen orientation in compact bone: I. A new method for the determination of the proportion of collagen parallel to the plane of compact bone sections. *Metabolic Bone Disease and Related Research*, **5**, 299–307.

Bray, D. (1970). Surface movements during the growth of single explanted neurons. *Proceedings of the National Academy of Sciences*, **65**, 905–10.

Brown, D. (1981). Contact stress distribution measurements in the human hip joint. *Proceedings of the Biomechanical Symposium, ASME*, ADM43.

Brown, R. A., Pedowitz, R. A., Kwan, M. K., Rydevik, B. L., Lee, S. U., Massie, J. *et al.* (1989). The effects of acute stretch upon conduction properties in the rabbit tibial nerve, *35th Annual Meeting, Orthopaedic Research Society*, 6–9 February 1989. Las Vegas, Nevada.

Brown, T. D. and Shaw, D. T. (1984) *In-vitro* contact stress distribution on the femoral codyles. *Journal of Orthopadic Research*, **2**, 190–9.

Buff, H.-U., Jones, L. C., and Hungerford, D. S. (1988). Experimental determination of forces transmitted through the patello-femoral joint. *Journal of Biomechanics*, **21**, 17–23.

Bullough, P. G., Goodfellow, J., and O'Connor, J. J. (1973). The relationship between degenerative changes and load-bearing in the human hip. *Journal of Bone and Joint Surgery*, **55B**, 746–58.

Burkhardt (1929). Ueber den Aufbau der menschlichen Osteone. *Verhandlungen der Anatomischen Gesellschaft*, **38**, 97–102.

Burr, D. B. and Martin, R. B. (1989). Errors in bone remodelling: toward a unified theory of metabolic bone disease. *American Journal of Anatomy*, **186**, 186–216.

Burstein, A. H., Zika, J. M., Heiple, K. G, and Klein, L. (1975). Contribution of collagen and mineral to the elastic–plastic properties of bone. *Journal of Bone and Joint Surgery*, **57A**, 956–61.

Carando, S., Portigliatti Barbos, M., Ascenzi, A., and Boyde, A. (1989). Orientation of collagen in human tibial and fibular shaft and possible correlation with mechanical properties. *Bone*, **10**, 139–42.

Carter, D. R., Smith, D. J., Spengler, D. M., Daly, C. H., and Frankel, V.H. (1980). Measurement and analysis of *in-vivo* bone strains on the canine radius and ulna. *Journal of Biomechanics*, **13**, 27–38.

Case, J. and Chilver, A. H. (1971). *Strength of materials and structures*. Edward Arnold, London.

Cochran, G. and Van B. (1972). Implantation of strain gauges on bone *in vivo*. *Journal of Biomechanics*, **5**, 119–23.

Collins, J. J. and O'Connor, J. J. (1991). Muscle-ligament interactions at the knee during walking. *Proceedings of the Institution of Mechanical Engineers Part H, Journal of Engineering in Medicine*, **205**, 11–18.

Currey, J. D. and Butler, G. (1975). The mechanical properties of bone tissue in children. *Journal of Bone and Joint Surgery*, **57A**, 810–14.

Currey, J. D. (1975). The effects of strain rate, reconstruction and mineral content on some mechanical properties of bovine bone. *Journal of Biomechanics*, **8**, 81–6.

Currey, J. D. (1988*a*). The effect of porosity and mineral content on the Young's modulus of elasticity of compact bone. *Journal of Biomechanics*, **21**, 131–9.

Currey, J. D. (1988*b*). Strain rate and mineral content in fracture models of bone. *Journal of Orthopaedic Research*, **6**, 32–8.

Daly, J. A. and Ertinghausen, G. (1972). Direct method for determining inorganic phosphate in serum with the centrifichem. *Clinical Chemistry*, **18**, 263–5.

Dambacher, M. A. and Rüegsegger, P. (1985). Nichtinvasive Untersuchungsmethoden bei Osteoporosen. *Therapeutische Umschau*, **42**, 6, 339–50.

Dixon, W. J. (ed) (1985). *BMDP Statistical Software*. University of California Press, Los Angeles, 1985.

Eanes, E. D., Termine, J. D., and Nylen, M.U. (1973). An electron microscopic study of the formation of amorphous calcium phosphate and its transformation to crystalline apatite. *Calcified Tissue Research*, **12**, 143–58.

Eberhardt, A. W., Lewis, J. L., and Keer, L. M. (1991). Normal contact of elastic spheres with two elastic layers as a model of joint articulation. *Journal of Biomechanical Engineering*, **113**, 410–17.

Ekeland, A., Engesæter, L. B., and Langeland, N. (1981). Mechanical properties of fractured and intact rat femora evaluated by bending, torsional and tensile tests. *Acta Orthopaedica Scandinavica*, **52**, 605–13.

Ekeland, A., Engesæter, L., and Langeland, N. (1982). Influence of age on mechanical properties of healing fractures and intact bones in rats. *Acta Orthopaedica Scandinavica*, **53**, 527–34.

Ekeland, A. and Underdal, T. (1983). Effects of salmon calcitonin on synthesis and mineralisation of collagen in rats. *Acta Orthopaedica Scandinavica*, **54**, 470–8.

Engel, W. K., Brook, M. H., and Nelson, P. G. (1966). Histological studies of denervated or tenotomized cat muscle. *Annals of the NY Academy of Sciences*, **138**, 160–85.

Engesæter, L. B., Ekeland, A., and Langeland. N. (1978). Methods for testing the mechanical properties of the rat femur. *Acta Orthopaedica Scandinavica*, **49**, 512–18.

Evans, F. G. (1953). Methods for studying the biomechanical significance of bone form. *American Journal of Physical Anthropology*, **11**, 413–35.

Evans, F. G. (1974). The mechanical properties of human cortical bone. *Accademia Nazionale dei Lincei*, **208**, 3–28.

Evans, F. G. and Lissner, H. R. (1948). 'Stresscoat' deformation studies of the femur under static vertical loading. *Anatomical Record*, **100**, 159–90.

Evans, F. G. and Vincentelli, R. (1974). Relations of the compressive properties of human cortical bone to histological structure and calcification. *Journal of Biomechanics*, **7**, 1–10.

Firschein, H. E. (1969). Collagen and mineral dynamics in bone. *Clinical Orthopaedics and Related Research*, **66**, 212–25.

Frasca, P., Harper, R. A., and Katz, J. L. (1977). Collagen fiber orientation in human secondary osteons. *Acta Anatomica*, **98**, 1–13.

Freeman, M. A. R. (1972). The pathogenesis of osteoarthrosis: an hypothesis. In *Modern trends in orthopaedics—6* (ed. A. G. Apley), p. 40. Butterworths, London.

Galardi, G., Comi, G., Lozza, L., Marchettini, P., Novarina, M., Facchini, R. *et al.* (1990). Peripheral nerve damage during limb lengthening. *Journal of Bone and Joint Surgery*, **72B**, 121–4.

Gershuni, D. H. and Kuei, S. C. (1984). Articular cartilage deformation following experimental synovitis in the rabbit hip. *Journal of Orthopadic Research*, **1**, 313–18.

Giraud-Guille, M. M. (1988). Twisted plywood architecture of collagen fibrils in human compact bone osteons. *Calcified Tissue International*, **42**, 167–80.

Gitelman, H. J. (1967). An improved automated procedure for the determination of calcium in biochemical specimen. *Analytical Biochemistry*, **18**, 521–31.

Goodfellow, J. and O'Connor, J. (1992). The anterior cruciate ligament in knee arthroplasty: a risk factor in unconstrained meniscal replacement. *Clinical Orthopaedics and Related Research*, **276**, 245–52.

Goodfellow, J., White, S. H., and Ludkowski, P. (1991). Anteromedial osteoarthritis of the knee. *Journal of Bone and Joint Surgery*, **73B**, 582–6.

Grynpas, M, D, Bonar, L. C., and Glimeher, M. J. (1984). Failure to detect an amorphous calcium-phosphate solid phase in bone mineral: A radial distribution function study. *Calcified Tissue International*, **36**, 291–301.

Haftek, J. (1970). Stretch injury of peripheral nerve. *Journal of Bone and Joint Surgery*, **52B**, 354–65.

Hayes, W. C., Keer, L. M., Herrmann, G., and Mockros, L. F. (1972). A mathematical analysis for indentation tests of articular cartilage. *Journal of Biomechanics*, **5**, 541–51.

Highet, W. B. and Holmes, W. (1943). Traction injuries to the lateral popliteal nerve and traction injuries to the peripheral nerves after suture. *British Journal of Surgery*, **30**, 212–33.

Highet, W. B. and Sanders, F. K. (1943). The effects of stretching nerves after suture. *British Journal of Surgery*, **30**, 355–69.

Hoch, D. H., Grodzinsky, A. J., Koob, T. J., Albert, M. L., and Eyre, D. R. (1983). Early changes in the material properties of rabbit articular cartilage after meniscectomy. *Journal of Orthopaedic Research*, **1**, 4–12.

Holmdahl, D. E. and Ingelmark, B. E. (1948). Der Bau des Gelenkknorpels unter verschiedenen funktionellen Verhaltnissen. Experimentelle Untersuchung an wachsenden Kaninchen. *Acta Anatomica (Basel)*, **6**, 309–75.

Hori, R. Y. and Mockros, L. F. (1976). Indentation tests of human articular cartilage. *Journal of Biomechanics*, **9**, 259–68.

Husby, O. S., Indrekvam, K., Gjerdet, N. R., Engesæter, L. B., and Langeland, N. (1988). *In vivo* strain measurements in rat femora. *Scandinavian Journal of Laboratory Animal Science*, **15**, 159–70.

Ilizarov, G. A. (1989). The tension-stress effect on the genesis and growth of tissues. *Clinical Orthopaedics and Related Research*, **238**, 249–81.

Jonsson, U., Netz, P., and Stromberg, L. (1984). Solid mechanics and strength of bone in young dogs. *Acta Orthopaedica Scandinavica*, **55**, 446–51.

Jurvelin, J., Kiviranta, I., Tammi, M., and Helminen, H. J. (1986*a*). Softening of canine articular cartilage after immobilization of the knee joint. *Clinical Orthopaedics and Related Research*, **207**, 246–52.

Jurvelin, J., Kiviranta, I., Tamnu, M., and Helminen, H. J. (1986*b*). Effect of physical exercise on indentation stiffness of articular cartilage in the canine knee. *International Journal of Sports Medicine*, **7**, 106–10.

Jurvelin, J., Kiviranta, I., Saamanen, A-M., Tammi, M., and Helminen, H. J. (1990). Indentation stiffness of young canine knee articular cartilage—Influence of strenuous joint loading. *Journal of Biomechanics*, **23**, 1939–46.

Jurvelin, J., Kiviranta, I., Arokoski, J., Tammi, M., and Helminen, H. J. (1987). Indentation study of the biomechanical properties of articular cartilage in the canine knee. *Engineering in Medicine*, **16**, 15–22.

Jurvelin, J., Säämänen, A-M., Arokoski, J., Helminen, H., Kiviranta, I., and Tammi, M. (1988). Biomechanical properties of the canine knee articular cartilage as related to matrix proteoglycans and collagen. *Engineering in Medicine*, **17**, 157–62.

Kawamura, B., Hosono, S., Takahashi, T., Yano, T., Kobayashi, Y., Shibata, N. *et al.* (1968). Limb lengthening by means of subcutaneous osteotomy. *Journal of Bone and Joint Surgery*, **50A**, 851–77.

Keller, T. S. and Spengler, D. M. (1982). *In vivo* strain gauge implantation in rats. *Journal of Biomechanics*, **15**, 911–17.

Keller, T. S., Spengler, D. M., and Garter, D. R. (1986). Geometric, elastic, and structural properties of maturing rat femora. *Journal of Orthopaedic Research*, **4**, 57–67.

Keller, T. S. and Spengler, D. M. (1989). Regulation of bone stress and strain in the immature and mature rat femur. *Journal of Biomechanics*, **22**, 1115–27.

Kempson, G. E. (1971). The determination of a creep modulus for articular cartilage from indentation tests on the human femoral head. *Journal of Biomechanics*, **4**, 239–50.

Kempson, G. (1979). Mechanical properties of articular cartilage. In *Adult Articular Cartilage* (2nd edn) (ed. M. A. R. Freeman), pp. 333–414. Pitman Medical, London.

Kempson, G. E., Muir, H., Swanson, S. A. V., and Freeman. M. A. R. (1970). Correlations between stiffness and the chemical constituents of cartilage on the human femoral head. *Biochimica et Biophysica Acta*, **215**, 70–7.

Kettelkamp, D. B. and Jacobs, A. W. (1972). Tibiofemoral contact area—Determination and implications. *Journal of Bone and Joint Surgery*, **54A**, 349–56.

Kiviranta, I., Jurvelin, J., Tammi, M., Säämänen, A-M., and Helminen, H. J. (1985). Microspectrophotometric quantitation of glycosaminoglycans in articular cartilage sections stained with Safranin O. *Histochemistry*, **82**, 249–55.

Kiviranta, I. (1987). *Joint loading influences on the articular cartilage of young dogs—quantitative histochemical studies on matrix carbohydrates.* Publications of the University of Kuopio, Medicine, Original reports.

Krause, W.-D. (1969). *Mikroskopishe Untersuchungen am Gelenkknorpel extrem funktionell belasteter Mause.* Thesis, Kolln, Gouder & Hansen.

Küntscher, G. (1935*a*). Die Bedeutung der Darstellung des Kraftflusses im Knochen für die Chirurgie. *Klinicheskaia Khirurgiia*, **182**, 489–551.

Küntscher, G. (1935*b*). Ueber Nachweis von Spannungsspitzen am menschlichen Knochengerüst. *Jahrbuch für Morphologie und mikroskopische Anatomie*, **74**, 427–44.

Küntscher, G. (1934). Die Darstellung des Kraftflusses im Knochen. *Zentralblatt für Chirurgie*, **61**, 2130–6.

Kwan, M. K., Rydevik, B. L., Myers, R. R., Garfin, S. R., Triggs, K., and Woo, S. L.-Y. (1988). Stretch injury of rabbit peripheral nerve; a biomechanical and histological study. *Orthopaedic Transactions*, **12**, 493–4.

Lai, W. M., Hou, J. S., and Mow, V. C. (1990). A triphasic theory for the swelling properties of hydrated charged soft biological tissues. In *Biomechanics of diarthrodial joints*, Vol. I (ed. V. C. Mow, A. Ratcliffe, and S. L-Y. Woo), pp. 283–312. Springer-Verlag, New York.

Lampert, K. L. (1971). The weight-bearing function of the fibula: a strain gauge study. *Journal of Bone and Joint Surgery*, **53A**, 507–13.

Lane, J. and Ralis, Z. A. (1983). Changes in dimensions of large cacellous bone specimens during histological preparation as measured on slabs from human femoral heads. *Calcified Tissue International*, **35**, 1–4.

Lane, J. M., Chisena, E., and Black, J. (1979). Experimental knee instability: early mechanical property changes in articular cartilage in a rabbit model. *Clinical Orthopaedics and Related Research*, **140**, 262–5.

Langenskiold, A., Michelsson, J. E., and Videman, T. (1979). Osteoarthritis of the knee in the rabbit produced by immobilization. *Acta Orthopaedica Scandinavica*, **50**, 1–14.

Lanier, R. R. (1946). Effects of exercise on knee-joints of inbred mice. *Anatomical Record*, **94**, 311–21.

Lanyon, L. E. (1987). Functional strain in bone tissue as an objective and controlling stimulus for adaptive bone remodelling. *Journal of Biomechanics*, **20**, 11–12, 1083–93.

Lanyon, L. E. and Smith, R. N. (1969). Measurements of bone strain in the walking animal. *Research in Veterinary Science*, **10**, 93–4.

Lanyon, L. E., Magee, P. T., and Baggott, D. G. (1979). The relationship of functional stress and strain to the processes of bone remodelling. An experimental study on the sheep radius. *Journal of Biomechanics*, **12**, 593–600.

Lees, S. and Davidson, C. L. (1977). The role of collagen in the elastic properties of calcified tissues. *Journal of Biomechanics*, **10**, 473–86.

Lundborg, G. and Rydevik, B. (1973). Effects of stretching the tibial nerve of the rabbit. *Journal of Bone and Joint Surgery*, **55B**, 390–401.

Manicourt, D. H. and Pita, J. C. (1988). Quantificaction and characterization of hyaluronic acid in different topographical areas of normal articular cartilafe from dogs. *Collagen and Related Research*, **8**, 39–47.

Mak, A. F., Lai, W. M., and Mow, V. C. (1987). Biphasic indentation of articular cartilage—I. Theoretical analysis. *Journal of Biomechanics*, **20**, 703–14.

Marotti, G. and Muglia, M. A. (1988). A scanning electron microscope study of human bony lamellae. Proposal for a new model of collagen lamellar organization. *Archivo Italiano oi Anatmia e oi Embriologia*, **93**, 163–75.

Maroudas, A. (1976). Balance between swelling pressure and collagen tension in normal and degenerate cartilage. *Nature*, **260**, 808–9.

Meachim, G. (1972). Articular cartilage lesions in osteoarthritis of the femoral head. *Journal of Pathology*, **107**, 199–210.

Meachim, G. and Bentley, G. (1980). Horizontal splitting in patellar articular cartilage. *Arthritis and Rheumatic Diseases*, **39**, 514–23.

Mezhenina, E. P., Roulla, E. A., Pecherskỳ, A. G., Babich, V. D., Shadina, E. L, and Mizhevic T. V. (1984). Methods of limb elongation with congenital inequality in children. *Journal of Paediatric Orthopaedics*, **4**, 201–7.

Milner, R. H. (1989). The effect of tissue expansion on peripheral nerves. *British Journal of Plastic Surgery*, **42**, 414–21.

Minns, R. J., Campbell, J. and Bremble, G. R. (1975). The bending stiffness of the human tibia. *Calcified Tissue Research*, **17**, 165–8.

Mitchell, S. W. (1872). *Injuries of nerves and their consequences*. J. B. Lippincott and Co., Philadelphia, Pennsylvania.

Mizrahi, J., Maroudas, A., Lanir, Y., Ziv, I., and Webber, T. J. (1986). The 'instantaneous' deformation of cartilage: Effects of collagen fiber orientation and osmotic stress. *Biorheology*, **23**, 311–30.

Monticelli, G. and Spinelli, R. (1981). Distraction epiphysiolysis as a method of limb lengthening. *Clinical Orthopaedics and Related Research*, **154**, 284–315.

Mow, V. C., Kuei, S. C., Lai, W. M., and Armstrong, C. G. (1980). Biphasic creep and stress relaxation of articular cartilage in compression: theory and experiment. *Journal of Biomechanical Engineering*, **102**, 73–84.

Mow, V. C., Hou, J. S., Owens, J. M., and Ratcliffe, A. (1990). Biphasic and quasilinear viscoelastic theories for hydrated soft tissues. In *Biomechanics of diarthrodial joints*, Vol. I (ed. V. C. Mow, A. Ratcliffe, and S. L.-Y. Woo), pp. 215–60. Springer-Verlag, New York.

Mow, V. C., Gibbs, M. C., Lai, W. M., Zhu, W. B., and Athanasiou, K. A. (1989). Biphasic indentation of articular cartilage—II. A numerical algorithm and an experimental study. *Journal of Biomechanics*, **22**, 853–61.

Myers, E. R., Zhu, W., and Mow, V. C. (1988). Viscoleastic properties of articular cartilage and meniscus. In *Collagen*, Vol II: *Biochemistry and biomechanics* (ed. M. E. Nimni), pp. 268–288. CRC Press, Boca Raton, Florida.

O'Connor, J., Shercliff, T., FitzPatrick, D., Bradley, J., Daniel, D., Biden, E., *et al.* (1990). Geometry of the knee. In *Knee ligaments: structure function injury and repair* (ed D. M. Daniel, W. H. Akeson, and J. J. O'Connor), pp. 163–200. Raven Press, New York.

O'Connor, J. (1992). Can muscle cocontraction protect knee ligaments after injury or repair? *Journal of Bone and Joint Surgery*, **74B**, in press.

Paley, D. (1988). Current techniques of limb lengthening. *Journal of Paediatric Orthopaedics*, **8**, 73–92.

Palmoski, M., Perricone, E., and Brandt, K. D. (1979). Developmentand reversal of a proteoglycan aggregation defect in normal canine knee cartilage after immobilization. *Arthritis and Rheumatism*, **22**, 508–17.

Palmoski, M. J., Colyer, R. A., and Brandt, K. D. (1980). Joint motion in the absence of normal loading does not maintain normal articular cartilage. *Arthritis and Rheumatism*, **23**, 325–34.

Parsons, J. R. and Black, J. (1977). The viscoelastic shear behavior of normal rabbit articular cartilage. *Journal of Biomechanics*, **10**, 21–9.

Parsons, J. R. and Black J (1987). Mechanical behavior of articular cartilage. Quantitative changes with enzymatic alteration of the proteoglycan fraction. *Bulletin of the Hospital for Joint Diseases Orthopaedic Institute*, **47**, 13–30.

Paul, J. P. (1966). The biomechanics of the hip-joint and its clinical relevance. *Proceedings of the Royal Society of Medicine*, **59**, 943–8.

Paul, J. P. (1971). Load actions on the human femur in walking and some resultant stresses. *Experimental Mechanics*, **11**, 121–5.

Pauwels, F. (1954). Die statische Bedeutung der Linea aspera. Vierter Beitrag zur funktionellen Anatomie und Kausalen Morphologie des Stützapparates. *Zeitschrift für Anatomie und Entwicklungs-Geschichte*, **117**, 497–503.

Pauwels, F. (1968). Beitrag zur funktionellen Anpassung der Corticalis der Röhrenknochen. Untersuchung an drei rachitisch deformierten Femora. 12. Beitrag zur funktionellen Anatomie und kausalen Morphologie des Stützapparates. *Zeitschrift für Anatomie und Entwicklungs-Geschichte*, **127**, 121–37.

Pauwels, F. (1979). Biomécanique de l'appareil moteur. *Contributions à l'étude de l'anatomie fonctionnelle*. Springer-Verlag, Berlin, Heidelberg, New York.

Portigliatti Barbos, M., Bianco, P., and Ascenzi, A. (1983). Distribution of osteonic and interstitial components in the human femoral shaft with reference to structure, calcification and mechanical properties. *Acta Anatomica*, **115**, 178–86.

Portigliatti Barbos, M., Bianco, P., Ascenzi, A., and Boyde, A. (1984). Collagen orientation in compact bone: II. Distribution of lamellae in the whole of the human femoral shaft with reference to its mechanical properties. *Metabolic Bone Disease and Related Research*, **5**, 309–15.

Portigliatti Barbos, M., Carando, S., Ascenzi, A., and Boyde, A. (1987). On the structural and biomechanical symmetry of human femurs. *Bone*, **8**, 165–9.

Radin, E. L. and Paul, I. L. (1971). Response of joints to impact loading: *in-vitro* wear. *Arthritis and Rheumatism*, **14**, 356–62.

Radin, E. L., Martin, B. M., Burr, D. B., Caterson, B., Boyd, R. D., and Goodwin, C. (1984). Effects of mechanical loading on the tissues of the rabbit knee. *Journal of Orthopaedic Research*, **2**, 221–34.

Radin, E. L., Whittle, M. W., Yang, K. H., Jefferson, R., Rodgers, M. M., Kish, V. L., *et al.* (1986). The heelstrike transient, its relationship with the angular velocity of the shank and the effects of quadriceps paralysis. In *Advances in bioengineering* (ed. S. A. Lantz and A. I. King), pp. 121–3. American Society of Mechanical Engineers, New York.

Radin, E. L. (1990). Factors influencing the progression of osteoarthrosis. In *Articular cartilage and joint function: basic science and arthroscopy* (ed. J. W. Ewing), pp. 301–9. Raven Press, New York.

Radin, E. L., Yang, K. H., Riegger, C., Kish, V. L., and O'Connor, J. J. (1991). Relationship between lower limb dynamics and knee joint pain. *Journal of Orthopaedic Research*, **9**, 398–405.

Radin, E. L., Ehrlich, M. G., Chemak, R., Abernethy, P., Paul, I. L., and Rose, R. M. (1978). Effect of repetitive impulsive loading on the knee joints of rabbits. *Clinical Orthopaedics and Related Research*, **131**, 288–93.

Riggs, C. M. (1990). *The relationship between mechanical function and microstructural properties of cortical bone in the race horse*. Ph.D. Thesis, London.

Roark, R. J. and Young, W. C. (1976). *Formulas for Stress and Strain* (5th edn), pp. 61–9, 89–112. McGraw-Hill Book Company, Singapore.

Rowe, R. W. D. (1974). Collagen fibre arrangement in intramuscular connective tissue. Changes associated with muscle shortneing and their possible relevance to raw meat toughness measurements. *Journal of Food Technology*, **9**, 501–9.

Ruff, C. B. (1981). *Structural changes in the lower limb bones with aging at Pecos Pueblo*. Ph.D. Thesis, Univ. Pennsylvania.

Ruff, C. B. and Hayes, W. C. (1983). Cross-sectional geometry of Pecos Pueblo femora and tibiae—A biomechanical investigation: I. Method and general patterns of variation. *Journal of Physical Anthropology*, **60**, 359–81.

Ryan, B. F., Joiner, B. L., and Ryan, T. A. Jr. (1985). *Minitab Handbook* (2nd edn). Duxbury Press, Boston.

Rybricki, E. F., Mills, E. J., Turner, A. S., and Simonsen, F. A. (1977). *In-vivo* analytical studies of forces and moments in equine long bones. *Journal of Biomechanics*, **10**, 701–5.

Sääf, J. (1950). Effects of exercise on adult articular cartilage. An experimental study of guinea-pigs with relevance to the continuous regeneration of adult cartilage. *Acta Orthopaedica Scandinavica (Supplement)*, **7**, 1–86.

Salter, R. B. and Field, P. (1960). The effects of continuous compression on living articular cartilage: an experimental investigation. *Journal of Bone and Joint Surgery*, **42A**, 31–49.

Seedhom, B. B. (1979). Transmission of the load in the knee joint with special reference to the role of menisci. *Engineering in Medicine*, **8**, 207–19.

-Seedhom, B. B. and Hargreaves, D. J. (1979). Transmission of the load in the knee joint with special reference to the role of menisci. Part II: experimental results, discussion and conclusions. *Engineering in Medicine*, **8**, 220–8.

Shrive, N. G., O'Connor, J. J., and Goodfellow, J. W. (1978). Load-bearing in the knee joint. *Clinical Orthopaedics and Related Research*, **131**, 279–87.

Simpson, A. H. R. W. and Kenwright, J. (1992). The response of nerve to stretch. *British Orthopaedic Research Society*, Spring Meeting, Nottingham.

Sjostrom, M., Wahlby, K., and Fugl-Meyer, A. (1978). Achilles tendon injury. Structure of rabbit soleus muscles after immobilisation at different positions. *Acta Chirugica Scandinavica*, **145**, 509–21.

Slowman, S. D. and Brandt, K. D. (1986). Composition and glycosaminoglycan metabolism of articular cartilage from habitually loaded and habitually unloaded sites. *Arthritis and Rheumatism*, **29**, 88–94.

Sneddon, I. N. (1951). *Integral transforms*. McGraw-Hill, New York.

Sood, S. C. (1971). A study of the effects of experimental immobilisation on rabbit articular cartilage. *Journal of Anatomy*, **108**, 497–507.

Spirt, A. S., Mak, A. F., and Wassell, R. P. (1989). Nonlinear viscoelastic properties of articular cartilage in shear. *Journal of Orthopaedic Research*, **7**, 43–9.

Stofft, E. and Graf, J. (1983). Rasterelektronmikroskopische Untersuchung des hyalinen Gelenkknorpels. *Acta Anat. (Basel)*, **116**, 114–25.

Sunderland, S. and Bradley, K. C. (1961). Stress-strain phenomena in human peripheral nerve trunks. *Brain*, **84**, 102–19.

Swann, A. C. and Seedhom B. B. (1986). The relationship between the stiffness of normal articular cartilage and the predominant acting stress levels. *Scandinavian Journal of Rheumatology (Supplement)*, **60**(80), 29.

Sweet, M. B. E., Thonar, E.J.-M.A., and Immelman, A. R (1977). Regional distribution of water and glycosaminoglycan in immature articular cartilage. *Biochimica et Biophysica Acta*, **500**, 173–86.

Termine, J. D. (1972). Mineral chemistry and skeletal biology. *Clinical Orthopaedics and Related Research*, **85**, 207–41.

Torzilli, P. A., Takebe, K., Burstein, A. H., and Heiple, K. G. (1981). Structural properties of immature canine bone. *Journal of Biomechanical Engineering*, **103**, 232–8.

Torzilli, P. A., Takebe, K., Burstein, A. H., Zika, J. M., and Heiple, K. G. (1982). The material properties of immature bone. *Journal of Biomechanical Engineering*, **104**, 12–20.

Urry, D. W. (1974). On the molecular basis for vascular calcification. *Perspectives in Biology and Medicine*, **18**, 68–84.

Vandenburgh, H. H. and Karlisch, P. (1989). Longitudinal growth of skeletal myotubes *in vitro* in a new horizontal mechanical cell stimulator. *In Vitro Cellular and Developmental Biology*, **25**, 607–16.

vander Sloten, J. and van der Perre, G. (1989). Trabecular structure compared to stress trajectories in the proximal femur and the calcaneus. *Journal of Biomedical Engineering*, **11**, 203–8.

Vasan, N. (1983). Effects of physical stress on the synthesis and degradation of cartilage matrix. *Connective Tissue Research*, **12**, 49–58.

Vincentelli, R. and Evans, F. G. (1971). Relations among mechanical properties, collagen fibers, and calcification in adult human cortical bone. *Journal of Biomechanics*, **4**, 193–201.

Vinz, H. (1970). Die Änderung der Festigkeitseigenschaften des kompakten Knochengewebes im Laufe der Altersentwicklung. *Gegenbauers Morphologisches Jahrbuch*, **115**, 257–72.

Vogel, H. G. (1979). Influence of maturation and aging on mechanical and biochemical parameters of rat bone. *Geriatrics*, **25**, 16–23.

Weber, W.E. and Weber, E. F. W. (1836). Mechanic der menschulichen Genwerkzeuge. *I. vol and atlas Göttungen, in der Dieterichschen Buchandlung.*

Weightman, B. and Kempson, G. E. (1979). Load carriage. In *Adult Articular Cartilage* (2nd edn), (ed. M. A. R. Freeman), pp. 291–331. Pitman Medical, London.

Williams, P. and Goldspink, G. (1978). Changes in sarcomere length and physiological properties in immobilized muscle. *Journal of Anatomy*, **127**, 459–68.

Wolff, J. (1892). *Das Gesetz der Transformation der Knochen*. Hirchwald, Berlin.

Woo, S. L-Y. and Akeson, W. H. (1987). Response of tendons and ligaments to joint loading and movements. In *Joint loading—Biology and health of articular structures* (ed. H. J. Helminen, I. Kiviranta, M. Tammi, A.-M. Säämänen, K. Paukkonen, and J. Jurvelin), pp. 287–315. Wright, Bristol.

Section 5—Cellular Remodelling

28 Mechanical properties of articular cartilage and the response of the chondrocytes to mechanical load

R. M. Aspden, T. Larsson, and D. Heinegård

Introduction

Articular cartilage provides a tough, resilient coating to the ends of bones where they articulate in synovial joints. Its primary functions are to keep contact stresses low by deforming sufficiently to distribute loads imposed on the joint while providing a low friction bearing surface. In adult humans it is typically about 1–4 mm thick and is therefore not thick enough to have any significant shock absorbing capabilities (Finlay and Repo 1979; Weightman and Kempson 1979).

The main components of the tissue are a fibrous protein, collagen, surrounded by a highly hydrated proteoglycan gel, water comprises about 70 per cent of the mass. Together these form a biological example of a fibrous composite material (Hukins and Aspden 1985). Collagen fibrils are formed from molecules of type II collagen, which are about 300 nm long and 1.5 nm in diameter, assembled in a distinctive staggered array to form fibrils about 20–100 nm wide. Their length is unknown, though from electron microscopy it can be deduced that it is probably several millimetres. Like ropes, these fibrils are strong in tension and can therefore reinforce the surrounding gel providing that they are oriented so as to be placed into tension when the tissue is loaded. This orientation has been measured using X-ray diffraction (Aspden and Hukins 1981; Kirby *et al.* 1988) and polarized light microscopy (Yarker *et al.* 1983), and the results used in an analysis of the expected stress and strain in the fibrils (Aspden and Hukins 1989). Together these show that the organization of the collagen and its mechanical strength are indeed appropriate to its functioning in this way.

The main proteoglycan in articular cartilage consists of a protein core, with a molecular weight of just over 200 000, to which are covalently bound a large number of polysaccharide side-chains of different types called glycosaminoglycans (Heinegård and Oldberg 1989). A number of these proteoglycans are attached by the amino terminus of the protein core to another linear polysaccharide, hyaluronan, to form a very large aggregate. The glycosaminoglycan side chains of the proteoglycan molecule have large numbers of sulphate and carboxyl groups on them which are negatively charged at physiological pH. These are contained within the tissue, and to balance their charges they attract a large number of positively charged, small, mobile counter ions such as sodium and calcium. The imbalance so created between the large numbers of these molecules inside compared with those outside the tissue leads to an osmotic effect, called the Donnan effect, which results in water being drawn into the tissue. The end result is a swelling pressure, of about 0.3 MPa in the unloaded state, trying to hydrate the tissue which is resisted by tension in the collagen fibrils (Maroudas 1976).

Connective tissues also contain a number of other matrix components, some of which are specific to articular cartilage (Heinegård and Oldberg 1989). The functions of these are unknown but it is likely that they are involved in matrix assembly and repair, and some may help to transfer stress from the mechanically weak proteoglycan gel to the collagen fibrils. The shear modulus of the proteoglycan gel is only about 10 Pa whereas that for cartilage is about 10 MPa (Myers *et al.* 1984) so that interactions between the proteoglycans and collagen are essential if the collagen is to provide effective reinforcing. These interactions are presently unknown and may be weak and nonspecific. Alterations in this stress transfer may be important in diseases such as osteoarthritis where evidence suggests that mechanical load may be a major factor in the breakdown of the cartilage.

The cells in cartilage, known as chondrocytes, are responsible for the formation and maintenance of the tissue and they can respond to alterations in the habitual mechanical load on the joint. Moderate exercise of dogs on a treadmill has been shown to increase the amount of proteoglycan compared with controls (Kiviranta *et al.* 1988) whereas immobilization of joints, or even motion of the joint without loading, results in depletion of the proteoglycans from the cartilage (Palmoski *et al.* 1979, 1980). This depletion is reversible after resumption of load bearing (Behrens *et al.* 1989; Jurvelin *et al.* 1989). The compressive stiffness of the femoral head in humans varies over the joint surface in a way that correlates strongly with the applied load (Kempson 1979). The superior surface, which is most heavily loaded with contact stresses up to about 18 MPa (Hodge *et al.* 1986), is stiffer than the cartilage at the joint margins or in the infero-foveal region, which experience relatively little loading. These *in-vivo* observations may be complemented by studies applying mechanical load to cartilage biopsies *in vivo* which provide a more controllable biochemical and physical environment. Hydrostatic pressure (Hall *et al.* 1991) and static load (Gray *et al.* 1988; Larsson *et al.* 1991) have been shown to reduce proteoglycan synthesis, whereas certain regimes of cyclic load result in an increase in proteoglycan production (Sah *et al.* 1989; Larsson *et al.* 1991). To investigate the relationship between mechanical load and the cellular response the authors have developed a novel mechanical testing system and apparatus to apply known mechanical loads to cartilage biopsies in culture.

Mechanical Testing

The response of the chondrocytes to mechanical load in altering the amount, and possibly the type, of matrix components produced presumably has the aim of adapting the mechanical properties of the tissue to the applied loading. In order to determine the effects of treatment on the cultured biopsy it is therefore essential to be able to measure the mechanical properties of the tissue in such a way as will characterize them most fully. Because cartilage is viscoelastic, simply measuring the compressive stiffness will not be representative. The mechanical properties depend on time and the rate of loading significantly affects the measured stiffness.

The thinness of articular cartilage, typically less than 1 mm in the bovine cartilage used, means that a testing machine has had to be built that can measure displacements to an accuracy of 1 μm (Aspden *et al.* 1991). This is computer controlled and uses a stepper motor to drive a ball screw so that a single step corresponds to 1 μm. The stepper motor can start immediately at stepping speeds of 40–1000 steps/s, which corresponds to a maximum strain rate of 1 mm/s for a 1-mm thick sample. Slower stepping rates are controlled directly by software. Load and displacement are monitored using a load cell and a linear variable differential transformer (lvdt), respectively, and these can be sampled and digitized at frequencies up to about 10 kHz. In practice the highest sampling frequency used is just less than 1 kHz as there is nothing to be gained by sampling faster than the stepping speed of the motor. Stress–strain, creep, and stress relaxation tests are all programmable from the software written to drive the machine. The sample is placed in a small well below the loading platen and kept immersed in phosphate buffered saline solution at pH 7.4. All tests are done in unconfined compression, which removes the need to ensure that the sample fits closely into the constraint and is closer to the physiological environment in that fluid flow in loaded cartilage *in situ* is predominantly laterally through the tissue (McCutchen 1983) and not through the joint surface, which is where the highest loads are.

To test whether the viscoelasticity of cartilage is linear or otherwise, a set of isochronal stress–strain curves may be derived from a series of creep tests at different stresses. This is important because linear viscoelastic behaviour allows interconversion between the compliance and stiffness moduli and there are many models for this type of behaviour (Ferry 1980). However, Fig. 28.1 shows that cartilage is markedly nonlinear at a wide range of different times. In tendon the origin of this nonlinearity is the crimped nature of the highly aligned collagen fibres (Hooley and Cohen 1979) in which straightening of the zigzag leads to different structures at different stresses. In cartilage the fibrils are not crimped, possibly because of the high internal pressure, and it is not clear what is causing the isochronal stiffness to increase with load though loss of water could be the primary cause. An additional consequence of this

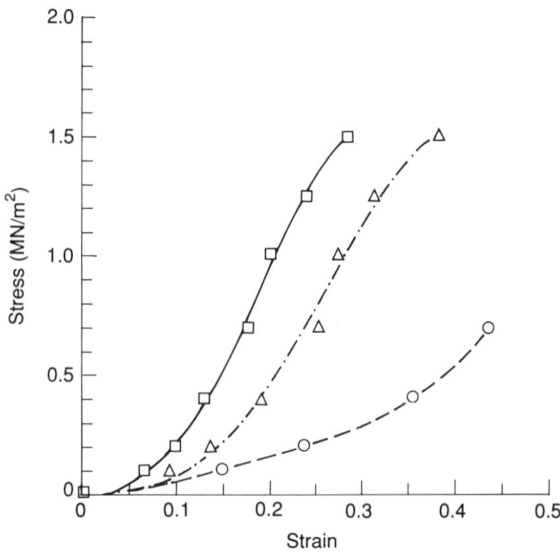

Fig. 28.1 Isochronal stress–strain curves from bovine articular cartilage at times of □ 2 s, △ 10 s, and ○ 100 s.

nonlinearity is that time–temperature superposition cannot reliably be used to form a master curve for compliance, say, from a series of curves at different temperatures. Again this has been well illustrated in tendon where the activation energy found by superposition was 50 per cent lower than that found by the more reliable temperature-jump method (Cohen *et al.* 1976).

Fourier-transform mechanical spectroscopy

Because of the time dependency of the mechanical properties of articular cartilage it is important in characterizing a tissue sample to measure the dynamic properties. There are a number of methods for doing this, depending on the frequency range to be covered, and these are all well described, for instance, by Ferry (1980). However, tests at frequencies below 1 Hz are difficult to perform, take a long time, and often necessitate different apparatus to tests at higher frequencies. In principle a stress–relaxation or creep test contains all the information described by the dynamic storage and loss moduli, which are obtained from a cyclic test at constant stress or strain amplitude. In the case of a linearly viscoelastic material the interconversion between the different kinds of test is mathematically exact. In a nonlinear material the descriptions may be seen as equivalent even though they may not be directly quantitatively comparable.

Consider, for example, a creep experiment in which a constant stress is applied in the form of a step function and the strain is monitored as a function of time. If $M(t)$ represents the time-dependent compliance which tends to a relaxed value of M_R in the limit as $t \to \infty$ then it is long established (Gross 1953; Ferry 1980) that the dynamic modulus

$$M^*(\omega) = M'(\omega) + M''(\omega)$$

is given by

$$M^*(\omega) = M_R + i\omega \int_0^\infty (M(t) - M_R)\exp(-i\omega t)\, dt.$$

This is a form of Fourier transform and has been little used in experimental work, probably largely because of the computational requirements. The advent of easily accessible, powerful computers makes the use of such computationally intensive methods more practical but the main drawback is that to make full use of fast Fourier-transform algorithms requires data at equally spaced time intervals. So if recordings are made at high frequency to detect the relaxations occurring at short times, then after the several hours, or even days, taken to complete the relaxation in some materials an impossibly large number of points would have been accumulated. To overcome this, a method has been described by which data can be recorded at high frequencies at the start of an experiment and the frequency progressively decreased by predetermined amounts as the experiment proceeds (Arridge 1984; Arridge and Barham 1986; Aspden 1991). Each data set at a given frequency is then Fourier transformed and the transforms combined to yield a representation of the loss and storage moduli over a large frequency range. This method has been called Fourier-transform mechanical spectroscopy (Arridge and Barham 1986).

This technique has been developed to provide a test for articular cartilage (Aspden *et al.* 1991). It is rapidly performed and measures the loss and storage moduli of a sample at frequencies from about 10^{-3} Hz to 250 Hz from one experiment lasting 18 min. Figure 28.2 shows the creep compliance as function of time measured at an applied stress of 0.4 MPa from the same sample of bovine articular cartilage as used above. This consists of four series each comprising 2048 points. The highest recording frequency was 950 Hz and this was reduced by a factor of eight between each of the subsequent sets. The first series was smoothed using cubic B-splines as the transform was very sensitive to noise in the data. These series are Fourier transformed in Fig. 28.3 to give the storage and loss compliances. The storage compliance only changes slowly over the range of frequencies measured, being 9.6×10^{-7}/Pa at 8.9×10^{-4} Hz and falling to 2.3×10^{-7}/Pa at 230 Hz. In contrast the loss compliance is 3.0×10^{-8}/Pa at 8.9×10^{-4} Hz and decreases rapidly to about 2×10^{-8}/Pa at walking frequencies of the order of 1 Hz, before falling still faster at higher frequencies. The gradient of the log compliance–log frequency

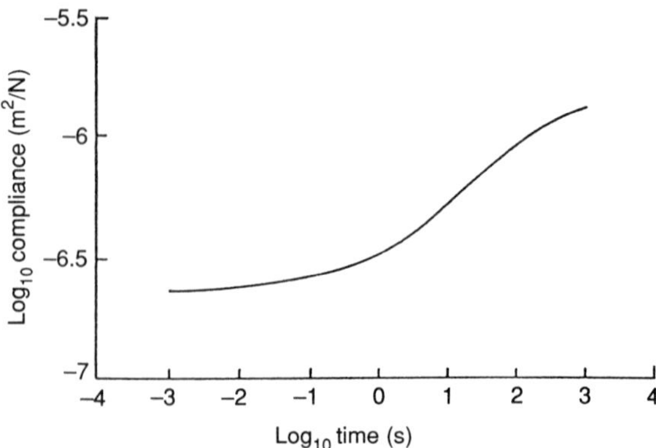

Fig. 28.2 Creep compliance as a function of time at an applied stress of 0.4 MN/m^2 for the same sample of bovine articular cartilage as in Fig. 28.1. Equilibrium has not been reached after 18 min.

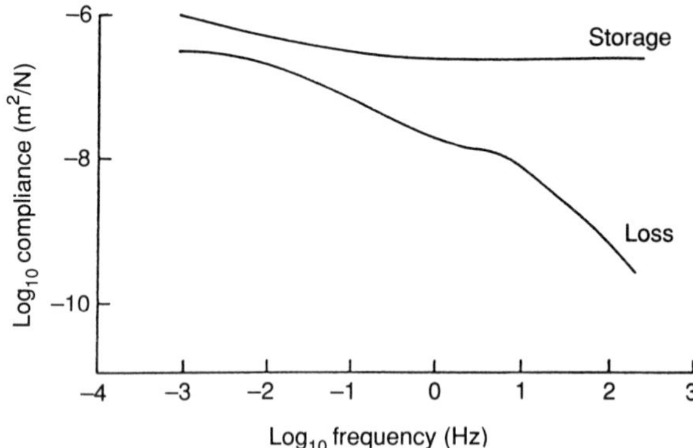

Fig. 28.3 Fourier transform of the compliance in Fig. 28.2 in four series to give the loss and storage compliances.

curve is -1 at high frequencies, which is predictable from the theory for an exponential form of relaxation in the tail of the transform. This does not mean that the creep curve could be fitted by an exponential. Combining a series of these should provide a reasonable approximation to the curve but so doing makes unwarranted assumptions regarding the nature of the relaxation processes. The advantage of Fourier-transforming the recorded data is that no such assumptions are needed and the method does not require recourse to abstract models based on springs and dashpots whose parameters are very difficult to relate to the composition and structure of the tissue.

Preliminary results suggest that the shape of the loss and storage compliances, and the loss tangent derived from them as the ratio of loss to storage, are sensitive to the applied stress and the site within the joint from which the sample was taken. Figure 28.4 shows the loss and storage compliances from the same sample as above at applied stresses of 0.1, 0.4, and 1.5 MN/m^2 and the loss tangents derived from these. The origin of the peak at around 10 Hz is unknown and current work is to determine whether it is an artefact, as it comes at overlap of the Fourier transforms with the sampling frequencies currently being used. However, different energy dissipation mechanisms operating at different frequencies could give rise to this form of curve. It is interesting that, irrespective of the exact shape of the curve at these frequencies, the loss tangent is an order of magnitude lower at walking frequencies of around 0.5–5 Hz than at the lowest frequencies measured. This suggests that the energy dissipated while walking or running is at most a few percent of the total imparted to the tissue, whereas at low frequencies it appears to be about 30 per cent.

Site variation is shown in Fig. 28.5 where three samples have been taken from different areas of the same joint. It has long been recognized that

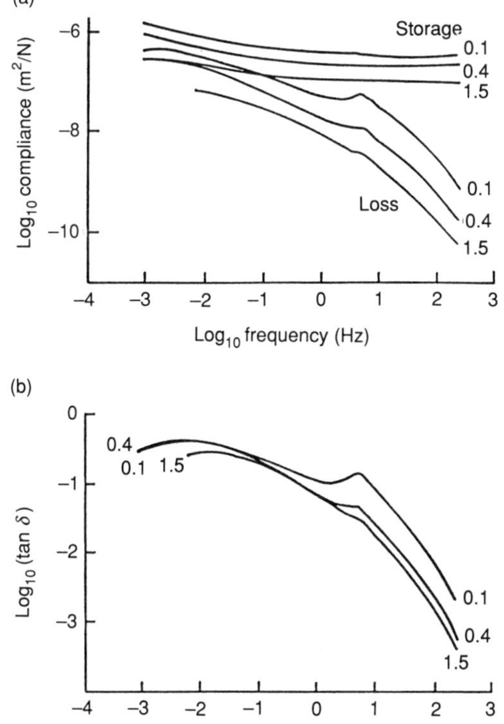

Fig. 28.4 (a) Loss and storage compliances and (b) the corresponding loss tangents from the same sample of bovine articular cartilage at different applied stresses.

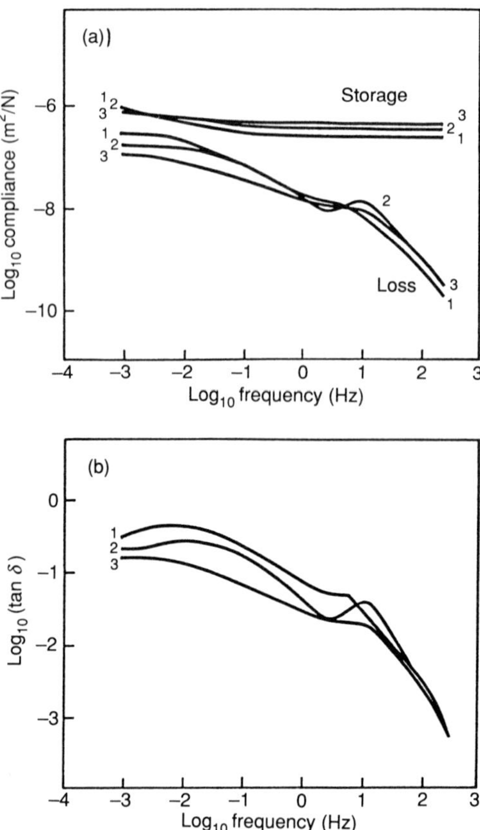

Fig. 28.5 (a) Loss and storage compliances and (b) the loss tangents derived from these for samples from three sites in a bovine carpo-metacarpal joint. Samples 1 and 2 are from sites along the edge of the joint, 3 is from the centre of the joint adjacent to 2.

habitually loaded areas of cartilage appear stiffer in a stress–strain test, or the so-called two-second modulus, than in a creep indentation test (Kempson 1979). However these data show the problems of a single frequency measurement when trying to characterize a material such as cartilage. For instance, because the curves can cross over, sample 1 appears most compliant at low frequencies but is the stiffest at higher frequencies. Unfortunately the detailed loading pattern of this bovine joint are not yet known, so a correlation with habitual load is not possible. However, these results illustrate that this method is capable of distinguishing different samples of cartilage and that it is important to have a wide frequency range to measure properly the material properties.

Using this technique it is now possible to measure the mechanical properties of an articular cartilage biopsy before and after culture, during which time the tissue can be subjected to mechanical load, treatment with enzymes to remove selected matrix components, or stimulated with growth factors or purified matrix components. In conjunction with biochemical tests it should be possible to determine more precisely the components and mechanisms involved in load transfer within the tissue and to begin to investigate ways in which these might fail.

Effects of load on cartilage cultures

Cartilage biopsies, similar to those used for mechanical testing, were obtained from the carpometacarpal joint of 6-month-old calves. Full-depth biopsies of 4-mm diameter were removed and cultured in 2 ml of Ham's F12 culture medium, buffered to pH 7.4 with 20 mM HEPES and supplemented by 50 μg ascorbic acid, 50 IU penicillin, and 50 μg streptomycin per ml. Loads were applied to the biopsies using a specially-constructed, pneumatically driven apparatus that enabled some samples to be subjected to a constant load while others had a stepwise cyclic load applied (Larsson *et al.* 1991). Three identical culture dishes were prepared, each containing up to 12 samples; one was left unloaded to act as a control, the second was placed under a static load calculated to produce a stress of 1 MPa on each biopsy, and the third was subjected to a cyclic load of the same magnitude. Cycle times were varied from 2+2 s (2 s load and 2 s unloaded) up to 60+60 s. Biosynthesis was studied by radiolabelling for 4 h at various times during an experiment using 50 μCi ^{3}H-leucine and 20 μCi $^{35}SO_4$ per ml of medium. Leucine is found in most proteins and commonly constitutes about 10 per cent of the amino acid composition while the glycosaminoglycans contain large numbers of sulphate groups which give them their high fixed charge density as previously described. Double labelling in this way therefore allows separate estimation of the synthesis of proteins and glycosaminoglycans. The amount of radioactivity incorporated is counted in a liquid scintillation counter calibrated against known sources and quoted as disintegrations per minute (dpm). In this set of experiments the biopsies were cultured over a period of 44 h in sets of four to enable some measurement of the variability. One set from each loading regime was radiolabelled immediately on starting the experiment, the second after 24 h, following which the load was removed from the remaining set which was finally labelled after 40 h (Larsson *et al.* 1991). This final unloaded period was to ensure that the biopsies were surviving and that the results obtained were not because of damage to the chondrocytes.

The synthetic response of the cells to the different loads may be seen in Figs 28.6 and 28.7 in which are shown results from 2+2 s and 60+60 s loading experiments. Other cycle times between these extremes gave intermediate results. There was an immediate response to both cyclic and static load as

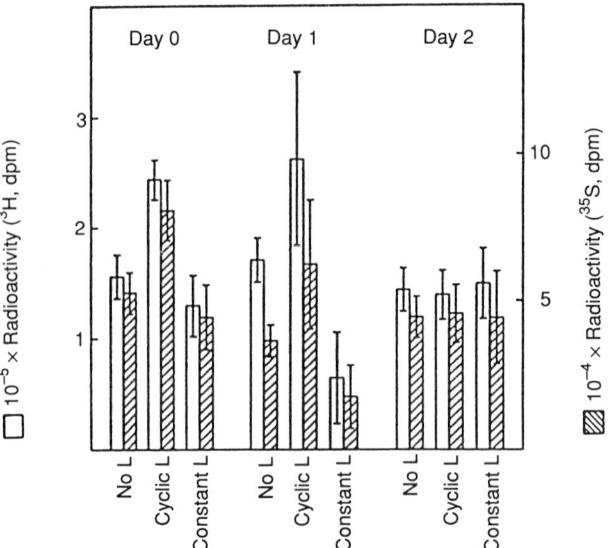

Fig. 28.6 Total incorporated radiolabel per mg wet weight of cartilage in unloaded, 2+2 s loading, and constantly loaded cartilage biopsies. See text for details.

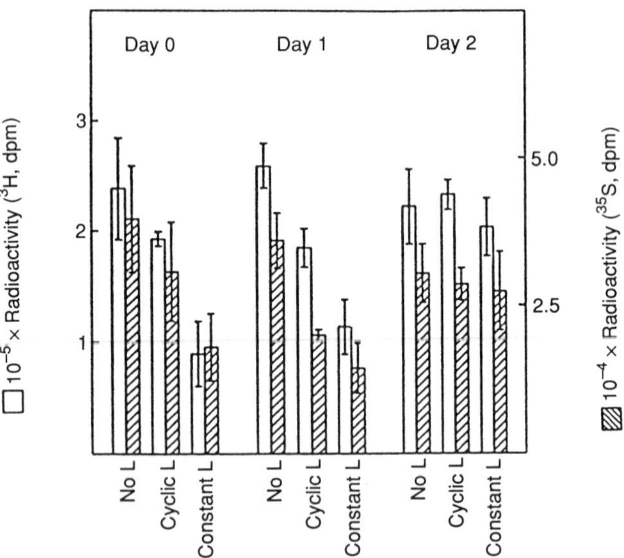

Fig. 28.7 Total incorporated radiolabel per mg wet weight of cartilage in unloaded, 60+60 s loading, and constantly loaded cartilage biopsies. See text for details.

demonstrated by the first 4-h radiolabel marked Day 0. The response to static load was consistently a depression of the synthesis of both protein and glycosaminoglycan by comparable amounts, though the magnitude of this varied between experiments. The effects of cyclic loading though were very dependent on the frequency; 2 + 2 s increasing the synthesis of both proteins and glycosaminoglycans by 56 per cent and 53 per cent respectively while 60 + 60 s depressed the synthesis of both to about 80 per cent of the unloaded control. This depression was not as great as that caused by the static load which in this same experiment was about 40 per cent of the unloaded value. Cyclic loads with a frequency of 10 + 10 s also increased the synthesis, though not as much, while 30 + 30 s slightly reduced it. After 24 h these same trends were maintained as shown by Day 1 in Figs 28.6 and 28.7: high frequency load caused an increased synthesis of both proteins and glycosaminoglycans, while low frequency and static loads inhibited synthesis. Once the load was removed the differences between the variously loaded groups disappeared, showing that not only were the cells still viable and that no irreversible changes had been induced by loading, but also that there was no memory of previous loading. In a few experiments there was an overshoot in production of glycosaminoglycans in those biopsies which had carried constant load, which may be indicative of a compensatory mechanism.

Comparison of the ratio of sulphate to leucine incorporation indicates whether synthesis of glycosaminoglycans or their sulphation is linked to the synthesis of proteins or is separately controllable. In the short cycle loading experiments there was no significant difference in this ratio compared with unloaded controls. As the cycle time increased up to 60 + 60 s, however, there was a significant decrease in the sulphate/leucine ratio compared with unloaded controls. This indicates that glycosaminoglycan production or sulphation of the molecules produced was inhibited more than protein synthesis. In some experiments an opposite effect was found in the constantly-loaded biopsies where the sulphate/leucine ratio increased, suggesting that protein synthesis was being inhibited more than glycosaminoglycan. While not being consistent, these results do suggest that separate regulation of the genes encoding proteins and proteoglycans is possible. It needs to be noted that the unloaded biopsies do not represent the conditions *in vivo*, as cartilage is normally subjected to load as a result of muscle and ligament tension, but are used here as a control.

It is now possible to study simultaneously the inter-relationships between mechanical load, mechanical properties of the tissue, and the control over these exerted by the chondrocytes using an *in-vitro* culture system, biochemical methods, and a simple mechanical test. Understanding the control mechanisms of the chondrocytes and the limits within which they function should lead to an understanding of factors that may cause tissue breakdown in diseases with an apparently mechanical aetiology such as osteoarthritis.

Acknowledgements

The authors thank the Welcome Trust of Great Britain for a fellowship for R.M.A. and gratefully acknowledge the financial support of the Swedish Medical Research Council, Folksams stiftelse, Gustaf V:s 80- åsfond, Kocks stiftelse, Österlunds stiftelse and the Medical Faculty, University of Lund. They also thank Runo Svensson and Lisbeth Camper for their skilful technical assistance.

Debate in the meeting

While the author (R.M.A.) made it clear that the Fourier-transform technique makes no assumption about the detailed structure of a material, the meeting considered it might be important that the conditions under which properties were being measured should bear some relationship to the mechanical situation in an animal. Separating cartilage from its bone surface may allow large amounts of lateral creep which do not occur in the animal joint. The author agreed that to test the material on a bone surface would be ideal but would present immense technical experimental difficulties, particularly in the repeatability of the measurements.

29 Functional adaptation of bone by strain-induced changes in lining cell association: an hypothesis

B. DERBYSHIRE

Introduction

Bone responds to changes in the range of forces that it customarily sustains by increasing or decreasing its wall thickness and/or by adjusting its orientation with respect to the direction of force (Wolff 1892). Thus, a long bone increases its cortical thickness in areas subject to unusually elevated stress, and cancellous bone trabeculae appear to be oriented along the directions of principal stresses (Lanyon 1974). Functional hypertrophy of bone may be observed in the preferred limbs of sportsmen (Jones *et al.* 1977), whereas atrophy is evident during prolonged periods of bed-rest (Krolner and Toft 1983) or space flight (Tilton *et al.* 1980). Despite much research over the past thirty years, the mechanism by which bone cells detect functional forces has yet to be elucidated. Force in the bone can only be detected by its displacement effect on the bone material, i.e. by bone strain. The osteogenic cells must, therefore, be responding to some sort of strain-related stimulus. Strain-gauge measurements of animal limb bones *in vivo* have revealed a fairly consistent level of peak functional strain (0.002–0.003) across different species, thus suggesting a common threshold strain beyond which bone starts to hypertrophy in order to maintain this threshold (Rubin 1984).

How does a change in strain magnitude affect the osteogenic cells? Several hypotheses have been put forward.

Previous hypotheses

Strain-generated electrical potentials

In the early fifties, it was demonstrated that transient electrical potentials could be measured in bone as it was deformed, and it was proposed that this electrical effect could be the stimulus for functional osteogenesis (Yasuda 1953). In dry bone, electrical potentials are generated mainly by the piezo-electric collagen fibres but in wet bone they are predominantly streaming potentials (Hastings 1988; Gross and Williams 1982) resulting from the expression of fluid from the charged surfaces of pores and channels in the bone.

The main evidence that strain-generated potentials (SGP) are associated with functional adaptation derives from the fact that exogenous electrical stimulation of bone induces osteogenesis (Brighton *et al.* 1979). Thus, bone formation occurs around an implanted cathode, and fracture repair is aided by electric or electromagnetic fields (Lavine and Grodzinsky 1987). However, there is no direct evidence that SGPs play any physiological role. Indeed, the ability of a pulsed or SGP-shaped current to stimulate bone formation is suspiciously limited compared to direct current stimulation (Treharne *et al.* 1979).

Mechano-chemical effects

Glegg and Leblond (1953) pointed out that when a crystal is compressed in a saturated solution, its solubility increases (Correns 1949). Accordingly, dissolution of bone hydroxyapatite crystals should occur in the compressed areas of bone and redeposition in the surrounding regions. Apparently unaware of this paper, Justus and Luft (1970) proposed the same mechanism but with the reverse effects. Compressive strain, they argued, would reduce the crystal solubility causing precipitation, whereas tensile strain would increase the solubility causing dissolution. Local changes in calcium concentration would then trigger a cellular response. Tests on hydroxyapatite-coated blades in bending did show an increase in solubility under tensile strain, but the theoretically expected correlation between stress magnitude and solubility could not be demonstrated.

Micro-damage repair

Greig (1931) proposed that functional stresses would produce miniature traumata in the bone leading to a local liberation of histamine, hyperaemia, and subsequent decalcification (Murray 1936). The locally high calcium concentration would then promote a repair response at the trauma sites.

In-vivo experiments by Chamay and Tchantz (1972) showed that repetitively over-loaded dog ulnae hypertrophied on the compressive sides at the sites of plastic slip lines. This led them to propose that functional adaptation was due to a local repair response. Micro-cracks were observed in the osteones of bones which had undergone 10 000 cycles of physiological strain (Burr *et al.* 1985). *In vivo*, these regions were resorbed which, they suggested, indicated a repair response.

Although it seems reasonable to assume that fatigue-related micro-damage would be 'cleaned up' by local remodelling (with no eventual loss or gain of bone), a correlation with functional adaptation would appear to be negated by the work of Lanyon and Rubin. They found that only a few load cycles per day, at peak physiological strain, were required to stimulate substantial bone formation (Rubin and Lanyon 1984; Rubin and Lanyon 1985). Such a small number of loading cycles would be insufficient to produce appreciable micro-damage (Lanyon 1987). Similar loading experiments also revealed a direct transformation of the periosteum from quiescence to active bone formation, unaccompanied by resorption or the presence of osteoclasts—thus suggesting no repair/remodelling response (Pead *et al.* 1988).

Physiological effects

Currey (1968) proposed that changes in bone strain could be detected by either local osteocytes or randomly-distributed nerves which would relay signals of the overall strain pattern to the surface osteogenic cells. Trueta (1968) also suggested that compression of the canaliculi might interfere with the exchange between osteocytes and surface cells, thus triggering osteogenesis.

Neural control of adaptation would seem to have been refuted by the demonstration that peripheral nerve resection of rabbit tibiae had no effect on functionally-stimulated osteogenesis (Hert *et al.* 1971).

Osteocytes communicate amongst themselves, and with the surface lining cells, via gap junctions between their processes (Doty 1981; Menton *et al.* 1984). Thus, the means for Currey's and Trueta's proposal does exist. High pressure may be generated in osteocyte lacunae and canaliculi when bone is compressed (Jendrucko *et al.* 1976; Piekarski and Munro 1977) and this may cause fluid flow and possibly generate streaming potentials.

Lanyon's group, who favour this osteocyte-relay hypothesis, have investigated the biochemical responses of osteocytes and lining cells after short periods of cyclic loading. Increases in RNA synthesis, glucose-6-phosphate-dehyrogenase activity, and prostaglandin production have been found *in vivo* or *in vitro* (Pead *et al.* 1988a; Skerry *et al.* 1989; El Haj *et al.* 1990; Rawlinson *et al.* 1991). However, because lining cells and osteocytes communicate with one another, it is difficult to discern from these tests which cell-type responds to strain and which is the mere recipient of intercellular signals.

Direct strain-cell effects

The most straightforward method of strain detection would be a direct sensing of bone strain by the surface osteogenic cells (lining cells, osteoblasts, or their precursors). Pauwels (1940–1980) proposed that cell distortion, of the order of one-thousandth of the cell length, might stimulate osteogenesis. More recently, Davidson *et al.* (1990) discovered mechano-sensitive ion channels in osteoblast-like cells and suggested that they might act as bone strain transducers.

The central fact to be appreciated when considering direct strain-cell mechanisms is that the strain level which stimulates adaptive hypertrophy is about 0.003 (Rubin and Lanyon 1985). Thus, a 30 μm-long cell, lying on the bone surface, would change in length by 0.09 μm. While it is conceivable that this change could be detected by the cell, this possibility rests on the assumption that the strain detector runs along the full length of the cell! A corresponding change in cell volume or pressure would be extremely small; too small, it would seem, to be detectable by the cell. Similarly, a cell membrane ion channel (diameter 1 μm, Sachs 1988) would barely change in area under such a strain. It is more likely, therefore, that mechano-sensitive ion channels are involved in osmoregulation or cell size regulation (Sachs 1988).

A new hypothesis

The above considerations involved the effect of direct bone surface strain on each individual surface cell. However, if the strain effect between the surface cells is considered, a more plausible possibility emerges. The cells covering all bone surfaces in the quiescent state are often referred to as bone lining cells. They are considered to be 'resting' osteoblasts or osteoblast precursors and they form a closely-packed 'pavement' of cells (Fig. 29.1) communicating via

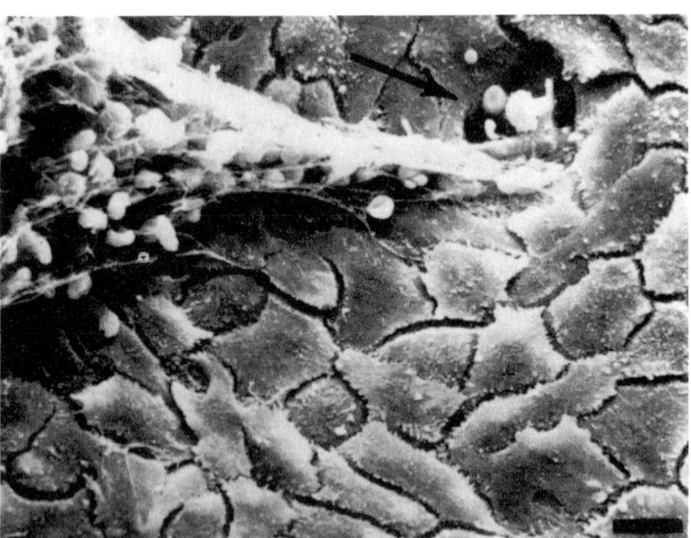

Fig. 29.1 Endosteal lining cells of the rat femur. Cell separation is exaggerated due to preparation artefact. Marrow cells remain attached at a vascular channel (arrow). Bar indicates 10 μm. (From Menton *et al.* 1984, copyright © Wiley–Liss 1984.)

gap junctions (Menton *et al.* 1984; Miller *et al.* 1989). Such quiescent surfaces can be stimulated to form bone by a brief period of cyclic loading (Pead *et al.* 1988*b*).

The 'standard' cell separation between most types of cell is about 20 nm. Thus, if bone were subjected to a compressive strain of 0.003, the cell membranes of any two adjacent lining cells would approach by only 0.06 nm (0.003×20) which is negligibly small. However, if cell shape or other effects are taken into account, this approach could be considerably amplified. It will be argued later that this change in proximity between lining cells has the propensity to initiate a cascade of cell responses leading to bone formation.

Figure 29.2(a) is a schematic diagram of two adjacent lining cells. Here, the peripheral cell shape entails that the point of attachment to the bone surface is a distance X from the cell periphery. Bone strain between the two cells, therefore, operates over a distance $2X$ (+ the 20 nm separation), i.e. between the points of cell attachment. If $X = 1$ μm (say), and the bone strain is 0.003, a change in separation of 6.06 nm (0.003×2.02) would be produced between adjacent cells. This is a substantial proportion of the 'standard' cell separation.

Lining cells project processes through canaliculi in the bone surface where they contact the processes of underlying osteocytes. Miller *et al.* (1989) suggested that these processes anchor the lining cells to the bone surface. The cells do not rest directly on the bone surface, however, but are separated from it by a thin layer of collagen fibres (Miller *et al.* 1989). Thus, in Fig. 29.2(b), each cell is anchored to the bone by the canalicular processes, and, lateral to these, the cell is considered to be unattached to the bone and resting on a 'carpet' of collagen fibres. Compressive strain, parallel to the bone surface, would then cause the anchor points of the adjacent cells to approach, while the

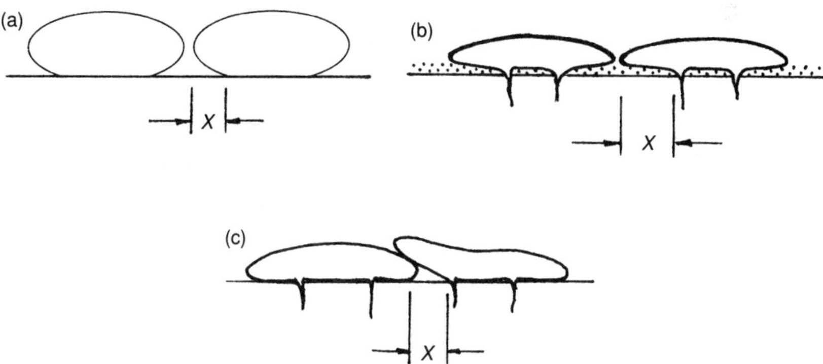

Fig .29.2 (a) Schematic diagram of two apposed lining cells on the bone surface. Cell separation 20 nm. (b) The two lining cells rest on a collagen fibre layer and are anchored to the bone by canalicular processes. (c) The effect of cell overlap.

periphery of each cell would glide over the shearing collagen fibres, thus bringing the adjacent membranes closer to each other.

It has been observed that some lining cells overlap slightly at their edges (Menton *et al.* 1984; Miller *et al.* 1989). Figure 29.2(c) shows how cell overlap requires that the periphery of one cell be separated from the bone surface over a distance X. Even if the whole lower surface of the underlying cell were attached to the bone, strain over the distance X would cause the surface of the overlapping cell to shear over its neighbour.

Possible effects of a change in proximity between lining cells

The cell surface is the means through which a cell senses its external environment. A strain-induced change in close proximity between neighbouring cells might, therefore, be expected to interfere with either the sensory apparatus or the environmental signals. The effects of a change in proximity between cells can be considered from three different levels: biophysical, biochemical, and biological.

A theoretical study by Gingell (1967, 1971) considered the biophysical effects of the approach of two cells. The plasma membrane of most cells is covered by a carbohydrate layer or coat which is negatively charged due to the sialic acid moieties within it. This charged surface results in the development of an ionic double layer in the intercellular fluid surrounding each cell. Gingell calculated that as two cells approached each other, their ionic double layers would interact, and the electrical potential at each cell membrane would increase. The change in membrane potential would depend upon the magnitude of variables such as coat thickness, coat separation, ionic strength of the fluid, charge density, and counter-ion penetrability of the coat. What is most interesting about his model is that neither the cell membranes nor even the cell coats need to make contact in order to produce this effect. Gingell suggested that an increase in membrane potential could affect the cell permeability and thereby act as a cell stimulus. Similar suggestions have been made to account for the exogenous electrical stimulation of bone cells (Brighton *et al.* 1979).

From a biochemical point of view, the cell membrane is the site of action of various biochemical stimuli. A change in proximity between cells could, conceivably, interfere with biochemical signals by causing the masking or exposure of receptors for these signals. For instance, the approach of charged surfaces might cause the redistribution of receptors—some receptors move around the membrane to the cathodal side of a cell in an electric field (Robinson 1985).

From a more general, biological, viewpoint, a change in proximity between lining cells might trigger the formation or disruption (Fig. 29.2(c)) of gap junctions between the cells. In other cell types, a change in number of gap junctions has been observed prior to differentiation (Caveney 1985; Sheridan and Atkinson 1985). With a change in separation between cells, the stress in the membrane around a pre-existing gap junction would be expected to change. It has been suggested that gap junctions may aggregate and coalesce in order to reduce such stresses (Kachar and Reese 1985). The increase in size of the gap junction might then provide the basis for cell stimulation.

Supporting evidence

In vitro *experiments*

During the last fifteen years, several groups have investigated the effect of force applied to the base supporting osteoblast-like cells cultured *in vitro*. Binderman's group have studied the effects of continuous tensile force applied to the base of a culture dish. Continuous strain, not exceeding 1 per cent, resulted in prostaglandin synthesis (PGE_2), an increase in cyclic adenosine monophosphate (cAMP), and DNA synthesis (Somjen *et al.* 1980). More recently, they have found that these responses may be triggered by the strain-induced activation of phospholipase A2 (Binderman *et al.* 1988). Other groups have also reported increases in PGE_2 synthesis (Yeh and Rodan 1984; Murray and Rushton 1990) and/or DNA synthesis in osteoblast-like cells similarly subjected to strain (Hasegawa *et al.* 1985; Buckley *et al.* 1988).

Although the maximum strain levels in most of these studies were larger than physiological strains, the strain distribution varied markedly in, for instance, a culture dish, and so the average strain might have been much lower. Furthermore, one needs to consider whether the type of strain (compressive, tensile, or shear) might be important in such tests. A large tensile strain might not be stimulatory in itself, but the corresponding compressive strain (contraction), perpendicular to it, might be stimulatory (even though it would be about one-third the magnitude). Another important factor is that the cells might move or reorientate to avoid strain. Buckley *et al.* (1988) found that osteoblast-like cells aligned themselves perpendicularly to the tensile strain direction. In cyclic strain tests (zero to maximum), the cells might re-attach to the substrate so that they undergo equal amounts of compression and tension (at half the original strain level).

In a different type of experiment, Reich *et al.* (1989) exposed osteoblasts on a cover-slip to fluid shear in a parallel-plate flow chamber. By testing different fluid viscosities and flow rates, they were able to show that increases in cAMP were not caused by streaming potentials but by the effect of fluid shear. A possible interpretation of this could be that the upstream cells were forced towards their downstream neighbours, increasing cell–cell proximity and thereby eliciting a response.

Cell density effects

In the early stages of embryonic limb formation, a condensation of mesenchymal cells occurs in the prospective chondrogenic limb bud. This condensation process causes a juxtaposition of the cells and an increase in the number of gap junctions, resulting, eventually, in chondrogenesis (Zimmermann *et al.* 1982). Similarly, when mesenchymal cells are cultured *in vitro*, plating density, sufficient to ensure close juxtaposition, is required before the cells will undergo chondrogenesis (Umansky 1966). High-density mesenchyme cultures elicit chondrogenesis whereas medium-density cultures generate a mineralized matrix, suggesting osteogenesis (Osdoby and Caplan 1979, 1980).

Acknowledgements

The main body of this hypothesis was proposed as part of a project at the University of Manchester Institute of Science and Technology (Derbyshire 1982) sponsored by the Science and Engineering Research Council.

The author thanks Dr J. Skorecki, UMIST, for his advice on this work. Figure 29.1 is reprinted with the kind permission of the authors, Menton *et al.* 1984, and the publisher Wiley–Liss, a division of John Wiley and Sons, Inc.

Debate in the meeting

Much work has been done by physiologists on vibration sensors which detect very low strains in the membrane. For example, insects sometimes have a hair which is levering a membrane to produce tiny motions in it of about 40 nm, but as the cell is attached to a rigid cytoskeleton all this strain takes place in one place, producing 40 nm dislocation between the cytoskeleton and the membrane. It has also been shown that in the stretching of muscle cells, the gap junction between cells provides a physiological control mechanism.

30 The stimulation of prostaglandin synthesis by micromovement in fracture healing

A. E. Goodship, R. Norrdin, and M. Francis

Introduction

The size and shape of bones are influenced both by the pre-determined genetic code and the response to prevailing hormonal and mechanical environment. In the 1890s both clinical and experimental observations on the adaptation of bone morphology to changes in the magnitude and direction of loading formed the basis of Wolff's Law of bone remodelling (Wolff 1892). With the development of *in-vivo* strain gauge techniques it has been possible to determine the response of living bone to changes in strain magnitude and direction in a quantitative manner (Lanyon 1974; Goodship *et al.* 1979; Lanyon 1987). Furthermore it has been shown that the mechanical stimulation of bone remodelling can be evoked by extremely short periods of cyclical deformation, using appropriate osteogenic strain regimes (Rubin and Lanyon 1987). This sensitivity to mechanical environment is seen not only in intact bone but also in the process of fracture healing.

Absolute interfragmentary stability with rigid internal fixation results in direct fracture healing, in which there is little or no periosteal callus formation and osteonal remodelling occurs across the fracture line. In this type of healing there is, in theory, no differentiation of tissue from haematoma to lamella bone, but direct bone formation following osteoclastic resorption associated with secondary osteon formation. Complete interfragmentary contact of the entire fracture line is probably not seen in clinical practice.

Using other methods of fracture repair such as bracing or external fixation, union occurs by indirect healing. In this pattern of fracture repair the fragments are stabilized by a rapid bridging of periosteal callus between the fragments. The amount and distribution of callus is related to the degree of stability afforded by the fixation device. With an increase in the stiffness and rigidity of an external fixation frame the extent of callus formation is reduced. A reduction in frame stiffness may result in prolific callus formation but also pin tract infections and a hypertrophic delayed union. However, modulation of indirect fracture healing can be attained by the daily application of short periods of controlled interfragmentary micromovement (Goodship and

Kenwright 1985). Application of these short periods of mechanical stimulation can override the inhibitory effect of a high rigidity frame. Although, it has also been shown that the pattern of repair is further influenced by the specific mechanical characteristics of the micromovement regime. Certain regimes were found to enhance healing, others resulted in inhibition, whereas some resulted in no different effect to that seen with the rigid frame (Kenwright and Goodship 1989). Micromovement regimes that stimulated fracture healing produced changes that were visible on radiological examination as periosteal callus formation within 2 weeks of application of the stimulus.

Transduction mechanisms involved in converting the mechanical signal to cellular activity are poorly understood. Prostaglandins, primarily of the E series, have been shown to be associated with modulation of bone remodelling. In addition, inhibitors of prostaglandin synthesis have been shown to delay the healing of bone (Norrdin *et al.* 1990). This study was designed to test the hypothesis that specific patterns of callus formation, associated with defined characteristics of mechanical stimulation, are associated with different levels of prostaglandin mediators at the fracture site.

Methods

In six groups of six skeletally mature Welsh Mule ewes a standard 3-mm mid-diaphyseal osteotomy was made in one tibia and stabilized with the uniaxial Oxford Mark II fixator. Three groups were allowed to heal for a period of 3 weeks, in the remaining three groups the healing period was extended to 5 weeks. The effect of two different micromovement regimes was compared with a rigid control in each experiment. The fixator was applied using a jig to ensure that frame geometry was identical in all groups. Two regimes of applied micromovement were used on the basis of previous studies. One regime had been shown to enhance healing, this comprized an initial axial displacement of 1 mm using a force of 200 N and a rate of displacement of 40 mm/s, and the other, with the same characteristics but a 2-mm axial displacement, inhibited the repair process.

At post-mortem the callus was snap frozen and sectioned on a heavy duty cryomicrotome. Serial sections were used alternately for prostaglandin E_2 and $F_2\alpha$ assay and histochemical staining for acid and alkaline phosphatases. The prostaglandins were measured by powdering approximately 1 g of the bone sample by crushing, followed by acidification with 0.34 M citric acid and extraction with cyclohexane : ethyl acetate (1 : 1; v : v). The lipid-containing supernatant was evaporated and the residue reconstituted in 1.0 ml phosphate buffered saline (pH 7.4). Radioimmuno assay was carried out on a double-blind basis and the code broken after analysis.

Results

Following reallocation of the assay data to experimental groups it was evident that values were consistent within groups with relatively low variances. The levels of both E_2 and $F_2\alpha$ were not significantly different between the contralateral intact control tibiae, in the groups subjected to micromovement and the rigidly fixed osteotomies at both time points. Prostaglandin E_2 and $F_2\alpha$ were significantly increased by the application of a short period (17 min) of micromovement at both 3 and 5 weeks compared with both rigidly fixed osteotomies and intact contralateral bones (Fig. 30.1). The difference in levels

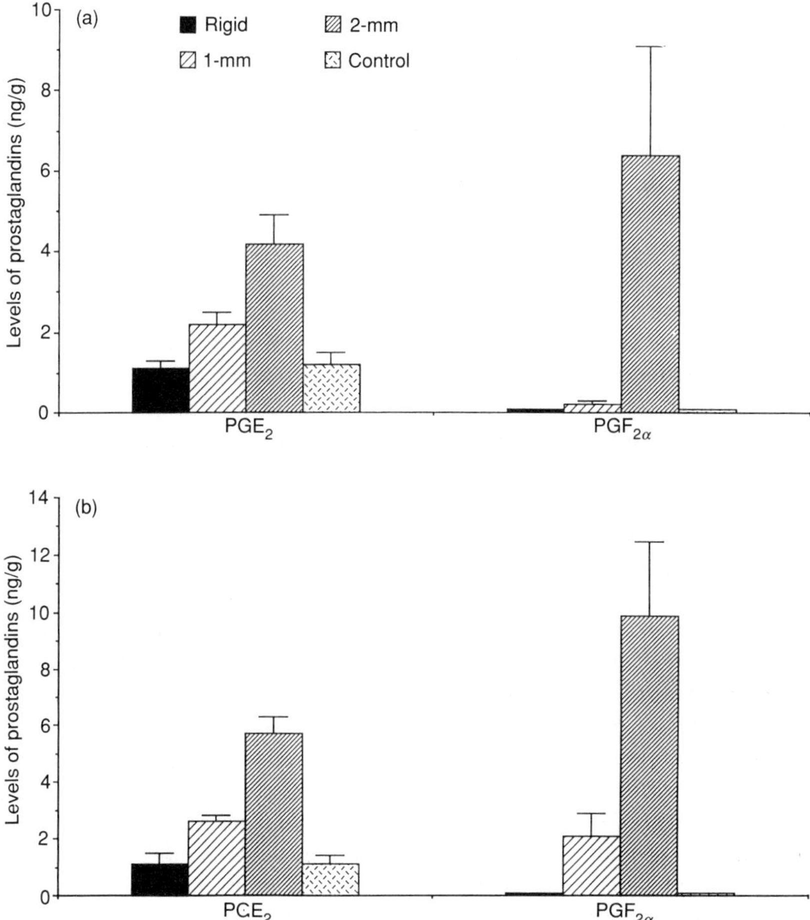

Fig. 30.1 Levels of prostaglandins in fracture callus undergoing regimes of micromovement; (a) at 3 weeks, (b) at 5 weeks.

of the two prostaglandins in relation to the mechanical stimulus of micromovement was seen as early as 3 weeks after osteotomy and 2 weeks after commencement of applied micromovement. The ratio of $F_2\alpha$ to E_2 was greater in the inhibitory micromovement regime at both timc points.

Discussion

These preliminary data suggest that mechanical stimulation may induce prostaglandin synthesis in fracture callus. Furthermore, the ratio of different types of prostaglandins is correlated with different mechanical characteristics of the applied micromovement in the early stages of fracture healing. This may suggest that the effect of mechanical stimulus on callus formation and remodelling could be mediated by prostaglandins. The role of prostaglandins in the control of bone formation and resorption is not fully understood. In a review by Norrdin *et al.* (1990) it has been suggested that the E series has the greatest activity in bone. Both formation and resorption of bone may be related to prostaglandin levels. There is also evidence of stimulation of prostaglandins by mechanical stress in cultured osteoblasts. Injection of PGE_2 has been shown to induce new bone formation on periosteal and endosteal surfaces. Furthermore, inhibitors of prostaglandin synthesis used during fracture healing lead to a delay in union. There is considerable circumstantial evidence to link prostaglandins with mechanically-related bone remodelling and repair. However, further work is required to determine whether these substances are inherent components of the transduction of mechanical stimuli to bone cell activity or merely coincidental products of the process of fracture healing.

Debate in the meeting

Is the prostaglandin that is being measured outside the cell really metabolic or physiological? Is it an exocronic factor, or, as most people think, an autocronic factor? (An exocronic factor is something which is producd by the cell with the purpose of targeting some other tissue or cell nearby. An autocronic factor is produced by the cell and influences the cell only.)—This is a preliminary study. One of the problems with prostaglandins is that there are too many of them. The authors have confirmed that the E_2 levels go up relative to other prostaglandins after a fracture and that the levels depend upon the biomechanical environment that the fracture has been under. Other studies suggest that the observed surge in E_2 indicates that, by chance, this was the right prostaglandin to measure. (Prostaglandin E_2 is normally measured simply because it was the first prostaglandin to be described and there are the best assays for it.) There are many other prostaglandins, however, which are being emitted by normal bone which is 'minding its own business'. The minute that bone is insulted, in whatever way, it suddenly produces a surge of prostaglandin E_2, i.e. it has switched, and it is the switch that is probably important rather than the absolute level. The other important observation to note is that, from other experiments, it seems that if you damage a bone on one side, the bones on the contralateral side and throughout the skeleton will respond. This is an important point for orthopaedics generally as well as for the interpretation of our results.

Prostaglandin E_2 seems to be there whether you are laying down bone or resorbing bone. If it is a switch, what is it switching to?—The switch depends upon the conditions at the time. There are also several different classifications of prostaglandin receptors so, although it may be possible to detect the release of prostaglandin, it will not necessarily be possible to know which receptors are active at any one time.

31 The effect of mechanical deformation on the distribution of potassium ions across the cell membrane of sutural cells

F. McDonald and W. J. B. Houston

Introduction

It has been recognized from very early times that bone growth can be modified by external forces. The cultural practices of head-binding of the children of South American Indians (cited by Poswillo 1989), or foot-binding in the children of the Chinese (cited by Veith 1980) extends far back into history. More specifically Galileo (1638) stated that there was a relationship between the loading of a bone and its form. The earliest scientific explanation offered for this relationship was by Wolff in 1892, who stated that the adult shape, structure and form of each bone is influenced by the mechanical circumstances to which it is subjected. In the following decades there was considerable controversy as to whether it was stress or strain that mediated the biological signal that led to bone remodelling. When considering bone remodelling, strain is outlined as the change in physical dimensions of either the bone structure or its cellular content Stress on the other hand is the force stored up within the structure to generate the change in shape.

These theories failed to explain how the physical change was detected by the cells that effected the bone remodelling. To do this it was proposed that the piezo-electric phenomenon, which can be detected when dry bone is strained, might be the signal detected by the cells (Bassett *et al.* 1965). This theory attracted considerable attention and several therapeutic procedures based upon electrical and electromagnetic phenomena were established (Bassett *et al.* 1974, 1976). This was in an attempt to improve fracture healing and repair in cases of nonunion of bony structures. The clinical results have been contentious. In some studies it appeared that there was an increase in the number of subjects producing bony repair (Brighton and Pollack 1985; Brighton 1982; Bruce *et al.* 1987). Other groups reported no evidence of increased repair (Papatheofanis and Papatheofanis 1989). A problem with these electrical theories is that in the living organism, electrical charges leak away into adjacent tissues and it is questionable whether the piezo-electric potentials could influence cells at the bone surface. Furthermore there are many other sources of electrical potential differences, such as nerves or striated muscles. These potentials are of such a magnitude that they overwhelm any possible electrical signals resulting from bone strain.

To investigate the effects of external loading at the cellular level, experiments were designed to assess the effects of membrane distortion on the ionic distribution across the cell membrane. The objectives of this study were:

1. To develop an experimental model for examining ion distribution across the membrane of bone-forming cells.
2. To examine the effects of physical distortion on ion distribution in these bone-forming cells.

Review of the literature

Electrical effects

The electrical events associated with many biological processes are related to their possible roles in the control of biological processes. The electrical effects of bone are broadly divided into the findings of dry bone (piezo-electric) and those of wet *in-vivo* bone (streaming potentials) (Bassett *et al.* 1965 and Andersen and Eriksson 1968 respectively).

Piezo-electric effects

Two types of piezo-electric effect are described: positive and negative. The positive effect is due to strains generated within the crystal lattice of a material leading to the production of a potential difference across the faces of that crystal. The negative effect is seen when an electric charge is passed across a molecule or crystal and leads to an inherent strain within that molecule (Isaacs 1987).

Both effects involve the organic molecules of collagen and the inorganic crystals of hydroxyapatite. The hypothesis that piezo-electric potential differences are the cause of bone remodelling was developed by Yasuda (1955) and Fukada and Yasuda (1957). The observed effect was assessed to be due to the distortion of long-chain molecules, particularly the crosslinkages within such molecules. The junction between the hydroxyapatite and the collagen has also been implicated as a site of charge production (Marino and Becker 1967). The ability of a symmetrical molecule such as collagen to generate charge can be questioned (Currey 1984*a*,*b*).

When a bone is deformed, an electronegative charge is produced on the concave surface produced by the deformation; while the production of a convex surface is linked with an electropositive charge (Bassett 1976). Placement of electrodes in bone leads to bone deposition around the charged cathode, and reportedly to bone loss around the anode (Bassett *et al.* 1965). A point not emphasized was that bone formed around the electrodes placed into bone when there was no potential difference across the electrodes and when no electrical circuit existed, thus excluding galvanic action. The technique of osseo-integrated implants is a further demonstration that the insertion of a metal within bone is sufficient *per se* to stimulate bone deposition.

Streaming potentials

An alternative source for the electrical occurrences described in the previous section is that derived from streaming potentials generated within the fluid of bone (Eriksson 1974). A streaming potential is an example of electric potential developed between two components by an electrolyte flowing between the solid surfaces. Physically it is generated by the movement of the electrostatic double layer at the fluid–bone interface. This electrical double layer arises because a net surface charge is gained by the bone surface when it is in contact with an ionic fluid (Anderson and Eriksson 1968).

In bone, such potentials could possibly arise in vascular channels, Haversian systems, canaliculi, microporosities of the structure, and because of blood flow and interstitial fluid movement. Differential movement of ions within a structure leads to a short-lived charge differential that behaves like a streaming potential (Pollack *et al.* 1984*a*). Neutrality is restored following equilibration of ion distribution. The external distribution of the electrical potential difference, observed between the two outer surfaces, is much smaller than that observed locally in the walls of the osteones. This has been attributed to the streaming potential (Pollack *et al.* 1984*b*).

The electromechanical theory has attracted much investigation. Microelectrodes were used to examine the distribution of stress generated potentials (SGP) within the structure of bone. SGP distributions were associated with Haversian systems and confined within the cement sheaths (Pienkowski and Pollack 1983).

SGPs are too high in wet bone and decay too rapidly for classical piezoelectricity to be solely responsible for the electrical effects observed (Johnson 1984).

Experiments conducted on both dry and wet bone in solutions of differing ionic strengths of potassium and sodium showed that the theory of piezoelectricity would explain the electromechanical behaviour of dry bone. In wet bone, the potential depended on the ionic concentration of the fluid and the differing results were due to the contributions made by streaming potentials (Williams *et al.* 1979). Stress generated potentials in fluid-filled bone varied with the fluid conductivity and fluid viscosity (Pollack *et al.* 1984*a*). It is possible to calculate the potential generated by the distortion of a fluid by the formula

$$V = \frac{\zeta \varepsilon \delta P}{4\pi\eta\sigma}$$

where

ζ = the electric potential difference between the slip plane and the bulk of moving liquid often known as the zeta potential (ZP),
V = the magnitude of the potential,
δP = the pressure difference that forces the liquid through the channel,
ε = the dielectric constant of the liquid,
η = the viscosity of the liquid, and
σ = the specific conductance.

The zeta potential of bone is dependent on the ion concentration of the fluid; i.e. interstitial fluid. At high concentrations the zeta potential reverses polarity (Williams *et al.* 1979). This suggests that the electrostatic double layer at the fluid-filled surface of bone matrix is an important region contributing to the electromechanical effects of bone.

An anatomical model for streaming potentials in osteones has been developed. This was based on the electromechanical theory and explained the behaviour of intra-osteonal potentials and the dependence of potentials on solution viscosity and conductivity. Time dependency of the electrochemical effects associated with ion distribution has been shown (Pollack *et al.* 1979). The slow decay of the signal in wet bone is not due to an electrical decay, as in the piezo-electric effect, but to the relaxation of fluid flow in the stressed bone. These are the characteristics of decay due to streaming potentials (Pollack *et al.* 1984*a*,*b*).

An electrokinetic model for the electrochemical effects of bone has been developed and shows that the fluid movement within the microporosity of bone is responsible for observed potentials whose origin is electrokinetic. The microporosity in bone, consisting of fluid spaces in and around mineral crystals encasing collagen fibrils, forms an enormous surface area. This area appears to dominate surface-related phenomena at low frequencies of loading (Salzstein *et al.* 1987; Salzstein and Pollack 1987). Emphasis was placed on the development of these electrical events around vascular channels. It was suggested that these channels must exist for streaming potentials to be generated. An inconsistent finding is that streaming potentials have also been shown in cartilage. Cartilage does not possess a Haversian system or canaliculi, nevertheless the currents generated and the characteristics that they showed were typical of streaming potentials. This was maintained when the surrounding ionic solution had its composition and pH varied (Eliot and Grodzinsky 1987). Vascular channels have been found within cartilage although they do not permeate to the same extent as bone channels. Their existence is temporary, appearing mainly in the first formed cartilages (Wilsman and van Sickle 1972). Owing to this temporary duration it is doubtful whether these channels play an important role in adult cartilage. To further attempt to explain this inconsistency, cartilage is said to hold the fluid between the extracellular matrix molecules. This does not have the relatively free flow of fluid as found in the Haversian system that is regarded as essential for the development of streaming potentials (Pollack *et al.* 1984*a*,*b*).

Both electrical effects suffer from a major criticism. The vascular system, periosteum, and tissue fluids are highly conductive and may drain away the change in potential difference (Jendrucko *et al.* 1977). It remains to be determined whether strain-induced potentials, however they are generated, are the irrelevant by-product of bone deformation or the vital link between it and the cellular population (Lanyon and Baggott 1976). If electrical effects are important it is appropriate to consider streaming potentials in an *in-vivo* situation rather than piezo-electric effects. This is because the events of dry bone while of interest may not reflect a biologically significant process in living bone.

Resting membrane potential

It is known that excitable cells (muscle and nerve) have a membrane potential at rest with the inner surface of the cell, negative to the outside. This is due to the unequal distribution of many ions, sodium, potassium, and chloride ions being the most prevalent The maintenance of the imbalance of ions is produced by the proteins found crossing the cell membrane and called pumps. The predominant pump is the sodium pump in which for every three sodium ions pumped out of the cell, two potassium ions are added to the intracellular contents. Gray *et al.* (1987) have described a class of vital dyes that bind selectively to the ions responsible for the membrane potential. Further reviews of these substances are given by Tsien (1989). Upon binding, fluorescence occurs and from standard solutions the concentrations of specific ions can be determined.

Materials and method

Standardized curves

Known concentrations of the ions were made up from 0 to 180 mM/l of water using either potassium or sodium chloride. Three millilitres of the solution were added to a petri dish. To these solutions were added either 10, 20, 50, or 100 µg of the sodium or potassium indicator dissolved in dimethyl sulphoxide. The dish was agitated and left for 5 min. Fluorescence was then recorded over 1, 2, and 5 min periods. Ten readings in total were achieved for each solution and the median and upper and lower limits plotted (Figs 31.1(a, b, c)). From

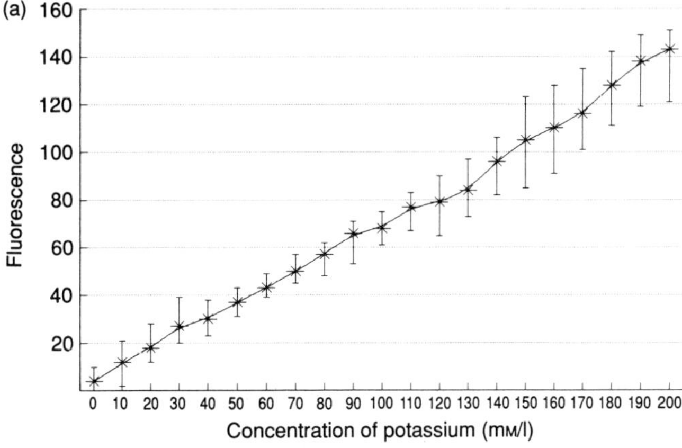

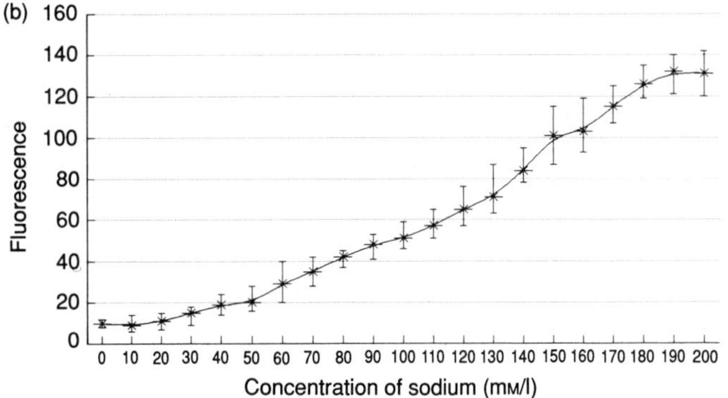

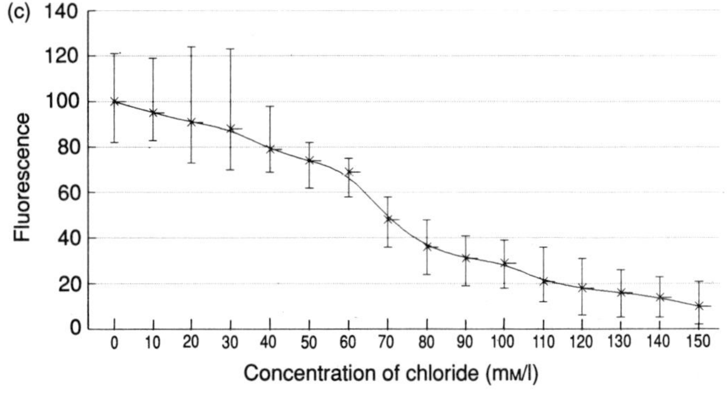

Upper limit Lower limit Median

Fig. 31.1 Relationship between fluorescence and ion concentration.

these plots minimal variation of the upper and lower limits was established using 50 μg of dye. A pilot study was undertaken using 3 ml minimum essential medium (MEM) and a rat calvarium similar to those used in the main study to establish the stability of the fluorescence of the *in-vitro* cells. Thirty fluorescent readings were taken at the start of the period and at 10, 30, 60, and 90 min. The readings were compared using Student's *t* and Mann–Whitney tests. Both showed no significant differences between intervals at the 95 per cent level. Further MEM was examined with 50–150 mM/l water and again 30 readings were taken. There was no significant difference between the fluorescence with MEM and the standardized curves for sodium and potassium.

The chloride ions were assessed with 1, 10, 20, and 50 mg of fluorescent indicator dissolved in water. The loading of the cells by this dye appeared stabilized at 20 mg per petri dish. Again a control was examined using MEM and a rat calvarium. From these standardizations it was decided to load the calvaria with 50 μg of dye for sodium and potassium and 20 mg for chloride ions. The chloride-binding fluorescent dye was water soluble and binding with chloride quenches the fluorescence of the dye. The dyes for sodium and potassium were dissolved in dimethylsulphoxide. The binding to sodium or potassium ions resulted in fluorescence. There was a level of auto-fluorescence but this was less than 1.5 per cent of the total fluorescence and so was ignored.

The system that was examined was the rat calvarium previously reported (Meikle *et al.* 1979). This has been used for the examination of biochemical changes following distortion of the cellular elements *in vitro*. This experimental system lends itself to organ culture and to the introduction of vital dyes. In addition it is possible to observe, by fluorescent microscopy, the quantities of dye bound to specific ions. The ions of interest are those likely to maintain the membrane potential of the osteocyte. This in turn could possibly influence intracellular processes when their concentrations are changed. Neonatal rat calvaria were removed after killing by an overdose of carbon dioxide. The calvaria were placed in MEM containing alanine and glucose. The pH was buffered to 7.3 with NaOH, and the solution filtered through a bacteria-free filter. The calvaria were randomly divided into three groups for examination. One group was used to examine changes in sodium ion distribution, a second for potassium ion distribution and a third for chloride ion distribution. The calvaria were cultured at room temperature (20°C) with the fluorescent marker specific to the particular ion in question. The fluorescent dyes used were:

(a) binding to sodium ions SBFI™;
(b) binding to potassium ions PBFI™;
(c) binding to chloride ions SPQ™.

The use of dimethyl sulphoxide acted to permeabilize the cell so that large molecules could gain entry. On return to the normal solution within the cell the molecules became trapped. After 30 min the calvaria were attached at one end of a Petri dish by cyanoacrylate adhesive. This was at a distance from the suture to prevent toxic effects of the adhesive on the cells (Fig. 31.2). At the opposite end a micromanipulator was attached, again using cyanoacrylate adhesive, and the Petri dish positioned on the microscope stage. The Petri dish contained the MEM and specific dye already used to culture the calvarium. Using polarized light, a cell was located and its suitability confirmed by using an ultraviolet filter and epifluorescence. For sodium ions the maximum amount of fluorescence was found externally to the cell while for potassium it was found mainly intracellularly. Once a suitable cell was located, the calvarial tissues were placed under tension by retracting the electrode manipulator. The fluorescence was then determined and the concentration of ion calculated from the standard curves.

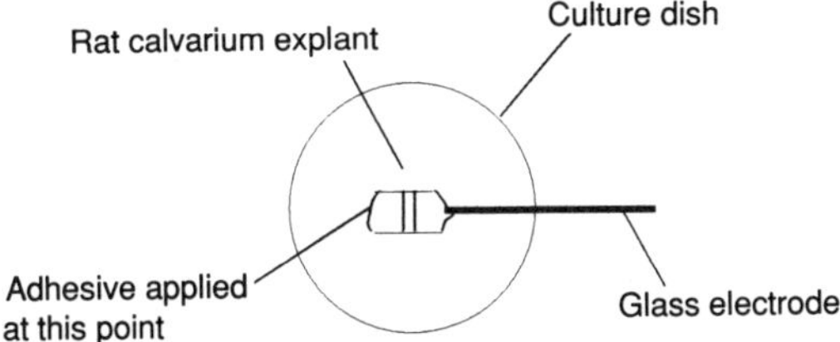

Fig. 31.2 Diagrammatic representation of manipulation of calvaria.

Potassium blocking agents

It appeared that the potassium distribution was affected mainly by physical distortion. To determine whether this change in K^+ distribution was the result of stretch activated channels, the K^+ channel blockers, quinine, and tetraethylammonium (TEA^+) were included in the media. TEA^+ was loaded at a rate of 1 and 5 mM while quinine was loaded at 10 μM, as used by Bokvist *et al.* (1990), in 10 preparations, 30 in total.

Results

Potassium ion concentration inside the cell, as determined by the fluorescence, decreased following physical distortion (Fig. 31.3). There was a non-normal distribution of the results for sodium and potassium ion distribution, consequently Mann–Whitney nonparametric tests were applied to the variations of ion distribution. The change in sodium and chloride ions, before and after stretching showed no significant difference (Figs 31.4 and 31.5), while

(a)

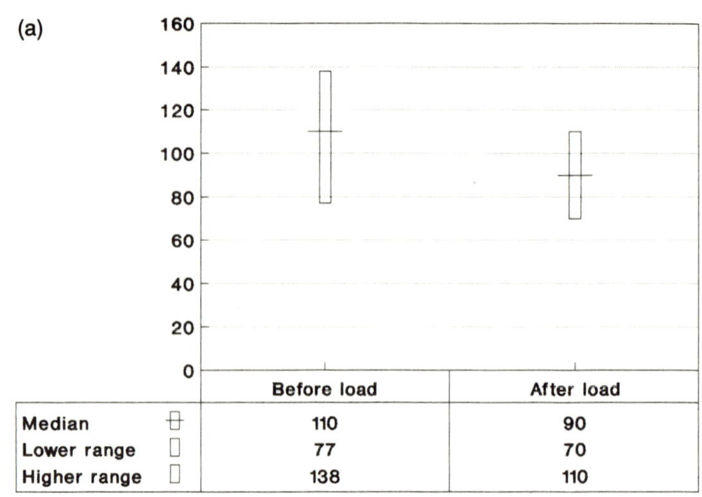

Intracellular

(b)

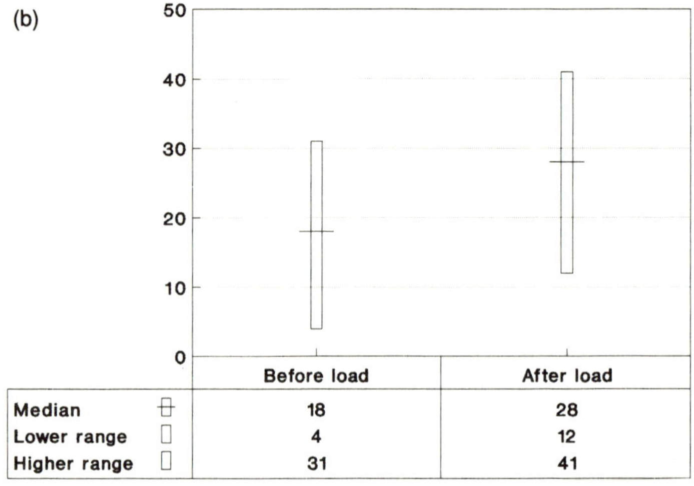

Extracellular

Fig. 31.3 Effect of load on potassium ion concentration inside and outside the cell.

the change in potassium distribution was significant at the 95 per cent level. The intracellular potassium change had a probability of $p<0.001$, while extracellular readings were $p<0.05$. Furthermore, it was noticed that the change in potassium permeability was not a gradual event. The response of the ion distribution demonstrated an all-or-none response. That is, until a displacement was applied to the calvarium, no change occurred. Once the

(a)

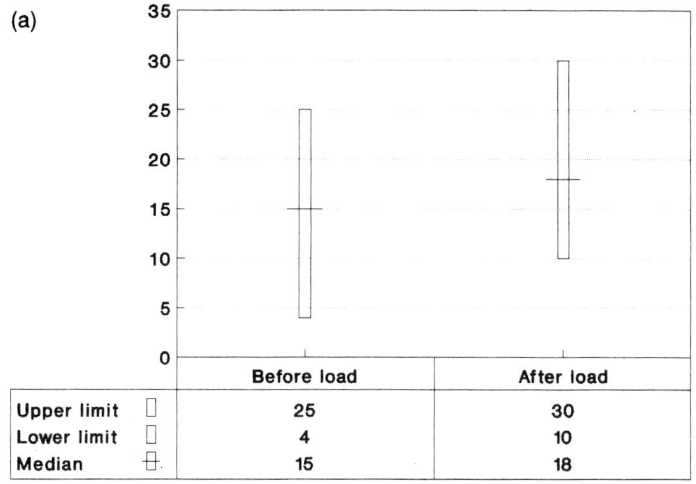

Intracellular

(b)

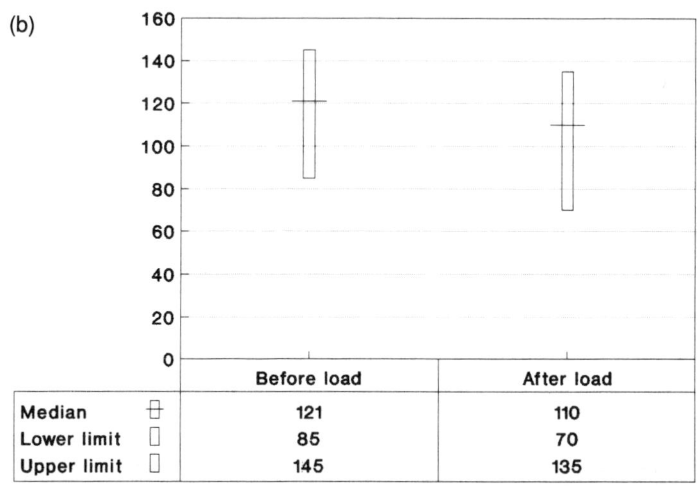

Extracellular

Fig. 31.4 Effect of load on sodium ion concentration inside and outside the cell.

extension was sufficient, a change in potassium-binding indicator fluorescence was detected that remained while the extension was maintained. Any attempt to increase the load did not change the fluorescence until the whole preparation failed. Upon removal, the fluorescence returned to its previous resting level. Of the cells that showed this phenomenon, the load/unload cycle could be repeated two to three times before the cell failed to respond.

(a)

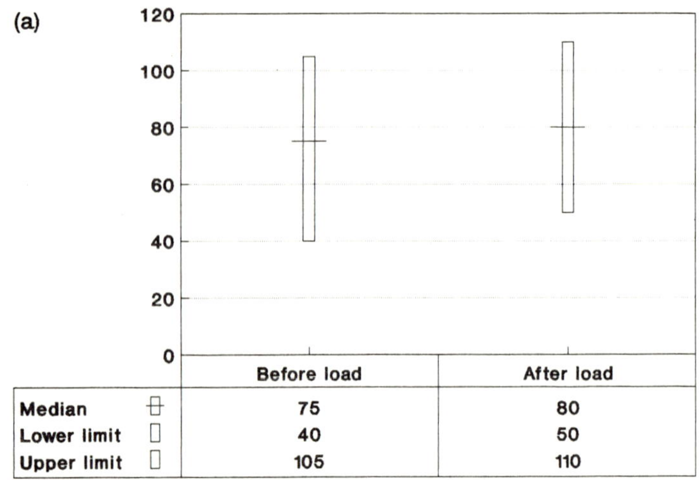

Intracellular

(b)

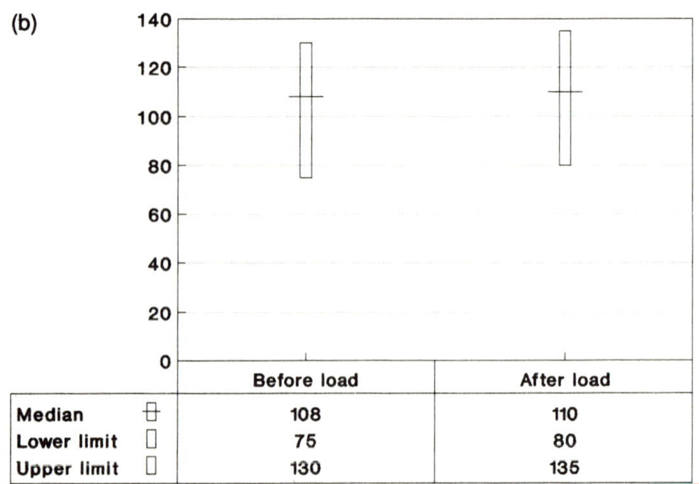

Extracellular

Fig. 31.5 Effect of load on chloride ion concentration inside and outside the cell.

Following the experimental period, the calvarium was marked close to the responsive cell with the tip of a microelectrode and fixed in formalin. The preparation was then stained with alkaline phosphatase to confirm that it was a bone-forming cell. Sodium and chloride ion fluorescence showed little variation, readings before and after stretching being statistically similar, i.e. no significant difference at the 95 per cent level. The variability of the chloride fluorescence made it difficult to validate.

Quinine and tetraethylammonium ions produced no effects on the change in potassium ion distribution before and after distortion.

Goldman equation

The potential across a membrane when several ions are present on each side is given by

$$E=\frac{RT}{F}\log_e\left(\frac{P_aA^+_{in}+P_bB^+_{in}+P_cC^+_{in}+\cdots+P_aX^-_{out}+P_bY^-_{out}+P_cZ^-_{out}+\cdots}{P_aA^+_{out}+P_bB^+_{out}+P_cC^+_{out}+\cdots+P_aX^-_{in}+P_bY^-_{in}+P_cZ^-_{in}+\cdots}\right)$$

in which

A^+, B^+, C^+ ... are monovalent cations,
X^-, Y^-, Z^- ... are monovalent anions,
P=permeability of the membrane to the ion indicated,
R=gas constant = 8.31 J/(mol K),
T=temperature in K (degrees absolute), and
F=Faraday constant = 96.50 C per equivalent.

This takes into account the relative permeability of the membrane to different ions and is applicable even in nonequilibrium conditions.

This equation has been modified, and used for the calculation of membrane potential by Chow *et al.* (1984), into the following format:

$$E=\frac{RT}{F}\log_e\left(\frac{[K^+]_0+(P_{Na}/P_K)\cdot[Na^+]_0}{[K^+]_i+(P_{Na}/P_K)\cdot[Na^+]_i}\right).$$

Values available for the expression P_{Na}/P_K are given by

Chow *et al.* (1984) =0.389
Civitelli *et al.* (1987) =0.317

Calculations of the change in membrane potential with these values before and after distortion are shown in Table 31.1.

Table 31.1 Change in membrane potential before and after distortion

P_{Na}/P_K	Before	After	Change
0.389	-17.4005 ± 1.212	-10.1245 ± 0.913	+7.276
0.317	-22.241 ± 0.986	-13.9928 ± 1.130	+8.2482

(Results are mean ± SD).

Many techniques have been used to measure the potential of osteoblast-like cells. There is a wide range varying from −4 mV (Jeansonne *et al.* 1978) to −41 mV (Dixon *et al.* 1984). The results from this experiment compare favourably, although the cells examined in some investigations were the transformed cells (Duncan and Misler 1989).

Discussion and conclusions

Until recently it has been difficult to determine the changes in membrane potential of bone cells. A pilot study was undertaken using glass ion-sensitive electrodes to examine fibroblasts before and after mechanical deformation. However, while insertion of the electrode was possible, only a short-lived electrical signal (10–15 mV change in potential lasting no more than 2 ms) was observed followed by complete loss of any potential. This implied cell death. It was consequently impossible to distinguish artefact from a response to strain. Previous reports have used patch clamps (Davidson *et al.* 1990) to distort the cells, a technique which reduces the damage to the cell. In an attempt to further limit the cellular damage, vital dyes have been used. These are capable of being cultured and introduced into cells with minimal effects on their metabolism (Tsien 1989).

Skerry *et al.* (1989) discussed the possibility that proteoglycan molecules crossed the phospholipid membrane and in so doing transferred the external distortion to the intracellular messengers such as cAMP. The implication of this work is that a major signal to modify the osteogenic cell's behaviour is the change in ionic distribution: specifically potassium ions. This change in distribution was an all-or-nothing response and could lead to a hypothesis involving recruitment of bone-forming cells. That is, the cells are only signalled into bone deposition when their potential has reached a threshold as determined by ion distribution. This would imply that strains, as they are distributed around a limb bone, could lead to a graded response. As the strain level increases, so more osteogenic cells are recruited into activity.

A significant finding was that any increase in strain produced an increase in potassium concentration external to the cell. Despite agitating the plate this did not appear to disperse. A possible explanation may be that the calvarium retained any ionic change preventing diffusion. There may however be an increase in ions local to the membrane. The variation in readings in standard solutions demonstrates the variability with known quantities of ion and this change extracellularly may be a variant of this variability. The probability was at the 95 per cent level for this but less than the 95 per cent level for the intracellular changes.

The distribution of ions responsible for the membrane potential would appear to explain the effect of electric current upon the biological response of bone cells in clinical situations. In many cells there is a relationship between membrane potential and internal behaviour. Here release of stored calcium is actuated by the change in membrane potential.

It has been postulated that the changes in potassium and calcium ion distribution are responsible for the regulation of vascular tone. This effect is mediated by the endothelial cells and varies as a response to haemodynamic stresses (Olsen *et al.* 1988; Lansman *et al.* 1987). With respect to osteoblasts and osteoblast-like cells, Duncan and Misler (1989) and Davidson *et al.* (1990) have demonstrated many types of ion channels, calcium, and potassium channels seeming to play a major role. The concept of stretch-activated channels in bone is not therefore new. This work has, however, focused on specific potassium channels. The inability of potassium blockers to inhibit the change implied that these channels were stretch-activated. This change in ion distribution would further demonstrate a change in membrane potential. This would agree with observations of excitable cells in that the trans-membrane changes may be the first stage in changing the intracellular events.

There appears to be a dilemma in the extrapolation of the existing theories of electric modulation of cellular behaviour. Known streaming potentials and piezo-electric potentials, being in the order of 1–6 mV, also change direction dependent upon the particular loading of the bone. This effectively makes the potential alternating in character. The therapeutic electrical stimuli apply potentials of 6 V in a single direction and it has been difficult to explain these two major differences (McDonald and Houston 1990). The soft tissues act as a current sink. This may well allow the 6 V to be reduced to the millivoltage range at the cellular level. The change of potential reported here demonstrates that this change exists in the millivolt range at the cellular level. This would clarify the situation concerning therapeutic use of currents. However, further work is required to establish the next possible stage in the cell signalling process.

Debate in the meeting

The volume of the extracellular material is vastly greater than the cells. Does this not affect precision of the dye technique?—The measurements are taken just at the periphery of the cell, although, because the cell is three-dimensional, it is difficult to guarantee that measurements are precisely adjacent to the membrane.

In the Goldman equation the membrane potential is set by the conductance and the inside-out ion concentration. The conductance changes when the channel opens, so it is the change in conductance which sets the membrane potential and not necessarily the number of ions flowing across the cell membrane. This means a large change in the number of ions is not required to generate a large change in membrane potential. Looking at potassium outflow across the cell, only 3000–4000 ions are needed to cross in a short time to create a large change in membrane potential, so this experimental method may not be the one of choice. Would a membrane potential dye be better?—The problem with that technique is to know which ionic imbalance is causing the observed change.

Concerning the time in which these changes took place, in order to differentiate dynamic and static changes, what was the time for which the strain had to be maintained before a change in potential was observed?—A lag existed in the signal from the spectrophotometer used to measure fluorescence that introduced a limited speed of response. There is evidence that calcium changes happen within milliseconds. The changes reported here began within a period of seconds to minutes, continued up to 30 min, and could then be maintained for a further 20 or 30 min.

32 Strain induced bone remodelling and prostaglandin E_2

D. W. MURRAY AND N. RUSHTON

Introduction

Bone remodels in order to withstand the mechanical stresses exerted upon it (Wolff's law). These stresses cause strains within the bone which may influence the resident bone cells. If the strains are outside the normal range then the cells may release mediators that stimulate remodelling. The aim of this investigation was to subject bone cells to strains both within and outside the normal range, and to determine how this effected the release of PGE_2, a possible mediator of bone remodelling.

Bone behaves elastically up to about 3 mstrain, and then plastically until it fails at about 30 mstrain (Dumbleton and Black 1975). The initial part of this plastic deformation, up to about 10 mstrain, is viscous, whereas larger strains are associated with microstructural damage which is soon repaired. The strains that occur in human long bones are similar to those that occur in animals, and may reach 2 or 3 mstrain during strenuous activity (Lanyon *et al.* 1975; Rubin 1984). The strain rates are of the order of 10 mstrain/s (Lanyon *et al.* 1975). The magnitude and direction of the strains varies within a bone and is altered by an endoprosthesis (Indong and Harris 1978). The physiological range of strain, which includes strains encountered with normal activity and those not associated with microstructural bone damage, therefore extends up to about 5 mstrain.

Bone remodelling depends partly on strain magnitude, with increased strains causing hypertrophy and decreased strains cause atrophy. Within a few months of increasing the loading on a bone, the bone has remodelled to such an extent that the strains, which initially rise, return to normal (Goodship *et al.* 1979). Bone remodelling also depends on the distribution of the strain and on the strain rate, with higher strain rates inducing more hypertrophy than lower ones (Rubin and Lanyon 1984; O'Conner *et al.* 1982). Bone remodelling is independent of the frequency of load cycles, provided that there are more than about 10 cycles per day (Rubin and Lanyon 1984).

PGE_2 is known to stimulate bone formation and resorption, both *in vivo* and *in vitro* (Raisz and Martin 1983; Shin and Norridin 1986). The PGE_2 released from bone cells has been shown to increase when they are subject to nonphysiological or biaxial strains (Somjen *et al.* 1980; Yeh and Rodan 1984;

Brighton *et al.* 1991). It is however not clear what effect cyclical uniaxial physiological strains have on PGE_2 release from bone cells. The bone cells that respond to strain are thought to be osteoblasts (Binderman *et al.* 1984).

Method

Bone cells were cultured in special wells, the bases of which were subjected to cyclical strains in a stretching machine (Murray and Rushton 1990). The PGE_2 released both during a control period prior to straining and during the straining period was determined, using a radioimmunoassay (New England Nuclear, Stevenage, UK). From these amounts the percentage increase in PGE_2 release was calculated. The cells therefore acted as their own controls.

A population of bone cells rich in osteoblasts was obtained using a method based on that of Wong and Cohn (1974). Calveria from 4-day-old mice, having had all fibrous tissue removed, were sequentially digested in 1 mg/ml collagenase (Sigma type 1, Sigma Ltd., Poole, UK) at 37°C. The cells released between 30 and 120 min were washed and then cultured in the stretching wells. After 3 days, when the cells were confluent, they were subjected to mechanical strain. They were cultured in HEPES buffered medium 199 with 5 per cent fetal bovine serum (FBS) and 1 per cent antibiotics (penicillin 1 U/ml and streptomycin 5 mg/ml) in a humidified atmosphere at 37°C. The cells were examined microscopically both before and after being subjected to strain. Phase contrast microscopy was used pre-strain. Post-strain the cells were either stained with May Grunwald and Giemsa stain, or they were stained for alkaline phosphatase using the azo-dye coupling method.

Special wells were constructed with transparent bases that could be stretched repeatedly and elastically. These were clamped into the stretching machine. The stretching machine was computer controlled and was able to apply any strain pattern. In these experiments the strain was applied repeatedly in the form of a ramped square wave for the duration of the test period. The strain rate was 10 mstrain/s. In different experiments, different strains and different cycle times were used.

For statistical analysis, t tests or analysis of variance (ANOVA) were used. A logarithmic transformation was used when percentage increases were compared statistically. On the graphs, means and standard errors are plotted; (a) signifies $p<0.05$ and (d) $p<0.001$.

Results

Time course of PGE_2 release

Two sets of wells containing bone cells cultured from the same population

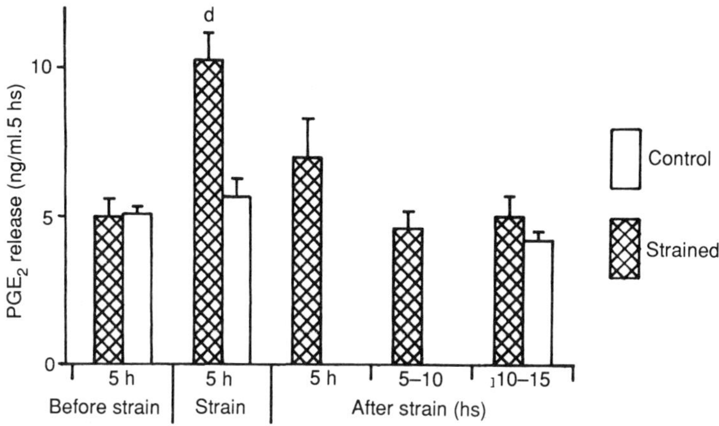

Fig. 32.1 Time course of PGE_2 release.

were treated in an identical manner, except that the test set was subjected to strain, and the control set was not. Every 5 h for a period of 25 h the medium was renewed and the PGE_2 production was assayed (Fig. 32.1). During the second 5 h period the test cells were strained repeatedly, by 28 mstrain, with a cycle time of 10 min. It was found that the PGE_2 release doubled when the cells were strained (a significant increase $p<0.001$). When the straining ceased, these cells continued to release an increased amount of PGE_2, but this gradually decreased until after about 5 h when it returned to the unstretched level and remained at that level. The PGE_2 released by the nonstrained control cells did not alter significantly during the 25-h period.

Strain magnitude

Bone cells were subjected to various strain magnitudes, with a constant cycle time of 10 min, and the percentage increase in PGE_2 release was determined (Fig. 32.2). The PGE_2 release was significantly different ($p<0.005$) at different strains. The experiment was repeated twice at 7 mstrain: no significant difference was found between the two sets of results, indicating that the results were repeatable. The graph comparing strain magnitude to PGE_2 release was biphasic. At 7 mstrain, and at $\geqslant$28 mstrain, the PGE_2 release was approximately twice the unstrained level, and these increases were highly significant ($p<0.001$, $p<0.001$, $p<0.001$, $p<0.001$). At 3 mstrain, and between 7 mstrain and 28 mstrain, the increases were small and not always significant. The increased release decreased from about 100 per cent at 1 and 28 mstrain to 20 per cent at 14 mstrain, decreases that were significant (both $p<0.01$).

At all strains tested, the PGE_2 release returned to the control levels by about 12 h after straining had ceased. The percentage difference in PGE_2 releases

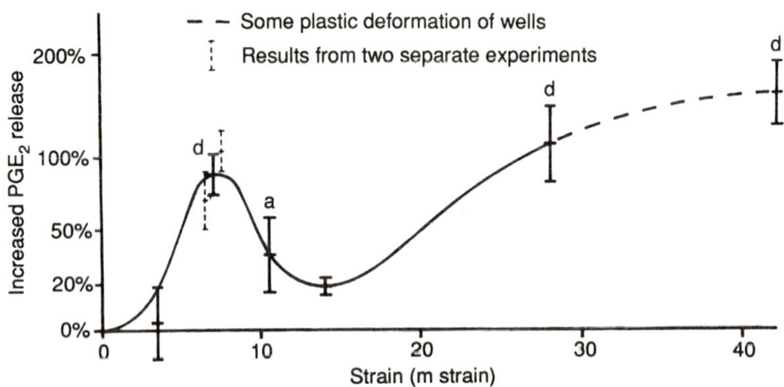

Fig. 32.2 The PGE_2 release at different strains.

between the post-strain levels (approximately 12 h after straining had ceased) and the pre-strain levels were: at 3 mstrain 4 per cent (SE 10), at 7 mstrain 4 per cent (SE 16), at 10 mstrain 19 per cent (SE 37), and at 28 mstrain 3 per cent (SE 6).

The influence of cycle time

Bone cells were subjected to strains of various magnitudes, with cycle times of 20, 100, or 600 s (Fig. 32.3). It was found that the PGE_2 release depended on the strain magnitude and was independent of cycle time.

Microscopy

Microscopy of both strained and nonstrained bone cells demonstrated that

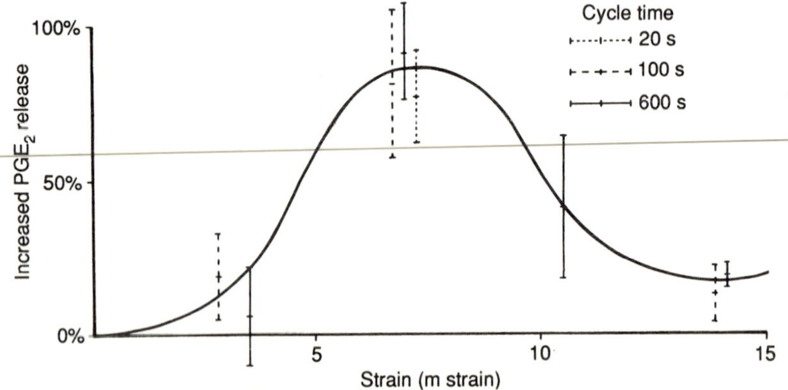

Fig. 32.3 The PGE_2 release at different cycle times.

strain did not cause a morphological change in the cells. A large proportion of cells stained positively for alkaline phosphatase confirming that they were probably osteoblasts.

Discussion

These results confirm the experimental findings of other workers: when bone cells are subject to large, nonphysiological, strains they release more PGE_2. This study has also shown that the PGE_2 release increases with physiological strain. The amount of PGE_2 released depends on the magnitude of the applied strain. For strain cycles of $\leqslant 10$ min, the PGE_2 release was independent of cycle time. As it takes approximately 5 h for the PGE_2 release to return to normal after straining, the PGE_2 release is likely to be independent of cycle time provided that this is less than a few hours. Other workers have shown *in vivo* that bone remodelling depends on strain magnitude, and is independent of cycle time provided that it is less than a few hours. The similarity of the *in-vivo* and the *in-vitro* results suggests that bone cells can detect strains directly and that they may control strain-induced bone remodelling by releasing PGE_2.

The relationship between the magnitude of the applied strain and PGE_2 release was biphasic. (It is unlikely that this was a statistical quirk as the dip that made the graph biphasic was statistically significant, $p < 0.01$.) The PGE_2 release was high with strains at the top of the physiological range and at strains that would be associated with microstructural damage. This suggests that PGE_2 may stimulate the osteogenesis that occurs with increased functional demands, and may have a role in initiating the remodelling that is necessary after microstructural damage. A possible explanation for the biphasic nature of the relationship is that there are two mechanisms that control the release of PGE_2. For physiological strains there may be a specialized receptor that controls the release of PGE_2. At very high strains the cell membrane disturbance may cause the increased release of PGE_2. If there is a specialized receptor that responds to physiological strains then it is likely to play a central role in strain-induced bone remodelling.

Debate in the meeting

Load was maintained on the cells for about 5 h. This is probably different from the physiological situation in which prostaglandin would presumably be removed. There is evidence that prostaglandin release rises and then diminishes during the first 5 min of load.

References to Section 5

Anderson, J.-C. and Eriksson C. (1968). Electrical properties of wet collagen. *Nature*, **218**, 166–8.

Arridge, R. G. C. (1984). Improvement of the high frequency content of fast Fourier transforms. *Journal of Physics D: Applied Physics*, **17**, 1101–5.

Arridge, R. G. C. and Barham, P. J. (1986). Fourier transform mechanical spectroscopy. *Journal of Physics D: Applied Physics*, **19**, L89–96.

Aspden, R. M. (1991). Aliasing effects in Fourier transforms of monotonically decaying functions and the calculation of viscoelastic moduli by combining transforms over different time periods. *Journal of Physics D: Applied Physics*, **24**, 803–8.

Aspden, R. M. and Hukins, D. W. L. (1981). Collagen organisation in articular cartilage determined by X-ray diffraction and its relationship to tissue function. *Proceedings of the Royal Society of London*, **B212**, 299–304.

Aspden, R. M. and Hukins, D. W. L. (1989). Stress in collagen fibrils of articular cartilage calculated from their measured orientations. *Matrix*, **9**, 486–8.

Aspden, R.M., Larsson, T., Svensson, R., and Heinegård, D. (1991). Computer controlled mechanical testing machine for small samples of biological viscoelastic materials. *Journal of Biomedical Engineering*, **13**, 521–5.

Bassett, C. A. L. (1976). Comment on theoretical evidence for the generation of high pressure in bone cells by R. J. Jendrucko *et al. Journal of Biomechanics*, **9**, 485.

Bassett, C. A. L., Pawluk, R. J., and Beckera, R. O. (1965). Effects of electrical current on bone *in vivo*. *Nature*, **204**, 652–4.

Bassett, C. A. L., Pawluk, R. J., and Pillam, A. A. (1974). Acceleration of fracture repair by electromagnetic fields.—A surgically non-invasive approach. *Annals of the New York Academy of Science*, **238**, 242–62.

Bassett, C. A. L., Pilla, A. A., and Pawluk, R. J. (1976). A non-operative salvage of surgically resistant pseudoarthroses and non-unions by pulsating electromagnetic fields. *Clinical Orthopaedics and Related Research*, **124**, 128–43

Behrens, F., Kraft, E. L. and Oegema, T. R. (1989). Biochemical changes in articular cartilage after joint immobilization by casting or external fixation. *Journal of Orthopaedic Research*, **7**, 335–43.

Binderman, I., Shimshoni, Z., and Somjen, D. (1984). Biochemical pathways involved in the translation of physical stimulus into biological message. *Calcified Tissue International*, **36**, S82–5.

Binderman, I., Zor, U., Kaye, A. M., Harell, A., and Somjen, D. (1988). The transduction of mechanical force into biochemical events in bone cells may involve activation of phospholipase A2. *Calcified Tissue International*, **42**, 261–6.

Bokvist, K., Rorsman, P., and Smith, P. A. (1990). Effects of external tetraethylammonium ions and quinine on delayed rectifying K^+ channels in mouse pancreatic beta-cells. *Journal of Physiology*, **423**, 311–25

Brighton, C. T. (1982). Use of constant direct current in the treatment of non-union. *Instruction Course Lecture*, **31**, 94–103

Brighton, C. T. and Pollack, S. R. (1985). Treatment of recalcitrant non union with a capacitatively coupled electric field. A preliminary report. *Journal of Bone and Joint Surgery*, **67A**, 577–85.

Brighton, C. T., Black, J., and Pollack, S. R. (ed.) (1979). *Electrical properties of bone and cartilage*. Grune and Stratton, New York.

Brighton, C. T., Strafford, B., Gross, S. B., Leatherwood, D. F., Williams, J. L., and Pollack, S. R. (1991). The proliferative and synthetic response of isolated calvarial bone cells of rats to cyclical biaxial mechanical strain. *Journal of Bone and Joint Surgery*, **73A**, 320–31.

Bruce, G. K., Howlett, C. R., and Huckstep, R. L. (1987). Effect of a static magnetic field on fracture healing in a rabbit radius. *Clinical Orthopaedics and Related Research*, **222**, 300–6.

Buckley, M. J., Banes, A. J., Levin, G., Sumpio, B. E., Solo, M., Jordan, R. *et al.* (1988). Osteoblasts increase their rate of division and align in response to cyclic mechanical tension *in vitro*. *Bone and Mineral*, **4**, 225–36.

Burr, D. B., Martin, R. B., Schaffler, M. B., and Radin, E. L. (1985). Bone remodelling in response to *in vivo* fatigue micro-damage. *Journal of Biomechanics*, **18**, 189–202.

Caveney, S. (1985). The role of gap junctions in development. *Annual Review of Physiology*, **47**, 19–35.

Chamay, A. and Tchantz, P. (1972). Mechanical influences in bone remodelling. Experimental research on Wolff's law. *Journal of Biomechanics*, **5**, 173–80.

Chow, S. Y., Chow, Y. C., Jee, W. S. S., and Woodbury, D. M. (1984). Electrophysiological properties of osteoblast-like cells from the cortical endosteal surface of rabbit long bones. *Calcified Tissue International*, **36**, 401–8.

Civitelli, R., Reid, I. R., Halstead, L. R., Avioli, L. V., and Hruska, K. A. (1987). Membrane potential and cation content of osteoblast-like cells (UMR 106) assessed by fluorescent dyes. *Journal of Cell Physiology*, **131**, 434–41.

Cohen, R. E., Hooley, C. J., and McCrum, N. G. (1976). Viscoelastic creep of collagenous tissue. *Journal of Biomechanics*, **9**, 175–84.

Correns, C. W. (1949). Growth and dissolution of crystals under linear pressure. *Faraday Discussions of the Chemical Society*, **5**, 267–71.

Currey, J. D. (1968). The adaptation of bone to stress. *Journal of Theoretical Biology*, **20**, 91–106.

Currey, J. D. (1984*a*). *The mechanical adaptation of bones*. Princetown University Press.

Currey, J. D. (1984*b*). What should bones be designed to do? *Calcified Tissue International*, **36**, S7-10.

Davidson, R. M., Tatakis, D. W., and Auerbach, A. L. (1990). Multiple forms of mechanosensitive ion channels in osteoblast-like cells. *Pflugers. Archiv. European Journal of Physiology*, **416**, 646–51.

Derbyshire, B. (1982), Ph.D. Thesis, University of Manchester Institute of Science and Technology.

Dixon, S. J., Aubin, J. E., and Dainty, J. (1984). Electrophysiology of a clonal osteoblast-like cell line: evidence for the existance of a Ca^{2+} activated K^+ conductance. *Journal of Membrane Biology*, **80**, 49–58.

Doty, S. B. (1981). Morphological evidence of gap junctions between bone cells. *Calcified Tissue International*, **33**, 509–12.

Dumbleton, J. H. and Black, J. (1975). Mechanical properties of tissues. In *An introduction to orthopaedic materials*, pp. 106–31. C. C. Thomas, Springfield, Illinois.

Duncan, R. and Misler, S. (1989). Voltage activated and stretch-activated Ba^{2+} conducting channels in an osteoblast-like cell line (UMR 106). *Federation of European Biochemical Societies Letters*, **251**, 17–21.

El Haj, A. J., Minter, S. L., Rawlinson, S. C. F., Suswillo, R., and Lanyon, L. E. (1990).

Cellular responses to mechanical loading *in vitro*. *Journal of Bone and Mineral Research*, **5**, 923–32.

Eliot, F. H. and Grodzinsky, A. J. (1987). Cartilage electromechanics 1. Electrokinetic transduction and the effects of electrolyte pH and ionic strength. *Journal of Biomechanics*, **20**, 615–27.

Eriksson, C. (1974). Streaming potentials and other water dependent effects in mineralized tissues. *Annals of the New York Academy of Science*, **238**, 321–38.

Ferry, J. D. (1980). *Viscoelastic properties of polymers* (3rd edn). Wiley, New York

Finlay, J. B. and Repo, R. U. (1979). Energy absorbing ability of articular cartilage during impact. *Medical and Biological Engineering and Computing*, **17**, 397–403.

Fukada, E. and Yasuda, I. (1957). On the piezoelectric effect of bone. *Journal of Physiological Society of Japan*, **12**, 1158–62.

Galileo, G. (1638). *Dialogues concerning the two sciences*. Grew and Salvio translation New York 1914, cited J. KIein Nulend, 1987.

Gingell, D. (1967). Membrane surface potential in relation to a possible mechanism for intercellular interactions and cellular responses: a physical basis. *Journal of Theoretical Biology*, **17**, 451–82.

Gingell, D. (1971). Cell membrane surface potential as a transducer. In *Membrane and ion transport* (ed. E. E. Bittar), pp. 317–57. Wiley, New York.

Glegg, R. E. and Leblond, C. P. (1953). Pressure as a possible cause of dissolution and redeposition of bone and tooth crystals. *Anatomical Record*, **72**, 97–115.

Goodship, A. E., Lanyon, L. E., and McFie, H. (1979). Functional adaptation of bone to increased stress. *Journal of Bone and Joint Surgery*, **61A**, 539–46.

Goodship, A. E. and Kenwright, J. (1985). The influence of induced micromovement upon healing of experimental tibial fractures. *Journal of Bone and Joint Surgery*, **67B**, 650–5.

Gray, M. L., Pizzanelli, A. M., Grodzinsky, A. J., and Lee, R. C. (1988). Mechanical and physicochemical determinants of the chondrocyte biosynthetic response. *Journal of Orthopaedic Research*, **6**, 777–92.

Gray, P. T. A., Standen, N. B., Whittaker, M. J., and the Company of Biologists (1987). Microelectrode techniques: The Plymouth Workshop Handbook (ed. N. B. Standen, P. T. A. Gray, and M. J. Whittaker). The Company of Biologists, Cambridge.

Greig, D. M. (1931). *Clinical observations on the surgical pathology of bone*. Oliver and Boyd, London.

Gross, B. (1953). *Mathemetical structure of the theories of viscoelasticity*. Herman, Paris.

Gross, D. and Williams, W. S. (1982). Streaming potential and the electromechanical response of physiologically-moist bone. *Journal of Biomechanics*, **15**, 277–95.

Guzelsu, N. and Walsh, W. R. (1990). Streaming potential of intact wet bone. *Journal of Biomechanics*, **23**, 673–85.

Hall, A. C., Urban, J. P. G., and Gehl, K. A. (1991). The effects of hydrostatic pressure on matrix synthesis in articular cartilage. *Journal of Orthopaedic Research*, **9**, 1–10.

Hasegawa, S., Sato, S., Saito, S., Suzuki, Y., and Brunette, D. D. M. (1985). Mechanical stretching increases the number of cultured bone cells synthesizing DNA and alters their pattern of protein synthesis. *Calcified Tissue International*, **37**, 431–6.

Hastings, G. W. (1988). Electrical effects in bone. *Journal of Biomedical Engineering*, **10**, 515–21.

Heinegård, D. and Oldberg, A. (1989). Structure and biology of cartilage and bone matrix noncollagenous macromolecules. *FASEB Journal*, **3**, 2042–51.

Hert, J., Skelenska, A., and Liskova, M. (1971). Reaction of bone to mechanical stimuli. Part 5. Effect of intermittent stress on the rabbit tibia after resection of the peripheral nerves. *Folia Morphologica*, **19**, 378–87.

Hodge, W. H., Fijan, R. S., Carlson, K. L., Burgess, R.G., Harris, W. H., and Mann, R. W. (1986). Contact pressures in the human hip joint measured *in vivo*. *Proceedings of the National Academy of Science USA*, **83**, 2879–83.

Hooley, C. J. and Cohen, R. E. (1979). A model for the creep behaviour of tendon. *International Journal of Biological Macromolecules*, **1**, 123–32.

Hukins, D. W. L. and Aspden, R. M. (1985). Composition and properties of connective tissues. *Trends in Biochemical Science*, **10**, 260–4.

Indong, O. H. and Harris, W. H. (1978). Proximal strain distribution in the loaded femur. *Journal of Bone and Joint Surgery*, **60A**, 75–85.

Isaacs, A. (1987). *Concise science dictionary*. Oxford Paperback, OUP, Oxford.

Jeansonne, B. G., Feagin, F. F., Shoemaker, R. L., and Rehm, W. S. (1978). Transmembrane potentials of osteoblasts. *Journal of Dental Research*, **57**, 361–4.

Jendrucko, R. J., Hyman, W. A., Newell, P. H., and Chakraborty, B. K. (1976). Theoretical evidence for the generation of high pressure in bone cells. *Journal of Biomechanics*, **9**, 87–91.

Jendrucko, R. J., Cheng, C. J., and Hyman, W. A. (1977). The distribution of induced electrical activity in bent long bone. *Journal of Biomechanics*, **10**, 493–503.

Johnson, M. W. (1984). Behaviour of fluids in stressed bone and cellular stimulation. *Calcified Tissue International*, **36**, S12–76.

Jones, H. H., Priest, J. D., and Hayes W. C. (1977). Humeral hypertrophy in response to exercise. *Journal of Bone and Joint Surgery*, **59A**, 204–8.

Jurvelin, J., Kiviranta, I., Säämänen, A.-M., Tammi, M., and Helminen, H. J. (1989). Partial restoration of immobilization-induced softening of canine articular cartilage after remobilization of the knee (stifle) joint. *Journal of Orthopaedic Research*, **7**, 352–8.

Justus, R. and Luft, J. H. (1970). A mechanical hypothesis for bone remodelling induced by mechanical stress. *Calcified Tissue Research*, **5**, 222–35.

Kachar, B. and Reese, T. S. (1985). Rapid formation of gap-junction-like structures induced by glycerol. *Anatatomical Record*, **213**, 7–15.

Kempson, G. E. (1979). In *Adult articular cartilage* (ed. M. A. R. Freeman), pp. 333–414. Pitman Medical, Tunbridge Wells.

Kenwright, J. and Goodship, A. E. (1989). Controlled mechanical stimulation in the treatment of tibial fractures. *Clinical Orthopaedics*, **241**, 36–47.

Kirby, M. C., Aspden, R. M., and Hukins, D. W. L. (1988). Determination of the orientation distribution function for collagen fibrils in a connective tissue site from a high-angle X-ray diffraction pattern. *Journal of Applied Crystallography*, **21**, 929–34.

Kiviranta, I., Tammi, M., Jurvelin, J., Säämänen, A.-M., and Helminen, H. J. (1988). Moderate running exercise augments glycosaminoglycans and thickness of articular cartilage in the knee of young beagle dogs. *Journal of Orthopaedic Research*, **6**, 188–95.

Krolner, B. and Toft, B. (1983). Vertebral bone loss: an unheeded side effect of therapeutic bed-rest. *Clinical Science*, **64**, 537–40.

Lansman, J. B., Hallam, T. J., and Rink, T. J. (1987). Stretch activated ion channels in vascular endothelial cells as mechano-transducers. *Nature*, **325**, 811–13.

Lanyon, L. E. (1974). Experimental support for the trajectorial theory of bone structure. *Journal of Bone and Joint Surgery*, **56B**, 160–6.

Lanyon, L. E. (1976). The measurement of bone strain *in vivo*. *Acta Orthopaedica Belgica*, **42**, 198–208.

Lanyon, L. E. (1987). Functional strain in bone tissue as an objective and controlling stimulus for adaptive bone remodelling. *Journal of Biomechanics*, **20**, 1083–93.

Lanyon, L. E. and Baggott, D. G. (1976). Mechanical function as an influence on the structure and form of bone. *Journal of Bone and Joint Surgery*, **58B**, 436–43.

Lanyon, L. E., Hampson, W. G. H., Goodship, A. E., and Shah, J. S. (1975). Bone deformation recorded *in vivo* from strain gauges attached to the human tibial shaft. *Acta Orthopaedica Scandinavica*, **46**, 256–68.

Larsson, T., Aspden, R. M., and Heinegård, D. (1991). Effects of mechanical load on cartilage matrix biosynthesis *in vitro*. *Matrix*, **11**, 388–94.

Lavine, L. S. and Grodzinsky, A. J. (1987). Electrical stimulation of repair of bone. *Journal of Bone and Joint Surgery*, **69A**, 626–30.

Marino, A. A. and Becker, R. O. (1967). Evidence for direct physical bonding between the collagen filament apatite crystals in bone. *Nature*, **213**, 697–8.

Maroudas, A. (1976). Balance between swelling pressure and collagen tension in normal and degenerate cartilage. *Nature*, **260**, 808–9.

McCutchen, C. W. (1983). Lubrication of and by articular cartilage. In *Cartilage*, Vol. 1, Biomedical aspects, (ed. B. K. Hall), pp. 87–107. Academic Press, New York.

McDonald, F. and Houston, W. J. B. (1990). An *in-vivo* assessment of muscular activity and the importance of electrical phenomena in bone remodelling. *Journal of Anatomy*, **172**, 165–75.

Meikle, M., Reynolds, J. J., Sellers, A., and Dingle, J. T. (1979). Rabbit cranial sutures *in vitro*; a new experimental model for studying the response of fibrous joints to mechanical stress. *Calcified Tissue International*, **28**, 137–44.

Meikle, M. C., Heath. J. K., Atkinson, S. J., Hembry, R. M., and Reynolds, J. J. (1986). Molecular biology of stressed connective tissues at sutures and hard tissues *in vitro*. In *The biology of tooth movement* (ed. L. A. Norton), C. J. Burstone, CRC Press, Florida.

Menton, D. N., Simmons, D. J., Chang, S. L., and Orr, B. Y. (1984). From bone lining cell to osteocyte: an SEM study. *Anatomical Record*, **209**, 29–39.

Miller, S. C., Saint-Georges, L., Bowman, B. M., and Jee, W. S. (1989). Bone lining cells: structure and function. *Scanning Microscopy*, **3**, 953–61.

Murray, P. D. F. (1936). *Bones: a study of the development and structure of the vertebrate skeleton*. Cambridge University Press, London.

Murray, D. W. and Rushton, N. (1990). The effect of strain on bone cell Prostaglandin E2 release: A new experimental method. *Calcified Tissue International*, **47**, 35–9.

Myers, R. R., Lai, W. M., and Mow, V. C. (1984). A continuum theory and an experiment for the ion-induced swelling behaviour of articular cartilage. *Journal of Biomechanical Engineering*, **106**, 151–8.

Norridin, R. W., Jee, W. S. S., and High, W. B. (1990). The role of prostaglandins in bone *in vivo*. *Prostaglandins Leukotrienes and Essential Fatty Acids*, **41**, 139–49.

O'Conner, J. A., Lanyon, L. A., and McFie, H. (1982). The influence of strain rate on adaptive bone remodelling. *Journal of Biomechanics*, **15**, 767–81.

Olsen, S. P., Clapham, D. E., and Davies, P. F. (1988). Haemodynamic shear stress activates a K^+ current in vascular endothelial cells. *Nature*, **331**, 168–70.

Osdoby, P. and Caplan, A. I. (1979). Osteogenesis in culture of limb mesenchymal cells. *Developmental Biology*, **73**, 84–102.

Osdoby, P. and Caplan, A. I. (1980). A scanning electron microscopic investigation of *in vitro* osteogenesis. *Calcified Tissue International*, **30**, 43–50.

Palmoski, M. J., Perricone, E., and Brandt, K. D. (1979). Development and reversal of a proteoglycan aggregation defect in normal canine knee cartilage after immobilization. *Arthritis and Rheumatism*, **22**, 508–17.

Palmoski, M. J., Colyer, R. A., and Brandt, K. D. (1980). Joint motion in the absence of normal loading does not maintain normal articular cartilage. *Arthritis and Rheumatism*, **23**, 325–34.

Papatheofanis, F. J. and Papatheofanis, B. J. (1989). Short term effect of exposure to intense magnetic fields on haematologic indices of bone metabolism. *Investigative Radiology*, **24**, 221–3.

Pauwells, F. (1940). Biomechanics of fracture healing. *Verhandlugen der Deutshen Orthopädishe Gesellschaft, 34th Kongreß.*

Pauwells, F. (1980). *Biomechanics of the locomotor apparatus*. Springer-Verlag, New York.

Pead, M. J., Skerry, T. M., and Lanyon, L. E. (1988*b*). Direct transformation from quiescence to bone formation in the adult periosteum following a single brief period of bone loading. *Journal of Bone and Mineral Research*, **3**, 647–56.

Pead, M. J., Suswillo, R., Skerry, T. M., Vedi, S., and Lanyon, L.E. (1988*a*). Increased 3H uridine levels in osteocytes following a single short period of dynamic loading *in vivo*. *Calcified Tissue International*, **43**, 92–6.

Piekarski, K. and Munro, M. (1977). Transport mechanisms operating between blood supply and osteocytes in long bone. *Nature*, **269**, 80–2.

Pienkowski, K. and Pollack, S. R. (1983). The origin of stress-generated potentials in fluid saturated bone. *Journal of Orthopaedic Research*, **1**, 30–41.

Pollack, S. R., Korostoff, E., Starkebaum, W, and Iannacone, W. (1979). Microelectrode studies of stress-generated potentials in bone. In *Electrical properties of bone and cartilage—experimental effects and clinical applications* (ed. C. T. Brighton, J. Black, and S. R. Pollack), pp. 69–81. Grune and Stratton, New York.

Pollack, S. R., Salzstein, R., and Pienkowski, K. (1984*a*). The electric double layer in bone and its influence on stress-generated potentials. *Calcified Tissue International*, **36**, S77–81.

Pollack, S. R., Petrov, N., Salzstein, R. A., Brankov, G., and Blagoeva, R. (1984*b*). An anatomical model for streaming potentials in osteons. *Journal of Biomechanics*, **17**, 627–36.

Poswillo, D. E. (1989). Myths, masks and mechanisms of facial deformity. *European Journal of Orthodontics*, **11**, 1–9.

Raisz, L. G. and Martin, T. J. (1983). Prostaglandins in bone and mineral metabolism. In *Bone and mineral research*, Vol. 2 (ed. W. A. Peck), pp. 286–310. Excerpta Medico, Amsterdam.

Rawlinson, S. C. F., El Haj, A. J., Minter, S. L., Tavares, I. A., Bennett, A., and Lanyon, L. E. (1991). *Journal of Bone and Mineral Research*, **6**, 1345–51.

Reich, K. M., Gay, C. V., and Frangos, J. A. (1989). Fluid shear stress as a mediator of the mechanical stimulation of osteoblasts. *Journal of Bone and Mineral Research*, **4**, S332.

Robinson, K. R. (1985). The response of cells to electrical fields: a review. *Journal of Cell Biology*, **101**, 2023–7.

Rubin, C. T. (1984). Skeletal strain and the functional significance of bone architecture. *Calcified Tissue International*, **36**, S11–18.

Rubin, C. T. and Lanyon, L.E. (1984). Regulation of bone formation by applied dynamic loads. *Journal of Bone and Joint Surgery*, **66A**, 397–402.

Rubin, C. T. and Lanyon, L. E. (1985). Regulation of bone mass by mechanical strain magnitude. *Calcified Tissue International*, **37**, 411–17.

Rubin, C. T. and Lanyon, L. E. (1987). Osteoregulatory nature of mechanical stimuli: function as a determinant for adaptive remodelling in bone. *Journal of Orthopaedic Research*, **5**, 300–10.

Sachs, F. (1988). Mechanical transduction in biological systems. *CRC Critical Reviews in Biomedical Engineering*, **16**, 141–69.

Sah, R. L.-Y., Kim, Y. J., Doong, J.-Y. H., Grodzinsky, A. J., Plaas, A. H. K., and Sandy, J. D. (1989). Biosynthetic response of cartilage explants to dynamic compression. *Journal of Orthopaedic Research*, **7**, 619–36.

Salzstein, R. A. and Pollack, S. R. (1987). E. M. potential in cortical bone II—experimental analysis. *Journal of Biomechanics*, **20**, 271–280

Salzstein, R. A., Pollack, S. R., Mak A. F. T., and Petkov, N. (1987). Electrochemical properties in cortical bone I—a continuum approach. *Journal of Biomechanics*, **20**, 261–70.

Sheridan, J. D. and Atkinson, M. M. (1985). Physiological roles of permeable junctions: some possibilities. *Annual Review of Physiology*, **47**, 337–53.

Shin, M. S. and Norridin, R. W. (1986). Effects of prostaglandins on regional remodelling changes during tibial healing in beagles: a histomorphometric study. *Calcified Tissue International*, **39**, 191–7.

Skerry, T. M., Bitensky, L., Chayen, J., and Lanyon, L. E. (1989). Early strain-related changes in enzyme activity in osteocytes following bone loading *in vivo*. *Journal of Bone and Mineral Research*, **4**, 783–8.

Somjen, D., Binderman, I., Berger, E., and Harell, A. (1980). Bone remodelling induced by physical strain is prostaglandin E2 mediated. *Biochimica et Biophysica Acta*, **627**, 91–100.

Tilton, F. E., Degioanni, T. T. C., and Schneider, V. S. (1980). Long term follow-up of Skylab Demineralization. *Aviation Space and Environmental Medicine*, **51**, 1209–13.

Treharne, R. W., Brighton, C. T., Korostoff, E., and Pollack, S. R. (1979). Application of direct, pulsed, and SGP-shaped currents to *in vitro* foetal rat tibiae. In *Electrical properties of bone and cartilage* (ed. C. T. Brighton, J. Black, and S. R. Pollack), pp. 169. Grune and Stratton, New York.

Trueta, J. (1968). *Studies of the development and decay of the human frame*. Heinemann, London.

Tsien, R. Y. (1989). Fluorescent indicators of ion concentrations. *Methods In Cell Biology*, **30**, 127–56.

Umansky, R. (1966). The effect of cell population density on the developmental fate of reaggregating mouse limb bud mesenchyme. *Developmental Biology*, **13**, 31–56.

Veith, I. (1980). The history of medicine dolls and foot binding in China. *Clio Medico*, **14**, 255–67.

Weightman, B. and Kempson, G. E. (1979). In *Adult articular cartilage* (ed. M. A. R. Freeman), pp. 291–331. Pitman Medical, Tunbridge Wells.

Williams, W. S., Johnson, M., and Gross, D. (1979). Piezoelectric description of strain-related potentials from whole bone and single osteons. In *Electrical properties of bone and cartilage—experimental effects and clinical applications* (ed. C. T. Brighton, J. Beck, and S. R. Pollack), pp. 83–9. Grune and Stratton, New York.

Wilsman, N. J. and van Sickle, D. C. (1972). Cartilage canals, their morphology and distribution. *Anatomical Record*, **173**, 79–94.

Wolff, J. (1892). *Das gesetz der transformation der knochen*. von August Hirsch-wald, Berlin. (As translated by Maquet, P. and Furlong, R. (1986). *The law of bone remodelling*. Springer-Verlag, Berlin.)

Wong, G. and Cohn, D. V. (1974). Separation of PTH and calcitonin sensitive cells from nonresponsive bone cells. *Nature*, **252**, 713–5.

Yarker, Y. E., Aspden, R. M., and Hukins, D. W. L. (1983). Birefringence of articular cartilage and the distribution of collagen fibre orientations. *Connective Tissue Research*, **11**, 207–13.

Yasuda, I. (1953). Piezoelectricity of living bone. *Journal of Kyoto Prefectural Medical University*, **53**, 325.

Yasuda, I. (1955). Fundamental aspects of fracture treatment. *Journal of Kyoto Medical Society*, **4**, 395–406.

Yeh, C. and Rodan, G. A. (1984). Tensile forces enhance prostaglandin E2 synthesis in osteoblastic cells grown on collagen ribbons. *Calcified Tissue International*, **36**, S67–71.

Zimmermann, B., Scharlach, E., and Kaatz, R. (1982). Cell contact and surface coat alterations of limb-bud mesenchymal cells during differentiation. *Journal of Embryology and Experimental Morphology*, **72**, 1–18.

Index